PROJECT MERCURY

America's First Manned Spaceflight Program

By TD Barnes

Copyright 2024

Table of Contents

Prologue

Project Mercury chronicles the inception, development, and execution of America's first manned spaceflight program. It captures the meticulous efforts of NASA, government agencies, and American industry, all working in unison to achieve mission success in humanity's initial step into space.

The journey began with an urgency fueled by the Cold War and the space race. The nation watched as engineers and scientists, driven by determination and fatigue, toiled tirelessly. The launchpads at Cape Canaveral became stages for history-making events, where the roar of rockets echoed against dawn skies, shaking the ground with their immense power.

In the annals of history, Project Mercury stands as a testament to humanity's insatiable curiosity and relentless pursuit of the stars. It marked the dawn of a new era in manned space exploration, characterized by its technical achievements and the invaluable experience it provided. This experience laid the groundwork for future space endeavors, executed with unprecedented quality and efficiency. Notably, Project Mercury was conceived and carried out for peaceful purposes, with every major event documented and broadcast to the public through comprehensive news and television coverage from Cape Canaveral.

The initial goal of Project Mercury was ambitious: to orbit a man in space and return him safely to Earth. Remarkably, this objective was achieved just three years after the prime contract was awarded, a feat particularly impressive compared to the typical five-year development timeline for conventional manned aircraft. This rapid progression through critical milestones underscored the unwavering dedication of thousands of individuals working behind the scenes.

The journey was not without its challenges. Each launch was a symphony of precision and coordination, a delicate dance of technology and human ingenuity. As the countdown to each manned mission echoed across the control room, the air was thick with anticipation and the aroma of coffee—a staple for the sleepless engineers and scientists. The roar of the launch vehicle's engines reverberated through the ground, a visceral reminder of the sheer power propelling humanity into the unknown.

Among the astronauts, the sense of history in the making was palpable. Encased in their suits, they experienced the claustrophobic confines of the Mercury spacecraft, a tiny capsule barely big enough to accommodate a single person. The astronauts' breaths, amplified by the close quarters, intermingled with the hum of life-support systems and the sporadic crackle of radio communications. Each mission tested human endurance and resilience as the astronauts faced the physical strains of space travel and the psychological pressures of isolation and confinement.

These astronauts, America's new breed of heroes, embodied the dreams and aspirations of a nation. Clad in their silvery spacesuits, they ventured into the cold void of space, experiencing the weightlessness that made their bodies feel simultaneously free and alien. The view from the tiny window of their capsules was a spectacle few had ever witnessed: the curvature of Earth, the deep blackness of space, and the distant, silent stars.

The Mercury Seven went on to pilot the six manned missions of the Mercury program from May 1961 to May 1963. These missions demonstrated that humans could survive and work in space. Each mission built on the successes and lessons of the previous ones, gradually expanding the capabilities of NASA's spaceflight program.

Alan Shepard, the first American in space, launched aboard Freedom 7 on May 5, 1961. His suborbital flight paved the way for

John Glenn's historic orbital flight aboard Friendship 7 on February 20, 1962, making Glenn the first American to orbit the Earth. These missions were not just technical achievements but also powerful symbols of American ingenuity and perseverance during the Cold War era.

Despite the challenges, the technical objectives of Project Mercury were met with remarkable success. The reliability of the spacecraft and launch vehicle systems allowed for eliminating certain qualification flights early in the program, streamlining the path to success. This efficiency enabled the expansion of Mercury's original scope, culminating in the final manned one-day mission that achieved 22 orbits and lasted 34 hours—a triumphant conclusion to a pioneering chapter in space exploration.

The invaluable experience gained from the Mercury spacecraft's design, development, and operation, as well as from managing such a complex program, has had a profound and lasting impact on subsequent projects like Gemini and Apollo. The lessons learned and the knowledge acquired during Project Mercury continue to shape the future of space exploration, ensuring that the spirit of innovation and discovery remains at the forefront of human endeavors in space.

Project Mercury was more than a series of technological triumphs; it was a major step in the national commitment to space research and exploration and to man's eternal struggle to fly. To fully appreciate this, one must contrast it with the achievements of the Wright Brothers sixty years earlier. Operating out of a modest bicycle shop with a handmade wind tunnel and a shoestring budget, the Wrights achieved controlled, powered flight. Their austere yet groundbreaking work and Professor Goddard's early rocket experiments stood in stark contrast to Mercury's highly complex, government-sponsored space program. This program involved thousands of individuals and hundreds of federal, industrial, and university activities, highlighting the new prominence of science and technology in our daily lives.

The evolution and achievements of Project Mercury exemplify a truly national effort in the advancement of knowledge and its application. From the drawing boards where engineers sketched initial designs to the laboratories where scientists tested materials and systems to the training facilities where astronauts prepared for the rigors of space, every step was a concerted effort. The American spirit of innovation and perseverance was alive and well, propelling the nation towards a new frontier.

The Project Mercury story must be examined in the full context of its scientific, engineering, and managerial features within the dynamic human environment of national and international life. The national commitment to Project Mercury and its successors underscores the potential accomplishments of science and technology and the response of a democratic society to the challenges of its day.

Chapter 1 - Space the Next Frontier

Project Mercury, America's first manned spaceflight program, is now a monumental achievement in the annals of science and technology. Spanning four years, eight months, and one week, Mercury was more than a series of missions; it represented the culmination of decades of rigorous research and application across various disciplines, including aerodynamics, rocket propulsion, celestial mechanics, aerospace medicine, and electronics. This audacious endeavor took humanity beyond Earth's atmosphere and into space orbit, confirming our potential for mobility in the universe. While Project Mercury laid the foundational groundwork, it was up to subsequent Projects, Gemini and Apollo, to fully realize this potential.

Major Events Leading to Project Mercury: March 1944 - December 1957

March 16, 1944: A Pivotal Seminar at NACA Headquarters

Amid World War II, a crucial seminar occurred at the National Advisory Committee for Aeronautics (NACA) headquarters in Washington, D.C. NACA personnel proposed developing a jet-propelled transonic research airplane. This bold idea would eventually give rise to the legendary "X" series of research aircraft. This seminar, attended by Air Force and Navy representatives, was charged with the energy of groundbreaking innovation. The proposal aimed to explore the transonic flight regime, where airflow around an aircraft approaches the speed of sound—a realm fraught with aerodynamic challenges requiring cutting-edge solutions.

Engineers and military representatives leaned forward in the grand, wood-paneled conference room, captivated by the ambitious vision. They fervently debated technical and logistical hurdles, recognizing the proposal's potential to revolutionize aviation and set the stage for supersonic flight exploration.

December 9, 1944: Inception of Supersonic Flight Testing

At Langley Aeronautical Laboratory, Langley Air Force Base, Virginia, experts convened to discuss forming an organization dedicated to studying the stability and maneuverability of high-speed weapons, particularly guided missiles. This meeting marked the inception of supersonic flight testing. By the spring of 1945, Congress provided supplemental funds, establishing the Auxiliary Flight Research Station (AFRS) on Wallops Island, Virginia.

Under the directorship of Robert R. Gilruth, the AFRS launched its first test vehicle, a small two-stage solid-fuel rocket, on July 4, 1945. The launch, set against the backdrop of Independence Day celebrations, was a hive of activity. Engineers and technicians made last-minute checks and adjustments as the countdown commenced. The successful launch, marked by a trail of smoke and fire, drew cheers from the assembled team and signified a significant milestone in the journey toward supersonic flight. The AFRS would play a crucial role in advancing America's aeronautical capabilities, paving the way for future military and civilian aviation breakthroughs.

March 1946: Establishment of Project RAND

In March 1946, the Army Air Forces established Project RAND, marking a strategic shift toward studying satellite applications and sparking the first conceptual discussions about artificial satellites. In the quiet, meticulously organized offices of the RAND Corporation, visionary scientists and military strategists explored the vast potential of space. Their meetings, filled with speculative yet rigorous debates, laid the groundwork for modern space exploration.

May 8, 1946: Navy's Directive on Earth Satellite Vehicles

On May 8, 1946, the Chief of Naval Operations directed the Navy's Bureau of Aeronautics to explore the possibilities of earth satellite vehicles. This directive propelled the Navy into the burgeoning space race. Engineers and researchers, filled with anticipation, delved into the complexities of orbital technology. Drafting tables were covered with intricate diagrams, and walls were lined with blueprints and equations as conversations centered around orbital mechanics and satellite stabilization.

The Navy's engineers and researchers, many veterans of World War II aviation projects, applied their extensive knowledge to this new frontier. They considered various propulsion systems, launch challenges, and materials needed to withstand the harsh conditions of space. This directive spurred technical innovation and fostered a sense of urgency and competition. Establishing a robust satellite program was essential for maintaining technological superiority and ensuring national security in an increasingly interconnected world.

During these formative years, the collective efforts culminated in Project Mercury, setting the stage for humanity's ventures into space and marking the beginning of an era where space exploration became a critical element of national pride and scientific progress. The foundation laid by Project Mercury was instrumental in shaping the future of America's space program, leading to the groundbreaking achievements of Projects Gemini and Apollo and beyond.

On May 12, 1946, Project RAND's "Preliminary Design of an Experimental World Circling Space Ship" report emerged as a landmark document in the annals of aerospace history. This detailed report provided the technical foundation for the feasibility of building and launching an artificial satellite, and its implications are profound and far-reaching across the military and scientific communities.

The report was meticulously dissected and discussed in laboratories and conference rooms nationwide. Engineers marveled at its insights, recognizing the practical applications of its theoretical underpinnings. On the other hand, military officials pondered the strategic advantages of orbital assets. The RAND report not only outlined the theoretical principles of satellite technology but also delved into practical considerations for construction and deployment, transforming the concept of an artificial satellite from a distant dream into a tangible, achievable goal.

These early steps in 1946 were pivotal, setting the stage for the United States' entry into the space race. The collaborative efforts of the Army Air Forces, the Navy, and Project RAND laid a robust foundation for future advancements in satellite technology. This groundwork ultimately led to the launch of the first artificial satellites, heralding a new era in both military and civilian aerospace endeavors.

1947: Breaking the Sound Barrier

On October 14, 1947, a clear day over Muroc Dry Lake, California, witnessed Captain Charles E. Yeager piloting the XS-1 rocket plane into the annals of history. The rocket's roar and the subsequent sonic boom marked humanity's first supersonic manned flight, breaking the sound barrier at 700 miles per hour (Mach 1.06) at an altitude of 43,000 feet. This achievement was a testament to the relentless pursuit of speed and altitude. As the clear blue sky was split by the thunderous noise of Yeager's historic flight, it proved that the barriers of speed could be shattered, setting a new standard for future aviation feats.

Coordinating Satellite Development

In October 1947, recognizing the proliferation of satellite study contracts, the Department of Defense assigned the coordination of this burgeoning field to the Committee on Guided Missiles of the Research and Development Board. This strategic move ensured a unified approach to satellite development, fostering collaboration and streamlining efforts across different military and scientific entities. The coordination provided by the Committee on Guided Missiles was crucial in aligning the scattered research and development activities into a cohesive strategy for space exploration.

These foundational efforts, marked by groundbreaking reports and historic flights, underscored the United States' commitment to advancing its aerospace capabilities. They laid the groundwork for the monumental achievements of Projects Mercury, Gemini, and Apollo, propelling humanity into the space age and establishing America as a leader in space exploration.

Strategic Shifts in Military Policy and Space Exploration

On January 15, 1948, General Hoyt S. Vandenberg, Vice Chief of Staff of the United States Air Force, sanctioned a pivotal policy advocating for the development of earth satellites. This strategic move underscored the military's growing interest in space as a new frontier for defense and exploration. Vandenberg's directive marked a significant shift in military policy, highlighting the potential of satellite technology for surveillance, communication, and national security.

Such a high-ranking official's approval signaled the Air Force's recognition of space as a critical domain for future military operations. The directive was more than just an endorsement; it was a call to action, urging the Air Force to invest resources and expertise into developing satellite technology. The potential applications were vast and transformative, promising to revolutionize how the military conducted operations and secured national interests.

The response to Vandenberg's directive was immediate and vigorous within the Air Force. Research teams and project managers began organizing meetings and setting up dedicated units to explore satellite technology. The corridors of the Air Force's research facilities buzzed with renewed purpose as engineers and scientists brainstormed innovative solutions to the challenges of spaceflight. They delved into topics such as orbital mechanics, launch vehicle design, and the miniaturization of electronic components to withstand the rigors of space.

This strategic policy shift also fostered collaborations with civilian agencies and private industry. The Air Force recognized that achieving success in space would require a concerted effort across multiple sectors. Partnerships were formed with leading aerospace companies and academic institutions, pooling knowledge and resources to accelerate the development of satellite technologies.

General Vandenberg's directive catalyzed a series of initiatives that eventually led to the United States establishing a strong foothold in space. It emphasized the importance of satellites for surveillance, providing the military with unprecedented capabilities to monitor global activities and gather intelligence. Communication satellites were envisioned to offer secure and reliable lines of communication across vast distances, critical for coordinating military operations.

The Use of Animals in Pioneering Space Travel

The journey to human spaceflight was paved with numerous experiments and missions involving animals. These early missions were crucial in understanding the biological impacts of space travel, ensuring the safety and success of subsequent manned missions. Various species, including primates, dogs, mice, and insects, were used to study the effects of microgravity, radiation, and the harsh conditions of space on living organisms.

Early Experiments and Species

Primates

Due to their genetic similarities to humans, primates were extensively used in early space missions. Their physiological and psychological responses provided valuable data for assessing the feasibility and safety of human spaceflight.

Albert Series: In the late 1940s and early 1950s, the United States launched a series of V-2 rockets carrying rhesus monkeys named Albert I, II, III, and IV. These missions aimed to study the effects of high-altitude flights and microgravity. Despite significant challenges, including the monkeys' deaths, these early missions provided critical insights into the physical stresses of space travel.

Ham and Enos: In the Mercury program, chimpanzees Ham and Enos played pivotal roles. Ham, launched on January 31, 1961, aboard Mercury-Redstone 2, became the first primate to perform tasks in space, demonstrating the functionality of life support systems and the spacecraft's ability to sustain life. Enos followed with a more complex mission on November 29, 1961, aboard Mercury-Atlas 5, completing two orbits around the Earth. His mission validated the environmental control systems in orbit, directly paving the way for John Glenn's historic flight.

Dogs

The Soviet space program famously used dogs in their early space missions, selecting them for their docile nature and ability to handle extended periods of confinement.

Laika: On November 3, 1957, Laika, a stray dog from Moscow, became the first animal to orbit the Earth aboard Sputnik 2. Although Laika did not survive the mission, her flight provided valuable data on the physiological effects of space travel and proved that living beings could endure space conditions.

Belka and Strelka: On August 19, 1960, dogs Belka and Strelka were launched aboard Sputnik 5. They became the first animals to return safely from orbit, providing critical insights into the effects of space travel on living organisms. Their successful mission demonstrated the feasibility of safe reentry and recovery, crucial for human spaceflight.

Rodents and Other Small Animals

Mice, rats, and other small animals were frequently used to study the biological effects of space travel, including radiation exposure, muscle atrophy, and bone density loss.

Mice and Rats: The United States and the Soviet Union both launched numerous missions involving rodents. These experiments provided extensive data on the biological effects of microgravity and space radiation on cellular processes, physiology, and behavior.

Frogs and Turtles: Frogs and turtles were also sent into space to study developmental biology and the effects of space travel on vestibular function (balance and orientation).

Insects and Microorganisms

Insects, such as fruit flies and beetles, and microorganisms, like bacteria and fungi, were sent into space to study genetic mutations, reproductive effects, and microbial behavior in microgravity.

Fruit Flies: The first living organisms sent into space were fruit flies, launched aboard a V-2 rocket by the United States in 1947. This mission aimed to study the effects of radiation exposure at high altitudes.

Bacteria and Fungi: Microbial studies were crucial for understanding how space conditions affect microbial growth, resistance, and behavior. These experiments helped develop protocols for planetary protection and understanding of potential health risks for astronauts.

Impact and Legacy

Using animals in space research was instrumental in advancing our understanding of space travel's physiological and biological challenges. These missions provided invaluable data that informed the design of life support systems, spacecraft, and mission protocols, ensuring the safety and success of human spaceflight.

Biological Insights: Animal missions revealed critical information about the effects of microgravity on muscle and bone health, radiation exposure, and the overall well-being of living organisms. This knowledge was essential for preparing humans for the rigors of space travel.

Safety and Feasibility: The data gathered from these missions helped engineers and scientists develop safer spacecraft and life support systems. The success of animal missions demonstrated that sending living beings into space and returning them safely was feasible, a crucial step towards human space exploration.

Ethical Considerations: While animal testing was vital in the early days of space exploration, it also raised ethical considerations regarding the treatment and welfare of animals. These missions sparked ongoing discussions about the ethics of using animals in scientific research and the need for humane treatment and care.

Uncrewed and Chimpanzee Flights

Project Mercury conducted 20 uncrewed flights using Little Joe, Redstone, and Atlas launch vehicles. These flights were essential for developing and testing the launch vehicles, launch escape system, spacecraft, and tracking network. These missions were crucial in validating the spacecraft's design and ensuring the success and safety of manned missions.

Testing and Trials: Uncrewed and Chimpanzee Missions

Before sending astronauts into space, NASA conducted numerous uncrewed and chimpanzee missions to rigorously test the Mercury spacecraft's systems, safety measures, and overall performance. These tests included a combination of boilerplate spacecraft, production models, and biological payloads to simulate real flight conditions and identify any potential issues.

One notable mission aimed to launch a specialized satellite using a Scout rocket equipped with Mercury communications components to test the ground tracking network. Unfortunately, the booster failed shortly after liftoff, underscoring the challenges and complexities of space exploration.

On June 11, 1948, White Sands, New Mexico's stark landscape, witnessed a somber milestone in space exploration. A V-2 rocket carrying a monkey named Albert was launched, aiming to study the effects of space travel on biological organisms. This mission marked one of the earliest attempts to understand how living creatures might endure the harsh conditions of space. Despite the promise of space travel, Albert succumbed to suffocation during the flight. The tragic outcome highlighted the immense challenges and risks of early spaceflight experiments.

The First Monkey Astronaut

In 1949, amidst the fervent developments in missile technology and space exploration, the White Sands Missile Range in New Mexico emerged as a pivotal launch site for the second V-2 Blossom mission. This vast desert expanse provided the ideal location for testing and launching rockets, setting the stage for a historic event.

On that significant day, the V-2 Blossom rocket stood tall on the launch pad, poised to catapult Albert II, a courageous rhesus monkey, into the uncharted realms of space.

As the countdown climaxed, the powerful engines ignited, propelling the rocket skyward and leaving a trail of exhaust fumes in its wake. The V-2 Blossom pierced through Earth's atmosphere, carrying Albert II within a specially designed container equipped with life support systems.

During this groundbreaking flight, Albert II experienced the weightlessness of space, becoming the first monkey to do so. This achievement underscored the incredible human ingenuity and the unwavering drive to explore the mysteries beyond our planet's boundaries. Albert II's pioneering journey laid the foundation for future space missions involving living organisms, deepening our comprehension of the effects of space travel on biological beings.

However, the mission was not without its challenges and heartbreak. Tragically, Albert II did not survive the flight due to a parachute failure during the capsule's descent. Despite this loss, the mission yielded invaluable data and insights that would play a crucial role in shaping the course of future manned space exploration.

Unfortunately, 1951 also witnessed a tragic event in the emerging Space Race. A rhesus monkey named Albert II became the first primate to venture into space aboard the V-2 Blossom rocket. However, a malfunction caused the parachutes to fail, resulting in the monkey's death upon impact. Subsequent missions involving Albert III and Albert IV also ended in similar tragedies. Despite these unfortunate outcomes, valuable biomedical data from Albert II's flight was successfully transmitted to ground control, contributing to the early efforts of space exploration.

Despite their tragic outcomes, these missions provided critical data that would inform and improve the safety of future human spaceflight endeavors. The data collected from Albert and Albert II's flights helped scientists understand the physiological stresses and environmental challenges living beings face in space. This information was invaluable in developing life support systems and safety protocols for future manned missions.

Genesis of the Atlas Rocket and the Dawn of a New Era

On July 13, 1946, Convair's MX-774 test vehicle, later known as the Atlas, roared to life for the first time. This test marked the genesis of a launch vehicle that would become pivotal in the Mercury program and a cornerstone of America's space exploration efforts. The successful ignition and flight of the MX-774 represented a significant advancement in rocket technology, demonstrating the feasibility of powerful rockets capable of reaching space.

The test site was abuzz with anticipation as engineers and technicians meticulously prepared for the launch. The MX-774 stood tall on the launch pad, a testament to the ingenuity and determination of the Convair team. As the countdown reached zero, the rocket's engines ignited, and it ascended with a thunderous roar, leaving a trail of smoke and flame. This moment was a technical achievement and a symbol of the burgeoning space age.

A team of scientists and engineers meticulously monitored the MX-774's flight. Their faces, lit by the glow of control panels and instrumentation, reflected hope and tension. The data collected during this test flight was crucial, providing insights into the performance of propulsion systems, structural integrity, and aerodynamic stability at high speeds.

This successful test was a major milestone, validating the design concepts and technological innovations that would underpin the development of the Atlas rocket. The MX-774 featured several groundbreaking technologies, including a gimballed engine for improved stability and a lightweight balloon-tank design that significantly reduced the vehicle's weight while maintaining strength.

These innovations were critical in achieving the high thrust-to-weight ratio for reaching orbit.

The success of the MX-774 test had far-reaching implications. It laid the groundwork for the Atlas series of rockets, which would become central to America's efforts in space exploration. The Atlas rocket would go on to launch satellites, crewed spacecraft, and intercontinental ballistic missiles, playing a crucial role in the space race and national defense.

In the context of Project Mercury, the Atlas rocket became the workhorse for sending American astronauts into orbit. Its powerful engines and reliable performance made it the ideal choice for the program, enabling the United States to make significant strides in manned spaceflight. The July 13 test of the MX-774 was more than just a successful launch; it was a pivotal moment that signaled America's entry into the space age.

Acknowledgment of the Earth Satellite Program

On December 29, 1948, Secretary of Defense James V. Forrestal delivered his inaugural report to President Harry Truman. Among the various defense initiatives outlined, Forrestal included a brief yet significant mention of the Earth satellite program, now coordinated by the Committee on Guided Missiles. This acknowledgment marked a pivotal moment, signifying the growing importance of satellite technology in national defense strategy.

Forrestal's report underscored the strategic value of satellites, recognizing their potential to revolutionize military operations, global communications, and scientific research. The brief mention in his report highlighted the evolving perspective within the U.S. government and military regarding space technology. It signaled that satellites could provide unparalleled advantages in surveillance, intelligence gathering, and secure communications, enhancing national security.

The Committee on Guided Missiles, tasked with overseeing the satellite program, had already begun to explore the practical applications of satellite technology. The committee's work involved rigorous research and development efforts to overcome the numerous challenges of launching and maintaining satellites in orbit. Their efforts were pivotal in transitioning the concept of artificial satellites from theoretical studies to actionable projects.

Forrestal's acknowledgment in the report also highlighted the broader implications of satellite technology for scientific research. Satellites promised to offer new ways to study the Earth's atmosphere, weather patterns, and even outer space, opening up unprecedented opportunities for scientific advancement.

The inclusion of the satellite program in Forrestal's report was a clear indication of its strategic priority. It reflected a growing consensus among military and scientific leaders that space was the next frontier for both defense and exploration. This early recognition and prioritization set the stage for subsequent investments and developments in space technology, eventually leading to the successful deployment of satellites and manned space missions.

The strategic value of satellites was particularly pronounced during the Cold War. The ability to monitor activities globally from space provided a significant advantage, contributing to the United States' efforts to maintain a technological edge over its adversaries. Forrestal's report thus not only highlighted the immediate military benefits of satellite technology but also underscored its long-term importance in shaping the future of defense and scientific inquiry.

The mention of the Earth satellite program in Secretary Forrestal's report to President Truman was a seminal moment. It reflected a shift in strategic priorities and paved the way for the United States' eventual leadership in space exploration and satellite technology.

Establishing the Atlantic Missile Range: A Cornerstone of American Space Exploration

On May 11, 1949, President Harry S. Truman signed a directive authorizing the creation of a missile test range, which would later evolve into the Atlantic Missile Range at Cape Canaveral, Florida. This significant decision set a geographic cornerstone for America's burgeoning space endeavors, establishing a vital site for missile and rocket testing that would become synonymous with the nation's space exploration efforts.

Establishing the missile test range was a strategic move, reflecting the growing recognition of the need for a dedicated facility to support developing and testing advanced missile technology. Cape Canaveral's location was ideal due to its proximity to the equator, allowing for more efficient launches into various orbital trajectories. Additionally, the vast expanse of the Atlantic Ocean provided a safe area for the testing of missiles and rockets, away from populated regions.

The creation of the Atlantic Missile Range marked the beginning of a new era in American aerospace research and development. The site quickly became a hub of activity, with engineers, scientists, and military personnel working tirelessly to push the boundaries of missile and rocket technology. The range was equipped with state-of-the-art tracking and telemetry systems, enabling precise monitoring of test flights and collecting critical data.

Cape Canaveral soon became the focal point for groundbreaking tests and launches. The site was crucial in developing the Redstone, Atlas, and Titan rockets, foundational to the United States 'missile program and early space exploration efforts. These rockets would eventually carry satellites, scientific instruments, and astronauts into space, marking significant milestones in the nation's quest to explore the final frontier.

The decision to establish the Atlantic Missile Range also underscored the United States 'commitment to maintaining technological superiority during the Cold War. The range provided a secure and controlled environment for the testing of intercontinental ballistic missiles (ICBMs), which were a critical component of the nation's defense strategy. The successful development and deployment of these missiles were essential for deterrence and maintaining the balance of power.

As the space race intensified, Cape Canaveral's role expanded beyond military applications. The site became the launch pad for some of the most iconic missions in space history, including the launches of the Mercury, Gemini, and Apollo programs. From here, astronauts first orbited the Earth and, eventually, the historic Apollo 11 mission began its journey to the Moon.

Revival of the Atlas Project: A Testament to Perseverance and Innovation

On January 16, 1951, the government resumed studies on the MX-774, now designated the Atlas, breathing new life into the project. This decision marked a significant revival for Convair, whose continued research and development, despite earlier setbacks, showcased the perseverance essential to eventual space success. The renewed focus on the Atlas project reflected a growing recognition of its potential as a powerful launch vehicle, crucial for both military applications and future space exploration missions.

The MX-774 had initially been shelved due to budget cuts and shifting priorities. However, the rapid advancements in missile technology and the intensifying Cold War tensions prompted a reevaluation of its potential. The Atlas rocket, with its innovative design features, such as a lightweight balloon tank and gimballed engines, offered the promise of significant advancements in both range and payload capacity.

Convair's engineers, who had never truly abandoned the project, eagerly resumed their work. Their offices and laboratories became hubs of renewed activity, with teams working tirelessly to refine and improve the Atlas design. The perseverance and dedication of these engineers were evident as they tackled the challenges that had previously hindered the project. Their commitment to overcoming obstacles was driven by a shared vision of pushing the boundaries of what was technologically possible.

The renewed focus on the Atlas project also highlighted its dual significance. Militarily, the Atlas was envisioned as a cornerstone of the United States' intercontinental ballistic missile (ICBM) capabilities. Its development was crucial for ensuring national security by providing a reliable and powerful deterrent against potential adversaries. The Atlas was expected to deliver nuclear warheads over vast distances, a capability that was becoming increasingly important in the geopolitical landscape of the time.

In parallel, the Atlas 'potential as a space exploration launch vehicle began to capture the imagination of scientists and engineers. The ability to carry heavier payloads into orbit opened up new possibilities for satellite deployment and manned space missions. The Atlas rocket would eventually become a critical component of the Mercury program, propelling America's first astronauts into space and marking significant milestones in the space race.

The government's decision to reinvest in the Atlas project was a testament to the foresight of military and scientific leaders who recognized this powerful launch vehicle's strategic and exploratory potential. It underscored the importance of innovation and persistence in the face of setbacks, highlighting the crucial role of continued research and development in achieving long-term success.

Historic Achievement in Biological Space Research

On September 20, 1951, the Western Hemisphere witnessed a historic milestone in biological space research. The first successful recovery of animals from a rocket flight was accomplished when a monkey and 11 mice survived an Aerobee launch to an altitude of 236,000 feet. This remarkable feat underscored significant advancements in the field and marked a pivotal moment in the journey toward human spaceflight.

The Aerobee rocket, designed for high-altitude research, was launched from the White Sands Proving Ground in New Mexico. As the rocket ascended, it carried its precious cargo— an assortment of biological specimens including a monkey and 11 mice—into the upper reaches of the atmosphere. The mission aimed to gather data on how these organisms would fare in the harsh near-space conditions, including exposure to cosmic radiation, extreme temperatures, and microgravity.

The animals' successful recovery was a testament to the meticulous planning and engineering behind the mission. The rocket's payload was equipped with life-support systems designed to ensure the animals' well-being during the flight. Upon reaching its peak altitude, the Aerobee's capsule separated and began its descent back to Earth, deploying parachutes to slow its fall and ensure a safe landing.

Ground crews eagerly awaited the capsule's return, and their anticipation turned to jubilation when they discovered that the monkey and all 11 mice had survived the journey unscathed. This achievement provided invaluable insights into the physiological effects of space travel on living organisms.

The data collected from the flight would help scientists understand how prolonged exposure to space conditions could impact biological systems, informing the design of life-support systems for future human missions.

The success of this Aerobee flight was more than just a technical triumph; it was a milestone that demonstrated the feasibility of sending living beings into space and bringing them back safely. It paved the way for more ambitious biological experiments and set the stage for human space exploration. The information gleaned from these early missions would prove critical in developing the Mercury and Apollo programs, ultimately leading to humans landing on the Moon.

Establishing the Atlantic Missile Range: A Cornerstone of American Space Exploration

On May 11, 1949, President Harry S. Truman signed a directive authorizing the creation of a missile test range, which would later evolve into the Atlantic Missile Range at Cape Canaveral, Florida. This significant decision set a geographic cornerstone for America's burgeoning space endeavors, establishing a vital site for missile and rocket testing that would become synonymous with the nation's space exploration efforts.

Establishing the missile test range was a strategic move, reflecting the growing recognition of the need for a dedicated facility to support developing and testing advanced missile technology. Cape Canaveral's location was ideal due to its proximity to the equator, allowing for more efficient launches into various orbital trajectories. Additionally, the vast expanse of the Atlantic Ocean provided a safe area for the testing of missiles and rockets, away from populated regions.

The creation of the Atlantic Missile Range marked the beginning of a new era in American aerospace research and development. The site quickly became a hub of activity, with engineers, scientists, and military personnel working tirelessly to push the boundaries of missile and rocket technology. The range was equipped with state-of-the-art tracking and telemetry systems, enabling precise monitoring of test flights and collecting critical data.

Cape Canaveral soon became the focal point for groundbreaking tests and launches. The site was crucial in developing the Redstone, Atlas, and Titan rockets, foundational to the United States 'missile program and early space exploration efforts. These rockets would eventually carry satellites, scientific instruments, and astronauts into space, marking significant milestones in the nation's quest to explore the final frontier.

The decision to establish the Atlantic Missile Range also underscored the United States 'commitment to maintaining technological superiority during the Cold War. The range provided a secure and controlled environment for the testing of intercontinental ballistic missiles (ICBMs), which were a critical component of the nation's defense strategy. The successful development and deployment of these missiles were essential for deterrence and maintaining the balance of power.

As the space race intensified, Cape Canaveral's role expanded beyond military applications. The site became the launch pad for some of the most iconic missions in space history, including the launches of the Mercury, Gemini, and Apollo programs. From here, astronauts first orbited the Earth and, eventually, the historic Apollo 11 mission began its journey to the Moon.

Revival of the Atlas Project: A Testament to Perseverance and Innovation

On January 16, 1951, the government resumed studies on the MX-774, now designated the Atlas, breathing new life into the project. This decision marked a significant revival for Convair, whose continued research and development, despite earlier setbacks, showcased the perseverance essential to eventual space success. The renewed focus on the Atlas project reflected a growing recognition of its potential as a powerful launch vehicle, crucial for both military applications and future space exploration missions.

The MX-774 had initially been shelved due to budget cuts and shifting priorities. However, the rapid advancements in missile technology and the intensifying Cold War tensions prompted a reevaluation of its potential. The Atlas rocket, with its innovative design features, such as a lightweight balloon tank and gimballed engines, offered the promise of significant advancements in both range and payload capacity.

Convair's engineers, who had never truly abandoned the project, eagerly resumed their work. Their offices and laboratories became hubs of renewed activity, with teams working tirelessly to refine and improve the Atlas design. The perseverance and dedication of these engineers were evident as they tackled the challenges that had previously hindered the project. Their commitment to overcoming obstacles was driven by a shared vision of pushing the boundaries of what was technologically possible.

The renewed focus on the Atlas project also highlighted its dual significance. Militarily, the Atlas was envisioned as a cornerstone of the United States' intercontinental ballistic missile (ICBM) capabilities. Its development was crucial for ensuring national security by providing a reliable and powerful deterrent against potential adversaries. The Atlas was expected to deliver nuclear warheads over vast distances, a capability that was becoming increasingly important in the geopolitical landscape of the time.

In parallel, the Atlas 'potential as a space exploration launch vehicle began to capture the imagination of scientists and engineers. The ability to carry heavier payloads into orbit opened up new possibilities for satellite deployment and manned space missions. The Atlas rocket would eventually become a critical component of the Mercury program, propelling America's first astronauts into space and marking significant milestones in the space race.

The government's decision to reinvest in the Atlas project was a testament to the foresight of military and scientific leaders who recognized this powerful launch vehicle's strategic and exploratory potential. It underscored the importance of innovation and persistence in the face of setbacks, highlighting the crucial role of continued research and development in achieving long-term success.

Acknowledgment of the Earth Satellite Program

On December 29, 1948, Secretary of Defense James V. Forrestal delivered his inaugural report to President Harry Truman. Among the various defense initiatives outlined, Forrestal included a brief yet significant mention of the Earth satellite program, now coordinated by the Committee on Guided Missiles. This acknowledgment marked a pivotal moment, signifying the growing importance of satellite technology in national defense strategy.

Forrestal's report underscored the strategic value of satellites, recognizing their potential to revolutionize military operations, global communications, and scientific research. The brief mention in his report highlighted the evolving perspective within the U.S. government and military regarding space technology. It signaled that satellites could provide unparalleled advantages in surveillance, intelligence gathering, and secure communications, enhancing national security.

The Committee on Guided Missiles, tasked with overseeing the satellite program, had already begun to explore the practical applications of satellite technology. The committee's work involved rigorous research and development efforts to overcome the numerous challenges of launching and maintaining satellites in orbit. Their efforts were pivotal in transitioning the concept of artificial satellites from theoretical studies to actionable projects.

Forrestal's acknowledgment in the report also highlighted the broader implications of satellite technology for scientific research. Satellites promised to offer new ways to study the Earth's atmosphere, weather patterns, and even outer space, opening up unprecedented opportunities for scientific advancement.

The inclusion of the satellite program in Forrestal's report was a clear indication of its strategic priority. It reflected a growing consensus among military and scientific leaders that space was the next frontier for defense and exploration. This early recognition and prioritization set the stage for subsequent investments and developments in space technology, eventually leading to the successful deployment of satellites and manned space missions.

The strategic value of satellites was particularly pronounced during the Cold War. The ability to monitor activities globally from space provided a significant advantage, contributing to the United States' efforts to maintain a technological edge over its adversaries. Forrestal's report thus not only highlighted the immediate military benefits of satellite technology and underscored its long-term importance in shaping the future of defense and scientific inquiry.

The mention of the Earth satellite program in Secretary Forrestal's report to President Truman was a seminal moment. It reflected a shift in strategic priorities and paved the way for the United States' eventual leadership in space exploration and satellite technology.

NACA's Vision for High-Altitude Flight and Early Space Exploration

On January 30, 1952, the National Advisory Committee for Aeronautics (NACA) released a pivotal report that outlined an ambitious vision for high-altitude manned and unmanned vehicle flights. This document laid the foundation for a transformative era in American space exploration, setting the stage for the nation's pioneering efforts in human spaceflight.

By June 24, 1952, the momentum generated by the report culminated in a decisive meeting where Robert J. Woods, an influential committee member, advocated passionately for NACA to take the lead in the nation's space flight research efforts. Woods, renowned for his persuasive eloquence and deep understanding of aeronautics, argued convincingly that NACA possessed the requisite expertise and infrastructure to spearhead these groundbreaking initiatives.

His advocacy resonated strongly within the committee, prompting swift action. Recognizing the potential and urgency of the endeavor, NACA swiftly formed a dedicated working group. Comprising forward-thinking engineers and scientists, this team endeavored to analyze existing data and rigorously conceptualize a manned test vehicle. Over the following two years, their relentless efforts bore fruit, bringing the dream of human spaceflight closer to reality.

The work environment during this period was charged with purpose and excitement. Engineers in their distinctive white lab coats meticulously worked over drafting tables, illuminated by the soft glow of desk lamps. They sketched designs, conducted calculations, and engaged in spirited debates. The air buzzed with the hum of slide rules and the rustle of blueprints, underscoring the intensity and dedication with which they pursued their goals. Late-night meetings were common as the team tackled the myriad challenges inherent in their pioneering work.

Central to their efforts was conceptualizing a manned test vehicle capable of safely transporting a human beyond Earth's atmosphere and back. This monumental task involved rigorous testing of materials, propulsion systems, and life-support mechanisms. Models were meticulously crafted and subjected to wind tunnel trials, while prototypes underwent subscale testing. Each step aimed to refine and validate the vehicle's design, ensuring it met the stringent demands of space travel.

The International Geophysical Year (IGY) and the Dawn of Space Exploration

Establishing the Special Committee for the International Polar Year on May 16, which later evolved into the International Geophysical Year (IGY), marked a pivotal milestone in global scientific research coordination. This ambitious initiative represented a concerted effort to advance understanding of Earth's physical properties through unprecedented international collaboration.

From July 1957 to December 1958, the IGY brought together scientists from 67 countries in a monumental effort to explore and gather data on a wide array of geophysical phenomena. Researchers constructed observatories in remote, often inhospitable locations such as Antarctica and the Arctic, where they braved extreme conditions to collect vital data.

Inside these makeshift laboratories, scientists calibrated instruments, analyzed samples, and meticulously recorded their findings. The collaborative spirit of the IGY fostered a culture of openness and data sharing that transcended national boundaries, laying a strong foundation for future scientific endeavors, including space exploration.

One of the most significant outcomes of the IGY was the discovery of the Van Allen radiation belts, regions of charged particles trapped by Earth's magnetic field. This breakthrough, facilitated by data from the first American satellite, Explorer 1, underscored the critical importance of space-based observations. It advanced our understanding of Earth's space environment and ignited global interest and investment in further space exploration endeavors.

The success of the IGY demonstrated the immense benefits of international cooperation in advancing scientific knowledge and tackling complex challenges. It set a precedent for future collaborative efforts in space exploration and solidified the United States' commitment to leadership in space science and technology.

These pivotal moments in history—from NACA's visionary reports on high-altitude flight to the collaborative achievements of the International Geophysical Year—were instrumental in shaping the trajectory of space exploration. They laid the groundwork for subsequent missions and discoveries that would ultimately lead humanity to the Moon and beyond, marking a testament to human ingenuity and perseverance in exploring the cosmos.

The Blunt Nose Principle: Revolutionizing Reentry Vehicle Design

On June 18, H. Julian Allen of the NACA Ames Aeronautical Laboratory introduced a revolutionary concept that would transform the design of reentry vehicles: the blunt nose principle. This innovative approach departed from traditional pointed designs by proposing a blunt nose shape, effectively mitigating the intense heat generated during reentry into Earth's atmosphere.

The physics behind Allen's concept were rooted in aerodynamics and thermodynamics. A blunt nose created a shockwave that stood off from the vehicle, allowing heat to dissipate over a larger surface area. This significantly reduced the vehicle's surface temperature, lowering thermal stress and potential damage. This breakthrough made reentry safer and more feasible for manned and unmanned missions.

The practical application of the blunt nose principle began with intercontinental ballistic missiles (ICBMs), where the reduction in thermal loads enhanced their reliability and effectiveness. Engineers and scientists conducted extensive wind tunnel tests and computer simulations to validate Allen's theory, confirming that blunt-shaped reentry vehicles could withstand reentry velocities far better than pointed designs.

The success of the blunt nose principle in military applications quickly attracted attention in the realm of space exploration. The Mercury spacecraft, designed to carry the first American astronauts into space, adopted Allen's innovative design to ensure safe reentry from orbit. Engineers at NACA (later NASA) refined and optimized the blunt nose concept, conducting rigorous tests on heat shield materials and configurations to maximize protective capabilities.

During the Mercury missions, the effectiveness of the blunt nose principle was put to the ultimate test. As the spacecraft reentered Earth's atmosphere, the heat shields performed flawlessly, safeguarding astronauts and ensuring their safe return. This success validated Allen's visionary thinking and the meticulous engineering efforts that went into developing and perfecting the blunt nose design.

The blunt nose principle quickly became a cornerstone of reentry vehicle design, extending its impact beyond Mercury to influence subsequent space missions such as Gemini, Apollo, and the Space Shuttle programs. This innovative concept made human spaceflight possible and ensured the safety of countless astronauts, marking a significant milestone in the history of space exploration.

Intensified Research Efforts at Langley Aeronautical Laboratory

On June 24, the NACA Committee on Aerodynamics issued a pivotal recommendation to intensify research efforts in both manned and unmanned high-altitude, high-speed flight. This directive sparked a concentrated research initiative at the Langley Aeronautical Laboratory, where a dedicated team of scientists and engineers tackled the formidable challenges posed by these conditions.

The Langley researchers identified aerodynamic heating and stability at high altitudes as critical focus areas. Aerodynamic heating, caused by air friction at high velocities, posed a significant threat to the integrity of aircraft and spacecraft. Meanwhile, maintaining stability in high altitudes's thinning atmosphere required innovative aerodynamic control solutions.

The laboratory buzzed with activity as interdisciplinary teams delved into these challenges. Engineers and scientists meticulously analyzed data, conducted extensive wind tunnel tests, and utilized emerging computational methods to model aerodynamic behaviors. The wind tunnel at Langley served as a crucible of innovation, where scale models of aircraft and spacecraft were subjected to extreme conditions simulating high-speed flight.

Addressing aerodynamic heating necessitated advances in materials science for heat-resistant alloys and ceramics. Concurrently, aeronautical engineers explored various design configurations to enhance heat dissipation and develop robust thermal protection systems for reentry vehicles.

Simultaneously, the team tackled stability issues by experimenting with different wing shapes, control surfaces, and aerodynamic configurations. The insights gained from these efforts were pivotal in refining supersonic and spacecraft designs, pushing the boundaries of technological capability in aviation and space travel.

The breakthroughs achieved at Langley laid the groundwork for significant advancements in aerospace engineering. Solutions to aerodynamic heating were instrumental in developing reentry vehicles that could safely return astronauts from orbit. Similarly, innovations in stability at high altitudes contributed to the design of supersonic jets and spaceplanes, shaping the future of aviation and space exploration.

These pioneering efforts underscored NACA's (and later NASA's) commitment to pushing the frontiers of aeronautical and aerospace research, ultimately enabling humanity to explore and navigate the challenges of space.

The Role of Centrifuge Training in Mercury Astronaut Preparation

In June, the Navy's human centrifuge at Johnsville, Pennsylvania, began operations, marking a significant advancement in preparing astronauts for human spaceflight. Renamed the Aviation Medical Acceleration Laboratory (AMAL), this state-of-the-art facility quickly became integral to the training regimen of the Mercury astronauts, known as the Mercury Seven.

Intense Training Regimen of the Mercury Seven

The training program for the Mercury Seven astronauts was meticulously designed to ensure they were thoroughly prepared for the unprecedented challenges of spaceflight. This comprehensive regimen included various simulations and exercises to test and enhance their physical and mental capabilities.

Centrifuge Simulations

The primary purpose of the centrifuge at AMAL was to simulate the intense G-forces that astronauts would experience during critical phases of spaceflight, such as launch, reentry, and other high-acceleration maneuvers. The centrifuge's design allowed astronauts to experience rapid accelerations and decelerations, mimicking the extreme conditions they would face in space.

The centrifuge's massive arm extended outward with a cockpit at its end, spinning at high speeds to create gravitational forces many times stronger than those on Earth. This grueling training was essential for building the physical and mental resilience needed to operate effectively under such extreme conditions.

Multi-Axis Spin-Test Inertia Facility (MASTIF)

At the Lewis Flight Propulsion Laboratory, astronauts trained in the Multi-Axis Spin-Test Inertia Facility (MASTIF), which simulated and prepared them to handle the disorienting effects of a spinning spacecraft. This training was crucial for emergencies, such as losing control during reentry, where maintaining orientation and control were critical. **Rocket Sled Testing**

Early tests involved dropping capsule models from rocket sleds to validate parachute deployment and escape systems. These tests provided essential data on the dynamics of rapid descent and the effectiveness of recovery systems under various conditions. They contributed significantly to developing safety protocols and emergency procedures crucial for ensuring the astronauts' safety during Mercury missions.

MASTIF at Lewis Research Center

G-Force Simulations

One of the most physically demanding aspects of astronaut training was enduring high g-force simulations in the centrifuge. Astronauts were subjected to accelerations that simulated the intense forces experienced during launch and reentry. This training was vital for acclimating them to the physical stresses of space travel, helping them develop techniques to maintain consciousness and functionality under extreme conditions.

Rigorous Training Sessions

Sessions at AMAL were rigorous and demanding. Astronauts experienced forces that pushed them into their seats, making movement difficult. Medical monitors tracked their vital signs while engineers and trainers observed their reactions and performance. Each session provided critical insights into how the human body responds to high g-forces, refining training protocols and safety measures based on collected data.

Conclusion

The centrifuge training at AMAL played a crucial role in preparing the Mercury astronauts for the challenges of spaceflight. By subjecting them to realistic simulations of space conditions, the facility ensured that astronauts were physically prepared and confident in handling the stresses of launch, reentry, and emergencies. This confidence was indispensable for the success of their missions, allowing them to focus on their tasks without being overwhelmed by the physical demands of space travel.

Egress training at Langley

Contributions of AMAL and NACA Langley/Ames Aeronautical Laboratories to Space Exploration

The Aviation Medical Acceleration Laboratory (AMAL) and the NACA Langley and Ames Aeronautical Laboratories played crucial roles in advancing human space exploration and missile technology during the early years of the Cold War. Here's a detailed overview of their contributions:

AMAL and Astronaut Training

High-G Training for Mercury Astronauts:

AMAL, formerly the Navy's human centrifuge at Johnsville, Pennsylvania, provided rigorous training in simulating the high gravitational forces (G-forces) astronauts would experience during spaceflight.

Astronauts underwent centrifuge sessions to acclimate to the physical stresses of launch, reentry, and emergency scenarios.

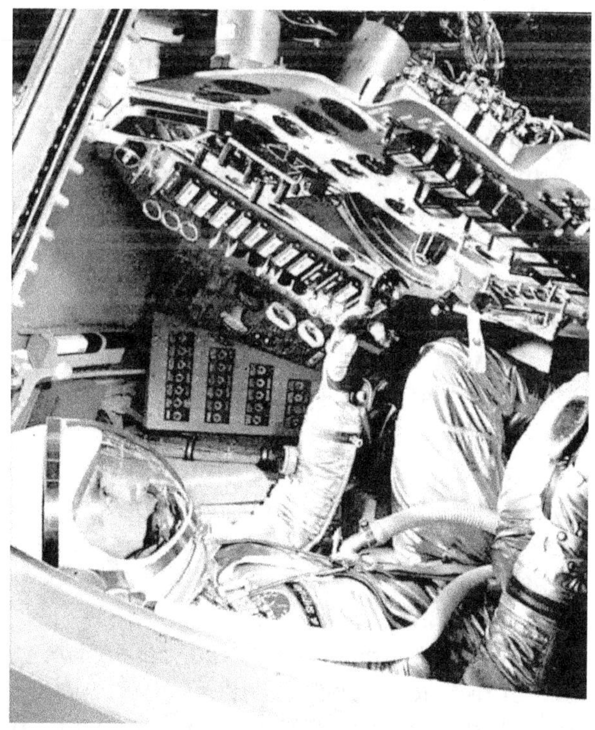

Flight trainer at Cape Canaveral

This training was vital for ensuring astronauts could maintain consciousness and functionality under extreme conditions, crucial for the success of Mercury missions.

Legacy and Influence:

The training protocols and knowledge gained at AMAL continued to shape astronaut preparation in subsequent space programs.

Insights into human endurance under high G-forces contributed to refining safety protocols and emergency procedures, enhancing astronaut safety and mission success.

NACA Langley and Ames Aeronautical Laboratories

Research on Reentry Aerodynamics (1952-1956):

During this period, NACA researchers at Langley and Ames focused on studying the aerodynamic characteristics of reentry vehicles.

Extensive wind tunnel tests, high-speed simulations, and theoretical analyses were conducted to understand vehicle stability and thermal protection under extreme conditions.

Impact on Project Mercury:

The findings from Langley and Ames were pivotal in designing the Mercury spacecraft, the United States 'first human spaceflight program.

Researchers applied their knowledge to develop spacecraft shapes and heat shield technologies capable of withstanding the intense heat of reentry, ensuring astronauts returned safely from orbit.

Advancements in Ablation Theory:

NACA's research contributed significantly to proving the effectiveness of ablation as a thermal protection method.

This was crucial for the Army's Jupiter missile development program. It demonstrated that ablative materials could safely dissipate heat during reentry, protecting the missile's nose cone.

ICBM Development:

Langley and Ames' aerodynamic data and materials research benefited the Air Force's intercontinental ballistic missile (ICBM) program.

Insights into reentry vehicle design helped create nose cones that endured high velocities and extreme temperatures during long-range ballistic trajectories.

Collaboration and Innovation:

Collaboration between NACA, the military, and industry partners accelerated innovation in aerospace technology.

Engineers and scientists collaborated across disciplines, sharing data and techniques to translate theoretical insights into practical applications.

Legacy of Knowledge:

The research conducted at Langley and Ames between 1952 and 1956 established a foundational knowledge base.

This knowledge addressed immediate challenges in aerospace technology and laid the groundwork for future advancements in space exploration and missile technology.

Conclusion

The efforts of AMAL and NACA Langley/Ames Aeronautical Laboratories between 1952 and 1956 were instrumental in advancing the United States 'capabilities in space exploration and missile technology. From pioneering astronaut training methods to groundbreaking research in reentry vehicle design, these institutions played key roles in shaping the course of aerospace innovation during a pivotal era of scientific and technological progress.

On July 30, 1953, C. E. Brown, W. J. O'Sullivan, Jr., and C. H. Zimmerman at the Langley Aeronautical Laboratory completed preliminary studies addressing the challenges of manned space flight and proposing a test vehicle to investigate these issues. Their work began in mid-1952 and explored the modification of the X-2 airplane to achieve greater speeds and altitudes, specifically aiming for altitudes around 200,000 feet. The idea was that such a vehicle could help resolve aerodynamic heating problems and simulate an environment with minimal atmospheric density, thus addressing many of the challenges associated with outer space flight.

However, there was a growing consensus among NACA scientists that the exploratory scope needed to be broadened. As early as July 1952, a resolution was presented, advocating for NACA to focus on problems of unmanned and manned flights at altitudes ranging from 50 miles to infinity and at speeds from Mach 10 to the velocity required to escape Earth's gravity. The Executive Committee of NACA adopted this ambitious resolution on July 14, 1952, setting a new direction for the organization's efforts.

A formal letter dated July 30, 1953, from NACA to the High-Speed Flight Research Station discussed these issues in detail. The letter outlined the challenges of high-speed, high-altitude flight and considered possible modifications to the X-2 airplane to extend its speed and altitude capabilities. This communication highlighted the importance of pushing the boundaries of current technology to achieve the objectives set forth by the resolution.

Just a few weeks later, on August 20, the Army successfully test-fired the first Redstone missile at Cape Canaveral, Florida. Research and development of the Redstone had begun in 1950, and it would later play a crucial role in the United States' space endeavors. The Redstone missile was used as a launch vehicle for manned suborbital flights and other development flights in Project Mercury, demonstrating the interconnection between missile technology and the emerging field of human spaceflight.

Technical Challenges

Reaching space required overcoming numerous technical challenges. Rockets needed to be powerful yet reliable, capable of withstanding extreme forces during launch and reentry. The spacecraft had to protect its occupants from the vacuum of space, radiation, and micrometeoroids. Reentry posed a particular danger, with temperatures exceeding 10,000 °F due to air compression.

The efforts at Langley Aeronautical Laboratory, combined with the developments in missile technology, underscored the rapid advancements in aerospace research during this period. These milestones addressed immediate technical challenges and set the stage for the United States 'future successes in space exploration. The pioneering work of scientists and engineers at NACA, alongside military collaborations, fueled a period of innovation that would lead to monumental achievements in human spaceflight.

Rocket Planes

In 1954, the Air Force pursued a parallel approach to human spaceflight with the Dyna-Soar project, which was canceled in 1963.

NASA later developed the reusable Space Shuttle, with the X-15 rocket plane being the first to enter space in 1963.

Dyna-Soar Project (Canceled in 1963):

The Dyna-Soar (Dynamic Soarer) project was initiated by the United States Air Force (USAF) as an experimental spaceplane program.

Its goal was to develop a reusable vehicle capable of performing a variety of military missions, including reconnaissance, bombing, and satellite deployment.

Despite early development and testing, the project was ultimately canceled in 1963 before a full-scale vehicle could be built and flown. This decision was influenced by changing military priorities and budget constraints.

X-15 Rocket Plane:

The X-15 program was conceived in 1954. It was a hypersonic rocket-powered aircraft operated by NASA and the United States Air Force. The goal of the X-15 program was to conduct research on high-speed flight and gather data on aerodynamics, structural heating, control systems, and other aspects of flight at altitudes and speeds that were beyond those feasible with conventional aircraft of the time.

In May 1954, the NACA made a pivotal decision by defining the characteristics of what would become the X-15 rocket aircraft. This marked a significant step forward in the quest for manned space flight, embodying the innovative spirit and engineering prowess that characterized NACA's contributions to aerospace technology.

The X-15 was conceived as a revolutionary vehicle designed to investigate the complex flight problems associated with high speeds and altitudes that were previously unattainable. Its development represented a crucial bridge between conventional atmospheric flight and the emerging domain of space travel. With its advanced design and cutting-edge technology, the X-15 was poised to push the boundaries of what was possible in aeronautics.

Engineers and scientists at NACA meticulously crafted the X-15's specifications to address the unique challenges of high-altitude, high-speed flight. The aircraft was equipped with a powerful rocket engine capable of propelling it to speeds exceeding Mach 6 and altitudes above 50 miles. This performance envelope allowed researchers to study the aerodynamic heating, stability, and control issues critical for future space missions.

The X-15 program also provided invaluable data on human factors in high-speed, high-altitude flight. Pilots who flew the X-15 faced extreme conditions, including intense G-forces and rapid transitions between atmospheric and near-space environments. Their experiences and the data collected from these flights informed the design of subsequent spacecraft and flight suits, contributing to the safety and success of manned space missions.

The development of the X-15 was a collaborative effort that involved NACA (which later became NASA), the United States Air Force, and various aerospace industry partners. This cooperation exemplified the synergistic approach necessary to overcome spaceflight's technical challenges. The program also benefited from the advancements in materials science, propulsion, and avionics that were emerging at the time.

The X-15 achieved numerous milestones throughout its operational life, setting records for altitude and speed for decades. The knowledge gained from its flights played a crucial role in developing the Mercury, Gemini, and Apollo programs, paving the way for human space exploration.

In essence, the X-15 rocket aircraft was not just a remarkable feat of engineering but also a symbol of the relentless pursuit of knowledge and innovation. It laid the foundation for the modern era of aerospace technology, demonstrating the potential of human ingenuity to transcend the limits of flight and venture into the uncharted realms of space.

The X-15 was a rocket-powered aircraft developed jointly by NASA, the Air Force, and the Navy.

It was designed to explore hypersonic flight and spaceflight at altitudes reaching the edge of space.

The X-15 made its first spaceflight on July 19, 1964, when pilot Joe Walker flew it to an altitude of 67 miles (108 km), surpassing the 50-mile boundary that defines the edge of space.

While the X-15 was not reusable in the same sense as later spacecraft like the Space Shuttle, it demonstrated key technologies and provided valuable data for the development of future spaceplanes and spacecraft.

NASA LLRV Program

The NASA LLRV (Lunar Landing Research Vehicle) program was a crucial initiative conducted in the 1960s to develop and test vehicles that simulated the lunar landing conditions anticipated for the Apollo missions. Here's an overview for your book:

Purpose and Objectives: Initiated by NASA in the early 1960s as part of the Apollo program, the LLRV aimed to develop and validate technologies and techniques for landing a spacecraft on the Moon. It was specifically designed to simulate the lunar environment and train astronauts in landing procedures.

Vehicle Design: The LLRV was a small, uncrewed aircraft resembling a single-engine jet trainer. It featured a framework of tubular aluminum alloy, with a rocket engine mounted vertically in the center to provide vertical thrust, simulating lunar descent and landing.

Flight Characteristics: The LLRV's design allowed it to hover and maneuver in a manner similar to what would be required for a lunar landing. It had a set of smaller thrusters around the vehicle's perimeter to control pitch, yaw, and roll during flight.

Training and Testing: The LLRV was used extensively at NASA's Flight Research Center (now Armstrong Flight Research Center) in California. Astronauts, including Neil Armstrong and other Apollo crew members, trained extensively in these vehicles to simulate lunar landing scenarios and practice abort procedures.

Successes and Challenges: The LLRV program successfully validated the concept of a lunar landing and refined techniques for controlling a spacecraft in the lunar environment. However, it also faced challenges, including several accidents during testing that led to the loss of vehicles.

Legacy: The knowledge and experience gained from the LLRV program were instrumental in the development of the Lunar Module (LM) that eventually landed astronauts on the Moon during the Apollo missions. The program demonstrated the feasibility of precision landing and provided critical insights into handling the unique challenges of lunar descent and landing.

NASA Lifting Body Program

NASA's Lifting Body Program was a pioneering effort in the 1960s and 1970s to explore the feasibility of using wingless aircraft with aerodynamic lift for atmospheric re-entry and controlled landing. Here are the key points you can include in your book:

Background and Goals: Initiated in the early 1960s, the Lifting Body Program aimed to develop vehicles that could return to Earth from space without relying on traditional capsule designs. The program sought to improve landing accuracy, reusability, and overall spacecraft safety.

Aircraft Designs: The program featured several distinctive lifting body designs, including the M2-F1, M2-F2, HL-10, X-24A, and X-24B. These vehicles were characterized by delta-shaped wings and fuselages that blended into the wing structure, eliminating the need for separate wings.

Flight Testing: Flight testing began with unpowered drop tests from B-52 aircraft to evaluate handling characteristics and aerodynamic performance during descent.

These tests provided crucial data on stability, control, and landing dynamics.

NASA Centers Involved: The program was primarily managed by NASA's Flight Research Center (now Armstrong Flight Research Center) in California, with significant contributions from other NASA centers and contractors.

Contributions to Space Shuttle Development: Findings from the Lifting Body Program greatly influenced the design and development of the Space Shuttle, particularly in shaping the orbiter's aerodynamics and approach to landing.

NASA's Space Shuttle Program:

NASA developed the Space Shuttle program based on the lessons learned from programs like Dyna-Soar and the X-15.

The Space Shuttle, officially named the Space Transportation System (STS), was designed as a reusable spacecraft capable of carrying astronauts and payloads to low Earth orbit (LEO).

The first Space Shuttle, Columbia, was launched on April 12, 1981, marking the beginning of an era of reusable space vehicles that would continue until the Shuttle fleet's retirement in 2011.

These developments illustrate the evolving approaches to human spaceflight during the Cold War era, with different agencies and programs pursuing various technological and operational concepts to advance space exploration and national defense capabilities.

On June 25, a critical meeting that included Dr. Wernher von Braun and other leading rocketry and space science experts occurred. During this meeting, it was concluded that a Redstone rocket equipped with a Loki cluster could successfully launch a satellite into a 200-mile orbit. This pivotal conclusion set the stage for the initiation of Project Orbiter, a groundbreaking endeavor that eventually led to the launch of Explorer I, America's first successful satellite.

The significance of this meeting extended beyond its technical conclusions. It highlighted the importance of collaboration between military and civilian scientists, illustrating how combined expertise could propel the United States into the space age. With his extensive rocketry background, Dr. von Braun played a crucial role in bridging the gap between the military applications of missile technology and its potential for scientific exploration.

Project Orbiter was an ambitious initiative to place a satellite into orbit using existing missile technology. The Redstone rocket, originally developed as a medium-range ballistic missile, was chosen for its reliability and performance. Adding the Loki cluster, a set of smaller rockets, provided the necessary boost to achieve the required orbital velocity.

The development process involved rigorous testing and refinement. Engineers and scientists worked tirelessly to modify and adapt the Redstone rocket for its new role. This included ensuring that the rocket could withstand the stresses of launch and the transition from the Earth's atmosphere into space. The collaborative effort saw contributions from various institutions, including the Army Ballistic Missile Agency and NACA, reflecting the multidisciplinary nature of the project.

On July 9, 1954, a pivotal meeting convened with members from NACA, the Air Force, and the Navy. Their objective: to address the critical need for a hypersonic research vehicle. This historic gathering marked the genesis of the ambitious X-15 project, a collaborative effort aimed at pushing the boundaries of aviation and space exploration.

The X-15 was designed to achieve unprecedented flight milestones, reaching speeds exceeding 4,000 miles per hour and soaring to altitudes as high as 314,750 feet. These achievements were not merely about setting records; they yielded essential data and insights crucial to the future of space exploration.

One primary goal of the X-15 program was to delve into the complexities of hypersonic aerodynamics. Researchers meticulously studied how air behaves at extreme velocities, providing fundamental insights into the aerodynamic challenges of near-space flight. This research was pivotal for designing future spacecraft, addressing critical issues such as thermal dynamics and structural integrity under extreme conditions.

Propulsion was another focal point of the X-15 project. Equipped with a robust rocket engine, the aircraft demonstrated unparalleled thrust capabilities, propelling it to the fringes of space. Engineers meticulously analyzed the engine's performance across various scenarios, refining their understanding of rocket propulsion for manned space missions. This knowledge formed the cornerstone for developing more advanced rocket technologies used in subsequent space programs.

Human factors also loomed large in the X-15's mission profile. Pilots who flew the X-15 encountered extreme G-forces, rapid accelerations, and transitions between atmospheric and near-space environments. The data collected on human performance and physiological responses under these conditions proved invaluable. It informed the design of life support systems, flight suits, and safety protocols critical for astronauts embarking on lunar missions and beyond.

The X-15 project exemplified interagency collaboration, drawing on the collective expertise and resources of NACA, the Air Force, and the Navy. This collaborative synergy ensured robust support and funding, enabling the program to achieve its ambitious objectives.

The legacy of the X-15 program remains profound. It generated a wealth of data that deepened our understanding of flight dynamics at hypersonic speeds and high altitudes, effectively bridging the gap between atmospheric flight and space travel. By addressing numerous challenges inherent to spaceflight, the X-15 laid the groundwork for subsequent missions, including Mercury, Gemini, and Apollo, marking a pivotal chapter in humanity's quest to explore the cosmos.

On August 7, a pivotal moment in space medicine unfolded as the Air Force School of Aviation Medicine at Randolph Field, Texas, welcomed its first space cabin simulator. This state-of-the-art facility represented a leap forward in understanding the intricacies of human adaptation to space environments, setting the stage for comprehensive research into the physiological and psychological challenges of space travel.

The space cabin simulator was designed to replicate the unique conditions astronauts would face during missions, including microgravity, confined spaces, and isolation. Within its controlled environment, researchers could meticulously study how the human body and mind responded to these unfamiliar settings.

A primary focus of the simulator was investigating the physiological impacts of spaceflight. Researchers closely monitored vital signs, muscle degradation, bone density changes, and cardiovascular responses in participants exposed to extended periods in simulated microgravity. This data proved essential in developing countermeasures to mitigate these effects, ensuring astronauts' health and performance could be sustained during prolonged missions.

Equally significant were the psychological studies conducted within the simulator. Researchers explored the psychological stressors of isolation, confinement, and the demands of space missions on mental well-being. Strategies such as communication protocols with mission control, recreational activities, and ergonomic cabin designs were evaluated to bolster astronauts' psychological resilience and overall mission effectiveness.

Moreover, the simulator served as a crucial training tool for astronauts. Simulating realistic space conditions allowed crew members to practice operational procedures, emergency responses, and daily routines in a safe environment. This hands-on training fostered confidence and preparedness, equipping astronauts to handle the challenges they would encounter during actual space missions.

The introduction of the space cabin simulator at Randolph Field marked a significant advancement in space medicine and astronaut preparation. It underscored the importance of rigorous research and training in ensuring the safety and success of human space exploration. Insights gleaned from the simulator directly informed the design of spacecraft cabins, life support systems, and operational protocols, enhancing overall mission safety and efficacy.

In summary, the debut of the space cabin simulator at the Air Force School of Aviation Medicine was a watershed moment in space exploration. Deepening our understanding of human factors in space travel enabled future astronauts to be better equipped physically and mentally for the challenges of spaceflight. This pioneering facility remains a cornerstone of space medicine, supporting ongoing efforts to expand humanity's presence beyond Earth's atmosphere with greater confidence and capability.

On October 14, a historic milestone in American rocketry unfolded at NACA's Pilotless Aircraft Research Division on Wallops Island with the launch of the first American four-stage rocket. This groundbreaking event marked a significant leap forward in rocket technology, showcasing the potential of multi-stage configurations to propel payloads to higher altitudes and beyond.

The four-stage rocket was a marvel of engineering, comprising successive propulsion units designed to ignite sequentially as each preceding stage exhausted its fuel and separated. This staging method optimized thrust and maximized efficiency by shedding weight at each stage. The successful launch validated the practical application of this intricate engineering concept, heralding a new era in atmospheric exploration and laying a solid foundation for future advancements in space travel.

Wallops Island test facility, 1961

During the launch at Wallops Island, each rocket stage performed flawlessly, igniting in succession to propel the payload to unprecedented altitudes. The meticulous planning and precise execution required for this multi-stage mission underscored the expertise and dedication of the engineers and scientists involved. The data gathered from this mission provided invaluable insights into the dynamics of multi-stage rocket flight, including critical aspects such as stage separation, ignition timing, and overall vehicle stability.

The success of the four-stage rocket launch marked a pivotal moment in the evolution of rocket staging technology. It provided compelling evidence that multi-stage rockets could effectively achieve the high altitudes necessary for space exploration, thereby influencing the design and development of subsequent launch vehicles. These advancements were crucial for enabling the deployment of satellites, scientific instruments, and eventually human crews into space with greater efficiency and reliability.

Furthermore, the Wallops Island launch underscored the importance of rigorous testing and experimentation in advancing rocketry. Lessons learned from this mission paved the way for enhancements in propulsion systems, structural integrity, and stage separation mechanisms, essential components for achieving sustained and dependable access to space.

In summary, the October 14 launch of the first American four-stage rocket by NACA's Pilotless Aircraft Research Division at Wallops Island was a landmark achievement in aerospace history. It not only demonstrated the feasibility and effectiveness of multi-stage rocketry but also provided critical data that propelled subsequent developments in space exploration technology. This successful mission exemplified the spirit of innovation and excellence driving America's journey toward broader horizons in human space exploration.

In March 1955, Dr. Alan T. Waterman, director of the National Science Foundation and a respected physicist, approached President Dwight D. Eisenhower with an ambitious proposal that would shape the future of space exploration. Dr. Waterman's vision centered on America's pivotal role in the upcoming International Geophysical Year (IGY), a global scientific initiative slated for July 1957 to December 1958. This unprecedented collaboration aimed to advance our understanding of Earth and its atmosphere through synchronized efforts across more than 60 countries.

At the heart of the IGY was the deployment of artificial satellites—an audacious concept for the time—to observe and gather data from outer space. Dr. Waterman's proposal emphasized that America's leadership in this international endeavor would bolster its scientific prestige and accelerate advancements in space technology. The plan outlined the launch of satellites that studied Earth's atmosphere, space environment, and celestial bodies, laying the groundwork for future space exploration.

President Eisenhower recognized the strategic significance of participating in the IGY and wholeheartedly endorsed Dr. Waterman's proposal. This decision marked a significant commitment by the United States to expand its capabilities in space science and technology. It spurred the mobilization of resources across governmental agencies, including the Department of Defense and the newly established Advanced Research Projects Agency (ARPA), to support satellite development and launch efforts.

As preparations for the IGY progressed, the United States embarked on an ambitious program to build and launch its first satellites. Notably, Explorer 1, launched in 1958, became the first American satellite in space and made groundbreaking discoveries regarding cosmic radiation and Earth's magnetic field. These early achievements provided crucial insights into space science and set the stage for subsequent space exploration endeavors.

Beyond scientific advancement, America's participation in the IGY carried significant geopolitical implications during the Cold War era. It demonstrated the nation's commitment to peaceful scientific exploration and international cooperation, contrasting with the Soviet Union's more secretive and militaristic approaches to space activities. This collaborative effort fostered unity among scientists worldwide and promoted a shared pursuit of knowledge that transcended political divides.

The comprehensive plan put forth by Dr. Waterman in 1955 marked a pivotal moment in the history of space exploration. It underscored America's dedication to scientific progress and global leadership, paving the way for the remarkable achievements that followed in the space age. The IGY satellite experiment expanded humanity's understanding of Earth and its environment, igniting a new era of space exploration and inspiring generations of scientists and engineers to push the boundaries of what was possible beyond our planet.

On July 29, 1955, President Dwight D. Eisenhower formally endorsed the International Geophysical Year (IGY) satellite proposal, marking a pivotal moment in the United States' commitment to space exploration. In a public announcement, Eisenhower declared that the United States would launch Earth satellites during the IGY period, spanning from July 1957 through December 1958. This endorsement was more than just a nod to scientific progress; it was a clear signal of the nation's dedication to advancing space science and contributing to global knowledge.

Eisenhower's endorsement came at a time when the world was on the cusp of a new era in scientific discovery. The IGY was conceived as a comprehensive international effort to study the Earth's geophysical properties, involving over 60 countries in a coordinated series of observations and experiments. The inclusion of satellite technology in these efforts was groundbreaking, as it promised to provide unprecedented insights into the Earth's atmosphere, magnetic field, and space environment.

The decision to launch satellites was bold and ambitious, reflecting the United States' desire to lead in space science. This commitment required significant resources and coordination across various branches of the government, including the Department of Defense and the newly formed Advanced Research Projects Agency (ARPA). It also necessitated collaboration with leading scientific institutions and the involvement of top engineers and researchers.

In the months following Eisenhower's announcement, preparations for the IGY satellite launches intensified. Scientists and engineers worked tirelessly to design and build the necessary technology. The first American satellite, Explorer 1, would eventually be launched in January 1958, providing valuable data on cosmic radiation and contributing to the discovery of the Van Allen radiation belts.

Eisenhower's endorsement of the IGY satellite proposal also had profound geopolitical implications. In the context of the Cold War, it was a statement of peaceful scientific ambition and international cooperation, contrasting with the Soviet Union's more militaristic approach to space exploration. This move helped to position the United States as a leader in the global scientific community, fostering a spirit of collaboration and mutual respect among nations.

The IGY satellite initiative advanced the space science field and laid the groundwork for future space exploration endeavors. It demonstrated the potential of international scientific cooperation and set the stage for the remarkable achievements of the space age. President Eisenhower's endorsement on July 29, 1955, thus stands as a testament to the United States' commitment to exploring the unknown and contributing to the collective scientific knowledge of the era.

On September 9, 1955, Project Vanguard officially commenced operations, signaling a new and focused direction for the United States' satellite development efforts. The Department of Defense authorized the Naval Research Laboratory to lead the Vanguard project by recognizing the strategic importance of participating in the International Geophysical Year (IGY) with a significant technological achievement. This decision followed a comprehensive review of various satellite proposals by the Stewart Committee, which ultimately recommended prioritizing Vanguard over other projects.

The Vanguard proposal aimed to launch a satellite into orbit during the IGY from July 1957 to December 1958. The project was chosen for its scientific potential and the feasibility of its technical approach. It marked a strategic pivot from the earlier Project Orbiter, a collaborative effort involving the Army and the Jet Propulsion Laboratory. The Stewart Committee's endorsement of Vanguard and the subsequent approval by Secretary of Defense Donald Quarles underscored the commitment to focusing resources and efforts on a single, robust project.

The selection of Project Vanguard was based on its innovative design and the capability to achieve a stable orbit. Unlike Project Orbiter, which relied on military rocket technology, Vanguard was envisioned as a purely scientific endeavor, emphasizing peaceful exploration and international cooperation. This distinction was important in the context of the Cold War, as it aligned with President Eisenhower's policy of promoting peaceful uses of outer space.

The Naval Research Laboratory (NRL) faced the significant challenge of developing the Vanguard rocket and satellite. This task involved pioneering rocketry, guidance systems, and satellite engineering work. The Vanguard team, composed of some of the nation's top scientists and engineers, faced numerous technical hurdles, from achieving the necessary launch velocity to ensuring the satellite's functionality once in orbit.

Project Vanguard's initiation reflected the broader aspirations of the United States during the IGY. It was about launching a satellite and demonstrating technological prowess and contributing valuable scientific data to the international community. The data collected by Vanguard would provide insights into the Earth's shape, gravitational field, and atmospheric density, enhancing our understanding of the planet and its environment.

Despite the ambitious goals and the prestige associated with being the first to launch a satellite, Project Vanguard faced intense competition from the Soviet Union, also making rapid advancements in space technology. This competitive atmosphere added urgency and pressure to the Vanguard team's efforts.

The decision to abandon Project Orbiter in favor of Vanguard marked a significant shift in the United States' approach to satellite development. It consolidated resources and streamlined efforts towards a singular, focused goal. Secretary Quarles' approval of this strategic pivot demonstrated a clear alignment within the Department of Defense and among civilian scientific leadership regarding the importance of the IGY and the potential impact of satellite technology.

In conclusion, the commencement of Project Vanguard on September 9, 1955, was a critical juncture in the United States' journey into space. It reflected a deliberate and strategic commitment to advancing scientific knowledge through international cooperation and peaceful exploration. The Vanguard project, focusing on innovation and collaboration, set the stage for America's future achievements in space exploration.

On October 2, 1955, the National Academy of Sciences took a decisive step to bolster the United States' efforts in space exploration by establishing a Technical Panel for the Earth Satellite Program. Chaired by Richard E. Porter, an eminent engineer and physicist, this panel coordinated and accelerated the nation's satellite research initiatives.

The formation of the Technical Panel was a response to the growing recognition of the complexities involved in developing and launching a satellite. The IGY satellite experiment required meticulous planning and seamless integration of various scientific and engineering disciplines. By bringing together leading experts in the field, the panel aimed to ensure that the United States would not only meet the ambitious goals of the IGY but also establish itself as a leader in the emerging field of space exploration.

Richard E. Porter, known for his aeronautics expertise and his missile technology contributions, was an ideal choice to lead this initiative. Under his guidance, the panel sought to harness the collective knowledge and capabilities of the nation's top scientists and engineers. The panel's mandate included reviewing existing research, identifying technological gaps, and recommending strategic investments in satellite technology.

Establishing the Technical Panel underscored the importance of structured and collaborative efforts in advancing satellite research. It facilitated better coordination between various governmental agencies, research institutions, and private sector partners involved in the Earth Satellite Program. This collaborative approach was essential in addressing the technical challenges of satellite design, propulsion, and communication systems.

One of the panel's first tasks was to ensure that Project Vanguard, the United States' primary satellite initiative, remained on track and received the necessary support. The panel's oversight helped streamline project management and foster innovation, ensuring the Vanguard team could overcome obstacles and meet critical milestones.

The Technical Panel also was vital in promoting the IGY's scientific objectives. By focusing on the potential discoveries and data that satellites could provide, the panel helped align the technical goals with broader scientific aspirations. This included understanding Earth's magnetic field, atmospheric composition, and space weather phenomena, which were crucial for future space missions.

Creating the Technical Panel for the Earth Satellite Program clearly indicated the United States' commitment to playing a leading role in the IGY and beyond. It highlighted the strategic importance of space exploration in the context of the Cold War, where scientific achievements were also seen as demonstrations of national prowess and technological superiority.

In conclusion, the National Academy of Sciences' decision to establish the Technical Panel for the Earth Satellite Program on October 2, 1955, was a pivotal moment in the history of American space exploration. Chaired by Richard E. Porter, the panel's work laid the groundwork for the successful development and deployment of satellites during the IGY, reinforcing the United States' position at the forefront of global scientific and technological advancement.

On February 1, 1956, the Army Ballistic Missile Agency (ABMA) was officially activated at Redstone Arsenal in Huntsville, Alabama, heralding a new era in American missile and space technology. Led by the visionary rocket engineer Wernher von Braun, who had pioneered Germany's V-2 rocket program during World War II, ABMA was tasked with developing advanced missile systems crucial to both military defense and scientific exploration.

At its inception, ABMA focused on the Redstone missile, initially designed as a short-range ballistic missile capable of delivering nuclear warheads. This versatile rocket quickly found application beyond its military role, becoming instrumental in America's early space efforts. Notably, a modified Redstone rocket launched Alan Shepard, America's first astronaut, on a historic suborbital flight in May 1961, marking a significant milestone in space exploration.

Building on the success of the Redstone, ABMA turned its attention to the Jupiter missile. This medium-range ballistic missile extended America's reach into space, providing the thrust necessary for heavier payloads and longer missions. The development of the Jupiter missile not only bolstered national defense capabilities but also solidified Huntsville's reputation as "Rocket City," a hub for cutting-edge rocket technology and innovation.

ABMA's establishment underscored the strategic importance of missile and rocket development during the Cold War. It exemplified America's commitment to advancing both national security and scientific progress through space exploration. The agency's achievements laid crucial groundwork for future endeavors, particularly the ambitious Apollo program that would eventually land humans on the Moon.

Meanwhile, in March 1956, the United States Air Force embarked on Project 7969, also known as the "Manned Ballistic Rocket Research System." This pioneering initiative aimed to achieve the recovery of a manned capsule from orbit, a critical step towards enabling human spaceflight. Project 7969 focused on developing recoverable satellites, essential for advancing the technology needed to safely sustain astronauts in space.

Central to Project 7969 was the development of life-support systems capable of sustaining human life in the vacuum of space. Engineers tackled challenges such as oxygen supply, temperature regulation, and radiation protection, pushing the boundaries of aerospace technology through rigorous testing and innovative solutions. The project's emphasis on ensuring astronaut safety during reentry into Earth's atmosphere led to the development of advanced heat shield technologies and aerodynamic designs.

The knowledge gained from Project 7969 directly informed the subsequent Mercury program, NASA's first manned spaceflight effort. Insights into capsule design, life-support systems, and reentry dynamics proved invaluable, shaping the spacecraft that would carry America's first astronauts into space. This integration between Project 7969 and Mercury underscored its pivotal role in advancing human space exploration capabilities.

Moreover, Project 7969 highlighted the United States' dedication to pushing the frontiers of space technology. It exemplified the collaborative efforts of military and civilian scientists and engineers, working in tandem to achieve breakthroughs that would pave the way for future missions. By addressing the formidable challenges of human spaceflight, Project 7969 set a precedent for developing safe and sustainable missions beyond Earth's atmosphere.

In summary, the activation of ABMA and the launch of Project 7969 in the mid-1950s marked pivotal moments in America's journey into space. These initiatives laid essential groundwork for developing critical missile and space technologies, advancing scientific exploration, and ultimately, setting the stage for historic achievements in human spaceflight.

On May 3, 1956, the United States Air Force unveiled plans for a substantial investment in the nation's missile production capabilities: a $41 million guided missile production facility in Sorrento, California, with Convair as the prime contractor. This facility was designated for developing and producing the Atlas launch vehicle, a pivotal asset that would become a cornerstone of the United States' burgeoning space exploration efforts.

The decision to build the facility in Sorrento represented a strategic expansion in the country's capacity to produce advanced missile and rocket technology. Convair, a division of General Dynamics, had already established itself as a leader in aerospace engineering, and the selection of Convair as the prime contractor underscored the importance of the Atlas program.

The Atlas launch vehicle was an intercontinental ballistic missile (ICBM) that was also adapted for space missions. Its design and development were crucial for both national defense and space exploration, highlighting the dual-use nature of missile technology during the Cold War era. The Atlas rocket's powerful engines and reliable performance would later be instrumental in launching satellites and human space missions, including the historic Mercury-Atlas missions that placed the first Americans in orbit.

Construction of the Sorrento facility marked a significant milestone in the United States' commitment to advancing its aerospace capabilities. The plant was envisioned as a state-of-the-art production center equipped with the latest technology and staffed by a skilled workforce dedicated to building the Atlas rockets. The facility's development was a massive undertaking involving extensive planning, engineering, and coordination.

The investment in the Sorrento facility also reflected the broader geopolitical context of the time. Amidst the intensifying space race with the Soviet Union, the United States recognized the need to accelerate its missile and rocket development programs. The new facility would enable the rapid production of Atlas launch vehicles, ensuring that the United States could keep pace with or surpass Soviet advancements in space technology.

The establishment of the guided missile production facility in Sorrento had far-reaching implications. It bolstered the local economy, creating jobs and stimulating technological innovation in the region. More importantly, it gave the United States the industrial capacity necessary to support an ambitious space exploration and missile development agenda.

The Atlas rocket would achieve numerous milestones in the following years, launching satellites, space probes, and astronauts. Its success demonstrated the effectiveness of the investments made in the Sorrento facility and underscored the strategic foresight of the Air Force and its partners.

In summary, the Air Force's announcement on May 3, 1956, to build a $41 million guided missile production facility in Sorrento, California, was a pivotal development in the history of American aerospace engineering. With Convair as the prime contractor, the facility played a crucial role in developing the Atlas launch vehicle, significantly enhancing the nation's missile and rocket production capabilities. This investment advanced the United States' defense posture and laid the groundwork for its historic achievements in space exploration.

On August 24, 1956, the National Advisory Committee for Aeronautics (NACA)'s Pilotless Aircraft Research Division achieved a historic milestone with the launch of the world's first five-stage, solid-fuel rocket test vehicle. This groundbreaking experiment reached an astonishing speed of Mach 15, advancing the understanding of hypersonic flight dynamics and providing invaluable data for developing future high-speed aircraft and spacecraft.

The success of this multi-stage rocket test was a significant leap in aerospace technology. It demonstrated the feasibility of using multiple stages in a rocket to achieve greater speeds and altitudes than ever before. Each stage of the rocket, powered by solid fuel, sequentially ignited and separated, propelling the vehicle to unprecedented velocities. This engineering marvel showcased the potential for reaching the hypersonic speed regime, defined as speeds greater than Mach 5.

The data gathered from this test were crucial for several reasons. Firstly, it provided insights into the behavior of materials and structures under extreme aerodynamic heating and stress conditions, which occur at hypersonic speeds. Engineers and scientists studied these effects to design more resilient and efficient airframes for future aircraft and spacecraft. The findings also contributed to understanding aerodynamic stability and control at high velocities, a critical factor for the safe operation of high-speed vehicles.

Moreover, this successful test highlighted the capabilities of solid-fuel rockets. Unlike liquid-fuel rockets, solid-fuel rockets are simpler and more reliable, with fewer moving parts and a greater ease of storage and handling. Demonstrating a five-stage solid-fuel rocket underscored the potential to develop more advanced and capable launch systems for military and civilian applications.

The implications of this test extended beyond immediate technological advancements. The success of the multi-stage solid-fuel rocket laid the groundwork for the future exploration of space. It provided a blueprint for designing rockets that could escape Earth's atmosphere and travel into space, paving the way for subsequent missions that would explore the outer reaches of our planet and beyond.

This achievement also contributed to the broader context of the Cold War era's technological race. As the United States and the Soviet Union competed for supremacy in space and missile technology, advancements like the Mach 15 rocket test underscored the United States' commitment to leading in aerospace innovation. The data and experience gained from this test would feed into various programs, including developing intercontinental ballistic missiles (ICBMs) and the nascent space program.

In conclusion, the launch of the world's first five-stage, solid-fuel rocket test vehicle by NACA's Pilotless Aircraft Research Division on August 24, 1956, marked a monumental step forward in aerospace engineering. Achieving a speed of Mach 15, the test provided crucial data on hypersonic flight dynamics informed the development of future high-speed aircraft and spacecraft and demonstrated the viability of multi-stage rockets. This success underscored the potential for reaching unprecedented speeds and altitudes, laying the foundation for future explorations into space and contributing to the technological advancements of the era.

In October of 1956, NACA scientists embarked on preliminary studies that would shape the future of hypersonic flight research.

These studies were initiated to explore the necessity for a successor to the X-15, a pioneering manned rocket research vehicle. The X-15 had already demonstrated remarkable capabilities in high-speed flight and had become a cornerstone of NACA's efforts to push the boundaries of aerospace technology.

The need for a follow-on vehicle to the X-15 stemmed from the desire to advance hypersonic flight research further. Hypersonic speeds, greater than Mach 5, presented unique challenges and opportunities for aerospace engineers and scientists. Research in this domain aimed to enhance understanding of aerodynamics, thermal management, materials science, and human factors at extreme velocities.

The preliminary studies conducted by NACA scientists in October 1956 laid the groundwork for future advancements in hypersonic technology. These studies assessed the technological feasibility, scientific objectives, and operational requirements for a next-generation manned rocket research vehicle. They identified key areas where improvements and innovations were needed to overcome the technical challenges posed by hypersonic flight.

The continued focus on hypersonic technology underscored the growing ambition and capability of the United States in advanced flight research. As the Cold War fueled competition between nations, achieving mastery in hypersonic flight was a scientific and engineering challenge and a strategic imperative. It represented a critical frontier in aerospace technology, with implications for both military superiority and civilian applications.

Moreover, the pursuit of a follow-on vehicle to the X-15 reflected NACA's commitment to maintaining leadership in aerospace research and development. The X-15 had already demonstrated its value as a platform for testing new technologies and pushing the limits of human and technological capability. A successor vehicle would build upon these achievements, leveraging lessons learned and incorporating new advancements in propulsion, materials, and flight control systems.

Ultimately, the preliminary studies initiated in October 1956 paved the way for the next generation of aerospace vehicles. They set in motion a series of developments that would culminate in the designing and constructing advanced hypersonic research platforms, contributing to significant advancements in aviation and space exploration.

In November of 1956, the Air Research and Development Command (ARDC) approached the National Advisory Committee for Aeronautics (NACA) with a proposal for collaboration on a new hypersonic research airplane project. This initiative began a cooperative effort between military and civilian sectors aimed at advancing aerospace technology to new heights. NACA promptly initiated feasibility studies in response, laying the groundwork for what would evolve into pioneering advancements in hypersonic and spaceflight technologies.

The partnership between ARDC and NACA underscored the synergy between military and civilian research efforts during the Cold War era. This collaboration was pivotal in driving forward the development of cutting-edge aerospace technologies, leveraging the expertise and resources of both sectors. It aimed to explore the frontiers of hypersonic flight and lay the groundwork for future space exploration capabilities.

Throughout 1956, NACA personnel concurrently engaged in studies to explore the feasibility of utilizing existing ballistic missile boosters for manned orbital spaceflight. This foundational research was crucial in shaping future space missions' early conceptual frameworks and technical requirements. By assessing the capabilities and limitations of missile technology for space exploration, NACA contributed critical insights that would inform the planning and execution of subsequent manned space missions.

The studies conducted by NACA personnel in 1956 provided valuable data on the performance characteristics of ballistic missile boosters when adapted for spaceflight. These insights were instrumental in evaluating propulsion systems, trajectory dynamics, and payload capacities necessary to achieve orbital velocities and sustain human presence in space. The research laid essential groundwork for developing launch vehicles and spacecraft, eventually leading to milestones such as the Mercury, Gemini, and Apollo programs.

Moreover, NACA's efforts in exploring the potential of ballistic missile boosters demonstrated foresight in repurposing existing military technology for peaceful scientific exploration. This dual-use approach accelerated progress in spaceflight capabilities and contributed to strategic advancements in national defense during the Cold War.

On January 14, 1957, the United States took a significant diplomatic step by proposing to the United Nations Assembly that outer space should be used exclusively for peaceful purposes. This proposal marked a seminal moment in developing international space policy, emphasizing the importance of global cooperation and peaceful exploration beyond Earth's atmosphere.

The U.S. initiative recognized that outer space represented a new frontier with immense potential for scientific discovery, technological advancement, and peaceful cooperation among nations. By advocating for the peaceful use of space, the proposal aimed to prevent the militarization of space and mitigate potential international tensions during the Cold War era.

The proposal highlighted the United States' commitment to promoting international norms and regulations governing space activities. It sought to establish a framework where space exploration could benefit all humanity, fostering collaboration in scientific research, satellite communication, and future space missions.

The proposal underscored the belief that space exploration should transcend geopolitical rivalries and contribute to humankind's collective progress. It laid the foundation for subsequent efforts to develop international agreements and treaties to ensure the peaceful and equitable use of outer space for all nations.

On June 11, a significant milestone in the development of rocket technology occurred at Cape Canaveral, Florida. The first launch attempt of the Atlas missile ended abruptly in an explosion shortly after takeoff. This incident underscored the formidable challenges of developing reliable launch vehicles, highlighting rocket technology's complexities and inherent risks.

The Atlas missile was critical to the United States' missile and space exploration programs. Designed as an intercontinental ballistic missile (ICBM) and later adapted for space missions, it was intended to deliver payloads into orbit and beyond. However, its maiden launch on June 11 exposed the intricate engineering required for a successful flight.

The explosion shortly after liftoff served as a stark reminder of the technical hurdles and uncertainties faced by aerospace engineers and scientists. Rocket propulsion, structural integrity, and aerodynamic stability were just a few of the myriad factors that had to align perfectly for a successful launch. The failure highlighted the need for rigorous testing, analysis, and iterative improvement in rocket design and manufacturing processes.

Moreover, the incident underscored the inherent risks associated with pushing the boundaries of aerospace technology. Rockets, by their nature, operate on the edge of what is physically possible, navigating extreme forces and environments during launch. The failure of the Atlas missile on June 11 underscored the dangers involved in exploring these frontiers and the dedication required to overcome setbacks in pursuit of scientific and technological progress.

In the aftermath of the explosion, engineers and scientists conducted thorough investigations to pinpoint the causes of the failure. Lessons learned from this setback would inform subsequent improvements in rocket design, testing protocols, and safety measures. The resilience and determination displayed in addressing the challenges highlighted by the failed launch would ultimately contribute to the eventual success of the Atlas program and the broader advancement of space exploration.

On June 20, significant advancements in aerospace research were made as two groups within the National Advisory Committee for Aeronautics (NACA) intensified their focus on high-speed, high-altitude aircraft performance and hypersonic flight and reentry issues. Their collaborative efforts marked a pivotal moment in advancing critical research essential for space exploration, deepening the understanding of the physical and engineering challenges associated with achieving and returning from space missions.

The first group concentrated on the performance of high-speed, high-altitude aircraft. This research was crucial for developing aircraft capable of reaching near space and conducting scientific observations, laying the groundwork for future spaceplanes and reconnaissance platforms. By exploring the aerodynamic, thermal, and propulsion challenges at extreme altitudes and velocities, NACA scientists and engineers contributed to the evolution of aircraft designs that could later inform spacecraft development.

Simultaneously, the second group delved into hypersonic flight and reentry complexities. This involved studying the dynamics of vehicles traveling at speeds greater than Mach 5 and reentering Earth's atmosphere from space. Understanding these dynamics was essential for designing spacecraft that could withstand the intense heat and forces encountered during reentry, ensuring the safe return of astronauts and scientific payloads.

The research conducted by these NACA groups was instrumental in addressing fundamental questions related to space exploration. They pioneered advances in materials science, thermal protection systems, and aerodynamic modeling that were essential for pushing the boundaries of human spaceflight. Insights gained from their work informed the development of spacecraft systems, propulsion technologies, and operational procedures that would later be integral to NASA's Mercury, Gemini, and Apollo missions.

Moreover, the collaborative efforts between these groups exemplified NACA's interdisciplinary approach to aerospace research. By integrating expertise from diverse fields such as aerodynamics, materials science, and propulsion, NACA was able to tackle complex challenges holistically, driving forward innovations that would shape the future of space exploration.

In July, the Langley Aeronautical Laboratory launched pioneering studies to harness solid-fuel upper stages to achieve payload orbit, laying the groundwork for the innovative Scout test-vehicle concept. These studies represented a critical step in exploring the viability and effectiveness of multi-stage rockets for launching satellites into orbit.

The initiative at Langley underscored a strategic shift towards leveraging solid-fuel technology in upper rocket stages. Solid-fuel propulsion offered distinct advantages in simplicity, reliability, and operational efficiency compared to liquid-fuel systems. By focusing on solid-fuel upper stages, researchers aimed to streamline launch operations, enhance payload capabilities, and

reduce costs associated with satellite deployment.

The development of the Scout test-vehicle concept was a direct outcome of these studies. The Scout program would later become synonymous with compact, cost-effective satellite launch capabilities, playing a pivotal role in the early years of space exploration. Its innovative use of solid-fuel upper stages set a precedent for future generations of launch vehicles, influencing design principles emphasizing efficiency and reliability.

Moreover, the studies initiated at Langley Aeronautical Laboratory exemplified NACA's commitment to advancing aerospace technology through systematic research and development. By exploring new propulsion strategies and operational concepts, researchers at Langley contributed to expanding the technological frontier of space exploration, laying the groundwork for future achievements in satellite communications, Earth observation, and scientific research from orbit.

During July-August, Alfred J. Eggers, Jr., working at the NACA Ames Aeronautical Laboratory, achieved a breakthrough by developing a semiballistic design for a manned reentry spacecraft. This innovative design represented a significant advancement in spacecraft technology, specifically focusing on ensuring the safe reentry of astronauts from space—an essential requirement for the future of human spaceflight.

Eggers' semiballistic spacecraft design incorporated critical principles to withstand the intense heat and forces encountered during atmospheric reentry. This included advances in thermal protection systems, aerodynamic stability, and structural integrity, essential for ensuring the safety and survival of astronauts returning from orbital missions.

On August 7, a pivotal demonstration of the ablative reentry principle occurred with the Jupiter-C test vehicle. This test successfully recovered a nose cone from a peak altitude exceeding 600 miles, validating the effectiveness of ablative materials in protecting spacecraft during reentry. President Eisenhower later showcased the recovered nose cone to the nation, emphasizing the significance of technological achievement in both national defense and the burgeoning field of space exploration.

The demonstration underscored the United States' capability to develop and deploy advanced aerospace technologies capable of operating at the frontiers of space. It highlighted the nation's commitment to pioneering efforts in reentry vehicle technology, which would prove instrumental in subsequent manned space missions and the development of space capsules like those used in the Mercury program.

On August 7, a pivotal event in aerospace history occurred when a Jupiter-C test vehicle successfully demonstrated the ablative reentry principle by recovering a nose cone from a peak altitude exceeding 600 miles. This achievement marked a significant advancement in space technology, showcasing the effectiveness of ablative materials in protecting spacecraft during the intense heat and friction of atmospheric reentry.

President Eisenhower recognized the importance of this technological milestone and personally displayed the recovered nose cone to the nation. His action underscored the United States' scientific and engineering prowess and the strategic significance of mastering reentry technology for national defense and space exploration ambitions.

The successful recovery of the nose cone from such a high altitude validated the United States' capability to develop and deploy advanced aerospace technologies. It bolstered confidence in the feasibility of future manned space missions, highlighting the progress made toward ensuring the safety and survival of astronauts returning from orbit.

On September 25, a setback occurred in the development of missile technology when the second Atlas launch vehicle was destroyed during a launch attempt at Cape Canaveral. This event underscored the persistent challenges inherent in refining and perfecting rocket technology.

The destruction of the Atlas launch vehicle highlighted the complexities involved in designing and testing rockets capable of reliably reaching orbit. Rocketry, by nature, demands precise engineering and flawless execution to navigate the extreme conditions of launch and ascent. Failures such as this one were integral to the iterative process of improving rocket designs, propulsion systems, and launch procedures.

Despite the setback, incidents like these gave engineers and scientists invaluable learning opportunities. Each failure yielded critical data and insights that contributed to refining subsequent iterations of the Atlas launch vehicle. These iterative improvements were essential for advancing missile capabilities and laying the groundwork for future successes in space exploration and satellite deployment.

Chapter 2 - The Space Race Begins

On October 4, the Soviet Union successfully launched Sputnik I, the first artificial Earth satellite. This historic achievement sparked widespread astonishment and urgency in the United States, catalyzing a swift escalation in American space research and exploration efforts.

Sputnik I's launch was not just a scientific milestone; it symbolized Soviet technological prowess and strategic capability. The satellite's successful orbit around the Earth sent shockwaves across the globe and triggered a heightened sense of competition between the superpowers, known as the space race. This geopolitical rivalry fueled a race to demonstrate superiority in space technology and exploration capabilities.

In response to Sputnik I, the United States mobilized its scientific and engineering communities, accelerating existing space programs and initiating new ones. The urgency to catch up and surpass Soviet achievements led to a surge in funding for aerospace research, educational reforms emphasizing science and technology, and a renewed national focus on space exploration.

The launch of Sputnik I fundamentally altered the trajectory of space exploration, pushing both nations to push the boundaries of scientific knowledge and technological innovation. It galvanized the United States to establish NASA (National Aeronautics and Space Administration) in 1958 and set ambitious goals for crewed space missions, culminating in the Apollo moon landing in 1969.

In the late 1950s, the world was gripped by the burgeoning Space Race, a fierce competition between the United States and the Soviet Union to achieve supremacy in space exploration. At the heart of America's efforts was Project Mercury, a groundbreaking program from 1958 to 1963. The primary goal of Project Mercury was ambitious: to send a man into Earth orbit and bring him back safely, ideally outpacing the Soviet Union. NASA, the newly established civilian space agency, assumed control of the project from the US Air Force, setting the stage for a series of historic milestones. During its tenure, Project Mercury conducted 20 uncrewed developmental flights, including missions with animals, and successfully completed six manned flights. Named after the Roman god Mercury, the project ultimately cost $2.68 billion when adjusted for inflation. The astronauts, famously known as the "Mercury Seven," each named their spacecraft with names ending in "7," symbolizing their unity and the spirit of exploration.

Early Days and Soviet Rivalry

The Space Race was officially ignited with the launch of the Soviet satellite Sputnik 1 in 1957, shocking the American public and revealing a perceived technological gap. The sight of the Soviet Union's satellite orbiting the Earth sparked a fervent national desire to catch up. In response, the US government expedited its space exploration efforts, placing them under NASA, a civilian agency designed to spearhead these ambitious missions. Following the successful launch of the Explorer 1 satellite in 1958, the next milestone became clear: manned spaceflight. This goal was set against the backdrop of Cold War tensions, where space exploration was not just about science but also about demonstrating national prowess.

The launch of the Soviet satellite Sputnik 1 on October 4, 1957, sparked the Space Race, a fierce Cold War competition between the United States and the Soviet Union. Shocked by the demonstration of Soviet technological superiority, the American public demanded a response. Initially, the USAF launched a project called Man in Space Soonest (MISS), but it faced significant challenges. President

Dwight D. Eisenhower's decision to create the National Aeronautics and Space Administration (NASA) in 1958 shifted the focus to a civilian-led space program.

On October 14, the American Rocket Society presented President Eisenhower with a proposal for establishing an Astronautical Research and Development Agency, akin to the National Advisory Committee for Aeronautics (NACA), but focused on space projects. This proposal underscored the increasing recognition of the need for structured and dedicated space research and development efforts.

The proposal reflected a growing consensus among aerospace experts and policymakers that the challenges and opportunities presented by space exploration required a specialized agency. Much like NACA had played a pivotal role in advancing aeronautical research and laying the groundwork for America's aviation industry, the proposed Astronautical Research and Development Agency aimed to spearhead efforts in space science, technology, and exploration.

By advocating for establishing such an agency, the American Rocket Society highlighted the strategic importance of coordinating and centralizing national efforts in space. This included fostering collaboration between government agencies, industry partners, and academic institutions to accelerate progress in spacecraft design, propulsion systems, orbital mechanics, and other critical areas of astronautics.

President Eisenhower's receipt of this proposal signaled his administration's recognition of the transformative potential of space exploration and the need for a unified approach to harnessing it. Ultimately, this proposal contributed to the establishment of NASA (National Aeronautics and Space Administration) in 1958, fulfilling the vision of a dedicated agency responsible for America's civilian space efforts.

From October 15 to 21, a pivotal "Round 3" conference occurred at the Ames Aeronautical Laboratory. This conference focused on the future direction beyond the X-15 program and laid the groundwork for the X-20 Dyna Soar project, a pioneering effort in manned satellite vehicle development.

Alfred J. Eggers, Jr., prominent at Ames, presented several visionary concepts during these discussions. His proposals marked a significant leap forward in satellite and space vehicle design, envisioning advanced capabilities for manned orbital missions. Eggers' contributions were instrumental in shaping the technological and strategic direction of the Dyna Soar project, emphasizing the integration of aerodynamic principles with spacecraft reentry and orbital maneuverability.

The discussions at the "Round 3" conference underscored the Ames Aeronautical Laboratory's role as a hub for innovative thinking and technical expertise in aerospace research. They reflected a concerted effort to expand the frontiers of space exploration, leveraging insights from previous programs like the X-15 to pioneer new concepts in manned spaceflight.

Ultimately, the conference's outcomes laid the foundation for the ambitious X-20 Dyna Soar project to develop a versatile manned spacecraft capable of orbital missions and reentry. Eggers' proposals contributed to advancing satellite vehicle technology and solidified America's commitment to advancing space capabilities during intense competition and rapid technological advancement in the space race.

On November 8, Secretary of Defense Neil McElroy instructed the Army to resume the Explorer Earth satellites project. This decision effectively revived the Orbiter project, initially excluded from the International Geophysical Year (IGY) satellite planning in 1955. The directive was a direct response to the Soviet Union's successful launch of Sputnik I, which sparked a sense of urgency in the United States to advance its own satellite technology.

McElroy's directive underscored the national imperative to accelerate American space exploration and satellite technology efforts. It reflected a strategic shift in prioritizing and funding initiatives to achieve scientific and technological milestones in space comparable to those of the Soviet Union.

The Explorer Earth satellites project aimed to deploy scientific satellites into Earth's orbit, enabling crucial research in various fields such as atmospheric science, space physics, and global environmental monitoring. By reinstating and supporting this project, the United States aimed not only to catch up with Soviet achievements but also to establish leadership in space science and technology.

On November 12-13, during a meeting of the NACA Subcommittee on Fluid Mechanics, a crucial emphasis was placed on the pivotal role of fluid mechanics in advancing space flight technologies. The subcommittee highlighted that many critical aspects of space exploration, particularly spacecraft reentry and high-speed flights, depended heavily on advancements in understanding gas flows.

The discussions underscored the necessity of intensifying research efforts in gas dynamics relevant to high-speed flights and spacecraft reentry. These efforts were essential for developing efficient and reliable reentry vehicles that could safely return from space missions. Fluid mechanics research was pivotal in optimizing the design and performance of spacecraft, ensuring they could withstand the extreme conditions encountered during atmospheric reentry.

The recommendations put forth by the subcommittee reflected NACA's commitment to leveraging scientific research and engineering expertise to overcome the challenges posed by space exploration. Researchers aimed to enhance future space missions' safety, efficiency, and capability by deepening the understanding of fluid dynamics in high-speed and high-temperature environments.

On November 19, Preston R. Bassett, a member of the NACA Committee on Aerodynamics, passionately advocated for NACA to adopt an ambitious space research technology program. His plea underscored the pressing need to push the boundaries of current aerospace capabilities in response to rapid international developments.

Bassett's call to action resonated with the heightened sense of urgency in the United States following significant advancements in Soviet space achievements, particularly the launch of Sputnik I. His proposal aimed to galvanize NACA into spearheading innovative research initiatives that could accelerate America's progress in space exploration and technology.

By urging NACA to embrace an aggressive space research technology program, Bassett emphasized the importance of proactive investment in cutting-edge aerospace technologies. This included advancing aerodynamics, propulsion systems, materials science, and spacecraft design— essential for achieving manned spaceflight and satellite technology breakthroughs.

His recommendation aligned with broader national efforts to bolster America's position in the space race, emphasizing the strategic importance of technological leadership and scientific innovation. By embracing Bassett's vision, NACA would play a pivotal role in laying the groundwork for future successes, including the establishment of NASA and landmark achievements such as the Apollo moon missions.

In summary, Preston R. Bassett's advocacy on November 19 reflected a pivotal moment in shaping America's approach to space exploration. It highlighted the imperative to adopt a forward-thinking space research agenda, setting the stage for transformative advancements that would define America's leadership in space technology for decades to come.

On November 21, the National Advisory Committee for Aeronautics (NACA) took a decisive step towards advancing space exploration by establishing a Special Committee on Space Technology. This committee, chaired by Dr. H. Guyford Stever of MIT, was tasked with identifying and addressing critical challenges necessary to realize practical space flight capabilities.

The formation of the Special Committee on Space Technology reflected NACA's proactive approach in response to the escalating demands of the space race, particularly in light of recent Soviet successes like the launch of Sputnik I. Under Dr. Stever's leadership, the committee aimed to leverage scientific expertise and technological innovation to propel American capabilities in space.

Simultaneously, the Rocket and Satellite Research Panel recommended the creation of a National Space Establishment under civilian leadership, advocating for a significant annual budget of $1 billion over ten years. This recommendation signaled a monumental strategic commitment by the United States to prioritize and fund the development of space exploration infrastructure and capabilities.

The proposed National Space Establishment would consolidate efforts across government agencies, industry partners, and research institutions, fostering collaboration and synergy in advancing space technology. This ambitious initiative aimed to accelerate progress in rocketry, satellite development, astronautics, and related fields, positioning the United States as a global leader in space exploration.

On November 21-22, a significant portion of the NACA Propulsion Conference was dedicated to exploring potential space propulsion systems, highlighting discussions on chemical, nuclear, and nuclear-electric rockets. The conference attendees deliberated on the capabilities and feasibility of each propulsion type for various space missions, focusing extensively on their potential applications, advantages, and challenges.

Among the discussed propulsion systems, chemical rockets emerged as the primary focus for near-term lunar missions and other initial space exploration endeavors. Experts agreed that chemical propulsion systems could support a round trip to the Moon and were deemed suitable for guiding future research priorities in space propulsion technology.

This conclusion underscored the pragmatic approach adopted by NACA in prioritizing achievable milestones in space exploration, emphasizing the reliability and maturity of existing chemical rocket technology for near-Earth and lunar missions. While nuclear and nuclear-electric propulsion systems held promise for longer-duration missions and interplanetary travel, their development and implementation were recognized as more complex and require further technological advancements.

By dedicating significant conference time to propulsion system discussions, NACA reaffirmed its commitment to advancing propulsion technologies essential for expanding human presence in space. The insights gained from these discussions informed strategic planning and research investment decisions, aligning with broader national goals to enhance America's space exploration capabilities during intense space race competition.

In November, Maxime A. Faget introduced a groundbreaking concept for achieving manned orbital flight using available ballistic missile technology. His proposal featured several innovative elements: utilizing existing ballistic missiles as launch vehicles, incorporating solid-fuel retrorockets to initiate reentry, and employing a non-lifting ballistic capsule shape for safe descent through Earth's atmosphere.

Faget's concept was strategically designed to expedite the development of manned spaceflight capabilities by leveraging existing resources and proven technologies. This approach was practical and considered the quickest and safest method for initiating human missions into space during the early stages of space exploration.

The concept profoundly influenced the design and development of the Mercury spacecraft program, which became NASA's first manned spaceflight program. By adopting Faget's recommendations, engineers and scientists streamlined efforts to create a spacecraft capable of carrying astronauts into Earth orbit and safely returning them to the planet's surface.

Faget's emphasis on simplicity, reliability, and safety in spacecraft design set a foundational precedent for subsequent space missions. His innovative approach demonstrated the feasibility of manned orbital flight using ballistic missile platforms, paving the way for America's entry into manned space exploration and contributing significantly to advancements in human spaceflight technology.

On December 4, the American Rocket Society publicly announced their proposal for an Astronautical Research and Development Agency, initially presented to President Eisenhower in October. This announcement marked a significant moment, emphasizing the growing consensus among scientists and aerospace experts regarding the necessity for a dedicated organization to oversee and coordinate space research and development efforts in the United States.

The proposal highlighted the complexities and challenges inherent in space exploration, particularly in the context of the burgeoning space race with the Soviet Union. It underscored the need for a centralized agency that could strategically plan, fund, and execute ambitious space missions while fostering collaboration between government agencies, industry partners, and academic institutions.

The American Rocket Society aimed to streamline and accelerate America's efforts in advancing space technology and exploration capabilities by advocating for an Astronautical Research and Development Agency. The proposed agency would play a pivotal role in harnessing national resources and expertise to maintain technological leadership in space, ensuring the United States remained at the forefront of global space endeavors.

The public announcement of this proposal reflected a pivotal moment in shaping American space policy, laying the groundwork for the eventual establishment of NASA (National Aeronautics and Space Administration) in 1958. This new agency would embody the vision outlined by the American Rocket Society, solidifying America's commitment to exploring space and pushing the boundaries of scientific and technological achievement in the decades to come.

On December 5, the Department of Defense announced a significant organizational shift in space research management: the creation of Advanced Research Projects Agency (ARPA). This announcement signaled the Department's intent to establish ARPA as a central agency dedicated to advancing defense-related space technologies and capabilities.

ARPA's mandate encompassed directing and coordinating cutting-edge research and development initiatives to enhance national security through space-based innovations. The agency was tasked with spearheading high-risk, high-reward projects that pushed the boundaries of scientific and technological feasibility, particularly in areas critical to defense and strategic interests.

The establishment of ARPA represented a strategic response to the evolving geopolitical landscape, especially in light of technological advancements demonstrated by global competitors. The Department of Defense aimed to foster rapid innovation and maintain technological superiority in space-based capabilities by consolidating and prioritizing defense-related space projects under a single agency.

This organizational shift underscored the growing recognition of space as a crucial national security and defense domain, highlighting the importance of integrated, forward-thinking approaches to space research and development. ARPA's creation laid the groundwork for future advancements in space technology, ultimately contributing to America's leadership in space exploration and defense-related initiatives.

On December 6, the first attempt of the International Geophysical Year (IGY) Vanguard launch, known as TV-3, ended in failure when the test satellite failed to reach orbit. This setback underscored the formidable technical challenges associated with satellite launches, particularly in the nascent stages of space exploration.

The failure of TV-3 highlighted the complexities involved in developing reliable launch systems capable of placing payloads into orbit. It underscored the critical need for rigorous testing, refinement, and iterative improvements in launch vehicle technology. The lessons learned from this initial failure prompted a reassessment of engineering approaches and operational procedures, guiding subsequent efforts to overcome technical hurdles in future launches.

Despite the disappointment of TV-3, the experience provided valuable insights that contributed to the evolution of space launch capabilities. It reinforced the determination of scientists, engineers, and aerospace professionals to persist in their quest for reliable and efficient methods of reaching space.

Ultimately, the failure of the TV-3 Vanguard launch underscored the resilience and determination of the space exploration community to learn from setbacks, refine technologies, and continue pushing the boundaries of what was achievable in space exploration.

On December 10, the Air Force took a significant step by establishing a Directorate of Astronautics to oversee its astronautical research programs, marking a proactive move in response to the escalating space race. However, this decision was short-lived as Secretary of the Air Force James H. Douglas rescinded the order later that month. He cited concerns that creating such a group prematurely, before the activation of the Advanced Research Projects Agency (ARPA), could lead to overlapping responsibilities and organizational inefficiencies.

Douglas' decision highlighted the dynamic and rapidly evolving nature of organizational structures within the U.S. military and government as they responded to the challenges and opportunities presented by space exploration. It underscored the complexities of coordinating and consolidating efforts across different agencies and departments to harness national resources and expertise in advancing space technologies effectively.

The Directorate of Astronautics order rescission reflected a strategic reassessment to ensure that future organizational frameworks would be well-aligned with broader national objectives and integrated with emerging initiatives like ARPA. This adaptive approach was essential in navigating the complexities of the space race and optimizing efforts to achieve technological milestones in space exploration.

Overall, the episode exemplified the government's commitment to refining its organizational strategies and structures to effectively meet the demands of a rapidly advancing space age, ultimately contributing to America's efforts to maintain leadership in space exploration and technology development.

On January 4, 1958, the atmosphere was charged with anticipation as members of the American Rocket Society and the Rocket and Satellite Research Panel convened. Their discussions resonated with a palpable sense of urgency and a shared vision of establishing a National Space Establishment. They passionately argued that this proposed body should operate independently of the Department of Defense, marking a pivotal shift towards civilian leadership in space exploration.

Amidst murmurs of agreement, the participants envisioned a future where space exploration would transcend military priorities, emphasizing peaceful and scientific endeavors. Their collective voices advocated for a civilian-led space program prioritizing exploration, discovery, and international collaboration over military applications.

The meeting underscored a growing consensus among scientists, engineers, and policymakers on formalizing and prioritizing space research and development within a civilian framework. It reflected a broader aspiration to harness space as a domain for peaceful cooperation and exploration, resonating with the aspirations of the International Geophysical Year and the evolving global perspectives on space.

Ultimately, the discussions on January 4, 1958, set the stage for transformative initiatives that would shape the future of American space policy, culminating in the establishment of NASA later that year. This pivotal moment marked a decisive step towards realizing the dream of a dedicated civilian space agency tasked with advancing humanity's understanding and capabilities in exploring outer space.

On January 10, under the clear Florida sky at Cape Canaveral, the fourth Atlas rocket soared into the heavens, marking a pivotal moment in America's nascent space program. With engineers and scientists watching anxiously, the powerful vehicle achieved a successful limited flight, showcasing significant progress and potential for future missions.

The successful launch of the Atlas rocket underscored its emerging role as a reliable and capable launch vehicle. It represented a crucial advancement in American efforts to explore and utilize space for scientific exploration and technological development. The achievement was a testament to the dedication and expertise of those involved in its development and launch, reaffirming confidence in America's ability to compete in the global space race.

The event on January 10 marked more than just a technical milestone; it symbolized a leap forward in the capabilities of American aerospace technology. It paved the way for ambitious future missions and cemented the Atlas rocket's reputation as a cornerstone of the nation's space exploration endeavors.

On January 12, President Eisenhower's demeanor in the Oval Office was steadfast and resolute as he addressed Soviet Premier Nikolai A. Bulganin. In his proposal advocating for the peaceful use of outer space, Eisenhower drew a parallel to his earlier efforts to channel atomic energy for constructive purposes. Despite the seriousness of his words, tension lingered in the room, tempered by the knowledge that Soviet authorities had already rebuffed similar overtures.

Eisenhower's stance underscored the United States' unwavering commitment to leveraging space for peaceful and scientific ends. His proposal aimed to establish a framework where space exploration and technology would serve as catalysts for international cooperation rather than sources of conflict. This vision aligned with broader efforts during the Cold War era to mitigate tensions through shared scientific endeavors and diplomatic dialogue.

On January 15, the Air Force was excited as it received eleven unsolicited industry proposals for Project 7969. In the background, observers from NACA, eager to contribute significantly, meticulously evaluated each submission. The atmosphere crackled with potential and competition, mirroring the intense enthusiasm and dedication to advancing manned spaceflight technology.

The influx of proposals underscored the growing interest and ambition within both military and civilian sectors to pioneer new frontiers in space exploration. Each submission represented a unique approach and vision for achieving the goals set forth by Project 7969, reflecting diverse perspectives on the technological challenges of manned spaceflight.

For NACA, the opportunity to assess these industry proposals signaled a pivotal moment in its involvement in astronautical research. As observers scrutinized each submission, they sought innovations to enhance spacecraft design, propulsion systems, and life-support technologies crucial for human space missions.

On January 16, the NACA assembly room was filled with determined faces as a resolution was adopted, emphasizing their responsibility in space technology research. Paul E. Purser and Maxime A. Faget's innovative solid-fuel launch vehicle design, later named Little Joe, was enthusiastically discussed. The Little Joe would be pivotal in testing full-scale spacecraft under extreme conditions, ensuring the safety and success of future manned missions. This marked an important step in developing robust testing protocols for spacecraft.

From January 29-31, Wright-Patterson Air Force Base in Ohio buzzed with activity during a pivotal conference. Inside, amidst the cold winter air contrasting with the heated discussions, concepts for manned orbital vehicles unfolded with ambitious fervor. Each proposal pushed the envelope further than the last, aiming to define the future of space exploration.

Maxime A. Faget's high-drag capsule design captured particular attention during the conference. This innovative concept promised a safe and controlled reentry, addressing a critical challenge in human spaceflight. With its emphasis on ensuring the safety of astronauts returning from orbit, Faget's design sparked enthusiasm and debate among aerospace experts and military officials alike.

The conference also saw presentations from major aerospace companies such as Northrop, Martin, and McDonnell, each unveiling their visionary concepts. Ideas ranged from boost-glide vehicles capable of extended orbital missions to elaborate space station designs envisioned as scientific research and exploration platforms.

On January 31, 1958, the air at Cape Canaveral was electric with anticipation. Engineers and scientists held their breath as the Jupiter-C missile stood poised on the launch pad, a testament to months of tireless effort and unyielding determination. As the countdown reached its final moments, the tension was palpable. Then, with a roar, the missile sprang to life, propelling Explorer I, America's first artificial satellite, into the heavens. The ground trembled, and a trail of fire and smoke marked its ascent into the night sky.

As the rocket climbed, the crowd below erupted into applause and cheers, their faces illuminated by the fiery trail. In the control room, technicians and engineers monitored their instruments with rapt attention, their hearts racing. Minutes stretched into an eternity until the crucial moment came—confirmation that Explorer I had successfully entered orbit. The room burst into jubilant celebration, a milestone that heralded America's entry into the space age.

While the jubilation at Cape Canaveral signaled a new chapter in space exploration, significant developments were unfolding at the Pentagon. Lieutenant General Donald Putt, a key figure in the Air Force's missile program, recognized the necessity of bridging the gap between military and civilian efforts in space research. In a historic move, he extended an olive branch to the National Advisory Committee for Aeronautics (NACA), inviting them to collaborate on the Air Force's manned ballistic rocket program.

This invitation was more than a gesture of goodwill; it marked a pivotal step towards unifying the nation's space endeavors. At the time, the United States' space efforts were fragmented, with various branches of the military and civilian organizations pursuing their own programs, often in isolation. Putt's initiative aimed to foster a spirit of cooperation and shared purpose, ensuring the wealth of expertise across different sectors could be harnessed more effectively.

The partnership between NACA and the Air Force symbolized a turning point. It promised to blend the cutting-edge technological advancements of the military with the innovative research and development expertise of civilian scientists. This collaboration was crucial for the ambitious goals of the nation's space program, including the development of manned spaceflight capabilities and the long-term vision of exploring beyond Earth's orbit.

In the following months, this newfound alliance began to take shape. Joint committees and working groups were established, bringing together the brightest minds from the military and civilian spheres. Meetings were held to align their objectives, share knowledge, and coordinate efforts. Laboratories buzzed with activity as researchers and engineers pooled their resources, working side by side to overcome the myriad technical challenges ahead.

This era of collaboration laid the groundwork for future achievements in space exploration. It set the stage for the creation of NASA, which would inherit the legacy of NACA and become the driving force behind America's space missions. The cooperative spirit fostered by Putt's initiative would prove invaluable as the nation faced the daunting challenges of space exploration, from landing on the Moon to launching interplanetary probes.

The Project Orbiter efforts culminated on January 31, 1958, when Explorer I was successfully launched into orbit. This historic event marked the United States' entry into the space race, following the earlier Soviet achievements with Sputnik. The successful deployment of Explorer I provided a wealth of scientific data, including the discovery of the Van Allen radiation belts, which was a significant milestone in our understanding of Earth's magnetosphere.

Project Orbiter's success and Explorer I's launch were testaments to the effectiveness of cooperative efforts between military and civilian sectors. It demonstrated how leveraging diverse expertise could overcome complex challenges and achieve groundbreaking results. This collaborative model would become a cornerstone of future space exploration endeavors, driving the United States' space program forward and fostering innovations that would shape history.

On February 6, 1958, the Senate chamber was abuzz with anticipation as lawmakers gathered to vote on a resolution that would profoundly impact the future of American space exploration. The resolution proposed the creation of a special Committee on Space and Astronautics, a dedicated legislative body tasked with overseeing the nation's burgeoning space activities.

As the senators cast their votes, the moment's gravity was palpable. This was more than a routine legislative.

In practical terms, this meant an intensified focus on several key areas:

The Committee on Space and Astronautics creation was met with widespread approval and enthusiasm. It represented a formal recognition of the importance of space exploration to national security, scientific advancement, and international prestige. The committee guided the legislative framework necessary to support America's ambitions in space, from funding allocations to regulatory oversight.

The committee members were appointed in the days following the Senate's decision, each bringing their unique expertise and vision to the table. The committee's mandate included evaluating the nation's current space capabilities, identifying areas for improvement, and recommending policies to support sustained progress in space exploration.

The committee's establishment also underscored the need for collaboration across various government agencies, academic institutions, and private sector entities involved in space research and development. It aimed to streamline efforts, eliminate redundancies, and ensure that resources were used efficiently to achieve the nation's space goals.

This legislative move clearly acknowledged that space exploration had become a critical arena of geopolitical competition and scientific inquiry. It set the stage for subsequent landmark decisions, including the establishment of NASA later that year, which would become the central agency for the United States' civilian space program.

The Senate's resolution and the formation of the Committee on Space and Astronautics were crucial steps in laying the groundwork for America's future achievements in space. They ensured that the nation's space activities would be conducted under structured and focused oversight, enabling the United States to pursue its ambitious goals with clarity and purpose.

On February 7, 1958, the United States took another decisive step in solidifying its space exploration and defense commitment. The Secretary of Defense issued a directive establishing the Advanced Research Projects Agency (ARPA). This new agency was designed to serve as a centralized entity overseeing outer space and antimissile projects, bringing coherence and direction to the nation's fragmented research efforts.

The formation of ARPA was seen as a critical yet temporary measure. It was intended to act as a bridge to a more permanent and comprehensive National Space Agency. The urgency of the Cold War context, marked by the Soviet Union's recent successes in space, necessitated swift and effective coordination of America's space and missile defense initiatives.

ARPA's mandate was clear: to streamline and accelerate the United States' space research and development efforts. The agency was tasked with cutting through bureaucratic red tape, fostering innovation, and ensuring that the country stayed ahead in the rapidly evolving arena of space technology and missile defense. This directive came when the need for a cohesive strategy in these domains was more pressing than ever.

In the weeks following the directive, ARPA quickly began to take shape. The agency attracted some of the brightest minds from various fields, including physics, engineering, and computer science. Laboratories and research centers across the country were mobilized, their projects now falling under the purview of ARPA's coordinated efforts. The goal was to harness the nation's collective expertise and resources to achieve significant technological advancements in a short period.

The establishment of ARPA marked a significant shift in the space and defense research approach. It signaled a move toward a more integrated and agile research environment where innovations could be rapidly developed and deployed. This approach was intended to provide the United States with a competitive edge, ensuring that it could respond swiftly to any advancements made by its adversaries.

While ARPA was designed as a temporary entity, its impact was immediate and profound. It laid the groundwork for the eventual creation of NASA later that year, which would inherit and expand upon ARPA's mission. The streamlined processes and accelerated pace of innovation established by ARPA became a model for future research and development initiatives within the United States government.

On February 10, 1958, the National Advisory Committee for Aeronautics (NACA) published a landmark study that underscored the urgent need for a rapid buildup in the United States' space technology capabilities. This comprehensive report laid out an ambitious expansion plan designed to catapult the nation to the forefront of space exploration.

The meticulous and far-reaching study addressed every facet of the burgeoning space program. It proposed an annual budget increase of $100 million, a significant sum aimed at accelerating research and development across multiple critical areas. Recognizing that financial investment alone was insufficient, the report also called for a substantial boost in personnel. The expansion would bring in a wave of scientists, engineers, and technical experts dedicated to advancing America's space endeavors.

The comprehensive program outlined in the study covered a wide range of technological and scientific domains. Propulsion systems were a primary focus, with goals set to develop more powerful and efficient rockets capable of carrying heavier payloads into orbit and beyond. The study emphasized the importance of advancements in materials science, which were crucial for building spacecraft that could withstand the harsh conditions of space travel and reentry into Earth's atmosphere.

Reentry techniques themselves were another critical area of the study. As the United States aimed to send humans into space, ensuring their safe return was paramount. The report detailed the need for rigorous research into heat shields and other technologies to protect astronauts during reentry's intense heat and friction.

Beyond these technical aspects, the study highlighted the necessity of expanding infrastructure. This included building state-of-the-art research facilities, launching sites, and testing grounds. These facilities would provide the environments for developing and testing new technologies, ensuring they met the rigorous standards required for space missions.

NACA's report did not merely outline technical requirements; it also articulated a vision of national leadership in space exploration. It positioned the United States as a competitor in the space race and a potential leader capable of pioneering new frontiers and setting global standards. The study's ambitious scope and detailed planning aimed to ensure that the United States could achieve significant milestones, from launching satellites to eventually sending humans to the Moon and beyond.

The publication of this study marked a turning point in the nation's approach to space exploration. It provided a clear, strategic roadmap for the rapid buildup of space technology capabilities, one that would guide the efforts of military and civilian agencies in the years to come. The recommendations within the report laid the groundwork for future successes and were instrumental in shaping the policies and priorities of the newly formed NASA later that year.

In sum, NACA's February 10 study was a bold declaration of intent. It called for significant investment and expansion, aiming to establish the United States as a dominant force in space exploration. The report's vision and recommendations would drive the nation's space program forward, inspiring innovations and achievements that captured the world's imagination.

On February 13, 1958, the inaugural meeting of the Special Committee on Space

Technology marked a pivotal moment in the United States' journey toward mastering space exploration. This gathering set the stage for a structured and systematic approach to addressing the complex challenges of space flight.

The meeting convened in a room filled with some of the brightest minds in aerospace, engineering, and science. A sense of purpose and urgency filled the air as the committee members settled into their seats. The recent successes of the Soviet space program and the launch of Explorer I had underscored the importance of coordinated and focused efforts in space research.

During this inaugural session, seven working groups were established, each tasked with tackling specific aspects of space research. These groups were meticulously organized to cover space exploration's broad and varied challenges, ensuring no critical area was overlooked.

The first working group focused on vehicular programs. This team was responsible for developing and refining spacecraft and rockets to create reliable vehicles capable of reaching and operating in space. Their work included advancements in propulsion systems, structural integrity, and fuel efficiency.

The second group concentrated on propulsion and power systems, vital for the spacecraft's sustained and controlled travel. They explored new propulsion technologies, including liquid and solid fuels, and innovative power sources that could support longer missions.

The third working group was dedicated to materials and structures. They investigated the properties of various materials that could withstand the extreme conditions of space, including temperature fluctuations, radiation, and the vacuum of space. Their findings would be crucial for building durable spacecraft and protective gear for astronauts.

The fourth group delved into reentry and recovery techniques. Their primary concern was ensuring that spacecraft and their occupants could safely return to Earth. This involved developing heat shields, parachute systems, and splashdown procedures to protect both equipment and human life during the intense reentry process.

The fifth working group focused on human factors and training. Understanding that space travel posed unique challenges to the human body and mind, this team worked on preparing astronauts for the physical and psychological demands of space missions. They developed rigorous training programs and studied the effects of microgravity on the human body.

The sixth group tackled life support and environmental systems. They were tasked with creating sustainable life support systems that could provide astronauts with air, water, and food in the harsh space environment. Their innovations would ensure that humans could survive and function effectively on long-duration missions.

The final working group was responsible for data and communications. Recognizing the importance of reliable communication between space missions and ground control, this team developed advanced telemetry systems, data handling techniques, and secure communication protocols to maintain contact with spacecraft in orbit.

Establishing these working groups marked the beginning of a coordinated and comprehensive effort to address the multifaceted challenges of space exploration. Each group operated with a clear mandate and a shared vision of advancing America's capabilities in space. The collaborative efforts of these groups laid the foundation for significant technological advancements and paved the way for the United States to achieve its ambitious space exploration goals.

The inaugural meeting of the Special Committee on Space Technology was more than just an organizational milestone; it was a declaration of intent. It signaled the start of a concerted effort to push the boundaries of human knowledge and capability, setting the United States on a path to become a leader in exploring the final frontier.

On February 14, 1958, the International Geophysical Year (IGY) Committee released a report further to galvanize the United States' commitment to space exploration. The IGY, an international scientific project that lasted from July 1, 1957, to December 31, 1958, had already achieved remarkable success, notably in earth sciences and space research. However, this new report emphasized a crucial insight: the long-term necessity of continued space research.

The IGY Committee's report highlighted that the benefits of space research extended well beyond the IGY itself. It detailed the potential for ongoing scientific discoveries and technological advancements arising from a sustained commitment to exploring and understanding outer space. The report predicted that space research would remain a vital study area, crucial for national interests and global scientific progress.

This forward-looking perspective underscored several key areas where space research was expected to yield significant benefits. It highlighted the importance of satellite technology for global communications, weather forecasting, and environmental monitoring. The report also emphasized the potential for breakthroughs in materials science, propulsion technology, and life sciences, all driven by the unique challenges posed by space exploration.

Furthermore, the IGY Committee's findings reinforced that space research could serve as a catalyst for educational and inspirational purposes. Space exploration could inspire new generations of scientists, engineers, and innovators by pushing the boundaries of what was possible. The report argued that maintaining a robust space program would advance technology and stimulate interest in STEM (science, technology, engineering, and mathematics)

fields, fostering a culture of innovation and discovery.

The IGY Committee's report profoundly impacted policymakers and the scientific community. It provided a compelling argument for why the United States needed to sustain and expand its space exploration efforts. The recognition of space research as a long-term necessity helped to secure broader support for initiatives like the establishment of NASA, which would soon become the central agency for coordinating and advancing the nation's space activities.

In the wake of the report, there was a renewed sense of urgency and purpose among those involved in space research and policy. The insights the IGY Committee provided helped shape the strategic direction of the United States' space program, ensuring that it was built on a foundation of continuous innovation and scientific inquiry.

The IGY Committee's commitment to sustained space exploration laid the groundwork for the many following achievements. It paved the way for historic milestones such as the Apollo moon landings, the development of the Space Shuttle, and the exploration of distant planets and moons. The report's predictions about the enduring importance of space research proved remarkably prescient, as the benefits of space exploration continue to resonate across numerous fields of science and technology today.

In essence, the IGY Committee's February 14, 1958 report was more than a call to action; it was a visionary document recognizing the limitless potential of space research. It solidified the nation's commitment to exploring the final frontier, ensuring that the pursuit of knowledge and innovation would remain at the forefront of America's scientific endeavors for decades.

On February 20, 1958, the evolution of America's aerospace ambitions was underscored by a symbolic yet significant change: the NACA Committee on Aerodynamics was renamed the Committee on Aircraft, Missile, and Spacecraft Aerodynamics. This renaming was more than a mere administrative adjustment; it marked a pivotal broadening of the committee's scope to formally include spacecraft and missile research, reflecting the intertwined futures of aviation and space travel.

The decision to rename the committee was driven by the rapid advancements and increasing overlap in aeronautics and astronautics. As the boundaries between atmospheric flight and space exploration blurred, it became clear that a comprehensive approach was essential to address the multifaceted challenges of modern aerospace engineering.

The newly named Committee on Aircraft, Missile, and Spacecraft Aerodynamics was tasked with an expanded mandate mirrored the field's growing complexity. Integrating spacecraft and missile aerodynamics into the committee's purview meant that experts would now tackle a wider array of technical challenges, from the intricacies of high-speed flight within Earth's atmosphere to the demands of reentry and the dynamics of space travel.

Propulsion Systems: Researchers worked on developing advanced propulsion technologies capable of supporting both high-altitude aircraft and space vehicles. This included refining rocket engines and exploring hybrid systems that could operate efficiently in both atmospheric and space environments.

Materials Science: The committee prioritized the development of materials that could withstand extreme conditions, such as the intense heat and pressure experienced during reentry. This research was crucial for the construction of both spacecraft and high-speed missiles.

Aerodynamic Testing: Wind tunnel tests and computational fluid dynamics simulations became even more critical as they were used to study the aerodynamic properties of various craft operating at different altitudes and speeds. The committee ensured that these tests covered the full spectrum of aerospace vehicles, from conventional aircraft to experimental spacecraft.

Reentry Techniques: The safe reentry of spacecraft into Earth's atmosphere posed significant technical challenges. The committee's expanded focus included developing robust reentry vehicles and heat shield technologies to protect crew and uncrewed missions.

Systems Integration: As aerospace projects became more complex, integrating various subsystems—such as guidance, navigation, control, and life support—became essential. The committee ensured these systems could function seamlessly in aircraft and spacecraft.

The renaming of the committee also reflected a broader cultural and strategic shift within NACA, soon to transition into NASA. It signified a recognition that the future of aviation could not be disentangled from the emerging field of space exploration. This holistic approach would prove crucial as the United States aimed to establish itself as a military and civilian aerospace technologies leader.

As the Committee on Aircraft, Missile, and Spacecraft Aerodynamics embarked on its expanded mission, it played a key role in numerous groundbreaking projects. It contributed to developing early space vehicles, including those used in the Mercury and Gemini programs. It laid the technical foundation for the Apollo missions that would eventually land humans on the Moon.

The committee's renaming was a declaration of intent, highlighting the interconnectedness of aircraft, missile, and spacecraft research. It recognized the comprehensive approach needed to tackle the challenges of modern aerospace engineering and underscored the United States' commitment to advancing its capabilities in the air and beyond. This strategic shift ensured the nation was well-prepared to lead in the new space exploration and technology era.

On February 27, 1958, the NACA High-Speed Flight Station was a hive of activity. The sound of papers shuffling and the murmur of discussions filled the room as Walter C. Williams, the station chief, addressed his colleagues with a tone of cautious concern. "Experience with the X-15 design indicates that many of the weight figures we've used for the drag or lift configurations of the reentry vehicle are too low," he stated. This revelation emphasized the need for more realistic weight assessments based on their work with the X-15. The revised weights were projected to be 2,300 pounds for the drag configuration and 2,500 pounds for the lifting configuration, marking a significant adjustment in their planning. The atmosphere was one of earnest determination as the team realized the critical nature of these adjustments for the success of future missions.

As these milestones unfolded, the groundwork for Project Mercury was laid meticulously. The collective efforts of engineers, scientists, and policymakers coalesced into a unified vision. From bustling conference rooms and laboratories to the tense moments of rocket launches, the period from January to October 1958 was marked by dynamic progress and unwavering determination. The air was thick with the scent of rocket fuel and the hum of innovation, heralding the dawn of America's journey into manned spaceflight.

Fervent discussions and strategic planning sessions occurred in conference rooms nationwide. Engineers sketched out blueprints and debated design specifications while scientists presented their latest research findings. Policymakers worked tirelessly to secure funding and navigate the bureaucratic challenges inherent in such an ambitious national endeavor. Each meeting was a step forward, bringing clarity and direction to the multifaceted challenges of sending a human into space.

Laboratories buzzed with activity, filled with the whirr of machinery and the clatter of tools. Technicians and researchers conducted experiments, testing materials and components to ensure they could withstand the harsh conditions of space. The development of life support systems, reentry techniques, and propulsion technologies was a priority, each breakthrough bringing Project Mercury closer to reality.

Rocket launch sites, such as Cape Canaveral, became the focal points of this extraordinary period. The thunderous roar of test launches echoed across the landscape, each one a testament to the progress being made. These launches were not without their tense moments—every successful lift-off was met with cheers and applause, while any setbacks only strengthened the resolve of those involved.

The vision for Project Mercury was clear: to send a man into space and safely return him to Earth. This goal required an unprecedented level of collaboration and innovation. The team had to develop a spacecraft capable of supporting human life and mastering the complexities of launch, orbit, and reentry. Each component had to be meticulously engineered and rigorously tested.

During these months, the dedication of those involved was unwavering. Long hours and sleepless nights were the norm as the team worked to overcome the myriad technical challenges. The spirit of innovation was palpable, driving them to push the boundaries of what was possible. They were not just building a spacecraft but forging a path to the stars.

By October 1958, significant strides had been made. The framework for Project Mercury was solidly in place, setting the stage for the United States to embark on its manned spaceflight program. The project symbolized the culmination of intense effort and the beginning of a new era in space exploration.

The progress during this period was not merely technical but also symbolic. It demonstrated America's commitment to leading the space race and exploring the unknown. The nation watched in anticipation, inspired by the possibility of human spaceflight.

In summary, the period from January to October 1958 was marked by dynamic progress and unwavering determination. The collective efforts of engineers, scientists, and policymakers coalesced into a unified vision, laying the meticulous groundwork for Project Mercury. This era was characterized by the scent of rocket fuel, the hum of innovation, and the dawn of America's journey into manned spaceflight. The achievements of these months set the stage for the following historic missions, propelling humanity into the cosmos and cementing America's place at the forefront of space exploration.

On March 10, the engineers at NACA convened to present and debate their latest reports on recoverable manned satellite configurations. The room buzzed with excitement and intensity, the air thick with the sound of technical discussions about the critical aspects of spacecraft design—motion dynamics, thermal protection, stabilization, and attitude control.

The team explored several innovative concepts, each offering a unique approach to ensuring a safe return to Earth. One prominent design was a blunt, high-drag, zero-lift vehicle. This straightforward and robust configuration relied on a parachute for the final deceleration stages. Its simplicity and reliability made it a strong contender for early missions.

Another intriguing concept was a winged vehicle designed to glide back to Earth. This design promised greater control and precision during reentry and landing, offering a more versatile approach to recovery. The engineers weighed the pros and cons of each design, considering the technical challenges and potential benefits.

These discussions were more than academic exercises; they laid the groundwork for America's first manned spaceflight program. The decisions made in that room would shape the future of space exploration, influencing the design and functionality of the spacecraft that would carry the first astronauts into orbit.

From March 10 to 12, the Air Force Ballistic Missile Division hosted a working conference in Los Angeles to support the "Man-in-Space Soonest" (MISS) program. General Bernard Schriever's opening remarks set a brisk pace for the event, emphasizing the urgency imparted by Roy Johnson of the Advanced Research Projects Agency. As Johnson pressed the Air Force for a concrete approach to putting a man in space, the attendees felt pressured to produce a viable proposal. Discussions revolved around three stages: an initial high-drag, no-lift spacecraft, followed by more sophisticated lifting vehicles, and ultimately, a long-range vision potentially culminating in a space station or lunar expedition. The sense of urgency and importance was palpable, with each participant acutely aware of their work's high stakes and monumental impact.

By March 12, NACA had completed a comprehensive outline for a manned satellite program. Researchers discussed orbits, propulsion systems, human factors, and materials in detailed presentations. The program was ambitious, encompassing full-scale mockups, simulators, unmanned flights, and eventually manned missions. Confidence was high that the Atlas launch vehicle could meet the requirements, with reentry heat management and parachute recovery as key components. The room hummed with shared purpose and optimism as they laid the groundwork for Project Mercury.

On March 17, the NACA Special Committee on Space Technology convened for its second meeting at Ames Aeronautical Laboratory. Reports from various working groups were presented, reflecting on past achievements and the increasing focus on astronautics. Between 1952 and 1956, only 10 percent of NACA's research had been related to space; now, it had surged to 30 percent. Discussions covered various topics, from fluid mechanics to spacecraft design, each critical to advancing America's space ambitions. The committee's discussions were both reflective and forward-looking, highlighting the dramatic shift in focus and the ambitious goals driving their research and development efforts. The period from January to October 1958 was one of unprecedented collaboration and innovation, setting the stage for the historic achievements that would follow in America's journey to the stars.

On March 18, 1958, NACA published a pivotal report titled "Preliminary Studies of Manned Satellites, Wingless Configuration, Non-Lifting," authored by Maxime A. Faget, Benjamin Garland, and James J. Buglia. This document laid the foundation for the development of Project Mercury, offering detailed insights into the design and feasibility of a wingless, non-lifting spacecraft configuration. This report became a cornerstone for America's manned spaceflight aspirations, emphasizing the practicality and advantages of a simpler, more robust design.

From March 18 to 20, a conference on high-speed aerodynamics at Ames Aeronautical Laboratory drew over 500 representatives from NACA, industry, military, and government agencies. The atmosphere was electric as 46 technical papers were presented, including manned space flight vehicle proposals. Maxime A. Faget's presentation on manned orbital satellites captivated the audience, highlighting the technical challenges and innovative solutions being explored. The discussions were intense and forward-thinking, reflecting the era's high stakes and rapid technological advancements.

Throughout March, Langley Aeronautical Laboratory was a hub of intense activity focused on refining manned satellite development plans. A dedicated working committee concluded that a ballistic-entry vehicle launched with an existing intercontinental ballistic missile propulsion system was feasible for the first manned satellite project. Key figures like Robert R. Gilruth, Clotaire Wood, and Hartley A. Soule played instrumental roles in shaping the design concepts. They advocated for a simple ballistic vehicle using existing missile technology and heat sink methods for reentry, emphasizing practicality and reliability.

On April 2, in a landmark move, President Eisenhower submitted a special message to Congress advocating for the creation of a civilian space agency with NACA at its core. This proposal aimed to centralize federal aeronautic and space activities under one cohesive organization, fostering a coordinated and focused effort in space exploration. The proposal underscored the national commitment to advancing space exploration and highlighted the need for a dedicated agency to lead these efforts. This initiative signaled the United States' determination to lead in the nascent space race.

During April, Maxime A. Faget and his team at NACA developed the concept of a contour couch to help astronauts withstand the intense g-forces experienced during acceleration and reentry. Recognizing the critical importance of astronaut safety, the team focused on creating a design that would distribute the forces more evenly across the body. Fabrication of test models began in May, and by July 30, the concept had been proven feasible. This innovation was a significant advancement in ensuring astronaut safety, demonstrating the team's dedication to addressing the human factors in space travel.

On June 5, after extensive collaboration with the Advanced Research Projects Agency, Maxime A. Faget presented a comprehensive report on recommendations for manned space flight to Dr. Hugh Dryden. The proposals, meticulously crafted by the panel, included using the Atlas launch vehicle and selecting astronauts from a pool of military volunteers. These recommendations were closely aligned with the emerging plans for Project Mercury, reflecting a strong consensus on the strategic direction for America's nascent manned space program.

A key recommendation was to use the Atlas launch vehicle, leveraging its proven capabilities to ensure a reliable launch system for the first manned missions. Equally strategic was the decision to select astronauts from military volunteers, ensuring that the individuals chosen would possess the discipline, training, and resilience required for the demanding conditions of space flight.

This alignment with Project Mercury's objectives indicated a unified vision among the key stakeholders involved in America's space endeavors. The report underscored the coordinated efforts and shared commitment to establishing a robust and effective manned space program. The meticulous planning and strategic choices made during this period were pivotal in laying the groundwork for the following historic achievements, marking a significant step forward in the nation's journey into space exploration.

Despite the absence of a designated manager for the project, the urgency to commence the program was palpable. The high stakes and rapid pace of developments underscored the nation's commitment to achieving a significant milestone in space exploration. The detailed planning and swift actions during this period reflected the intense drive to position the United States as a leader in the space race.

Faget's report and the panel's recommendations highlighted the coordinated efforts between various agencies and the military. This collaboration was crucial in addressing manned space flight's technical and logistical challenges. The proposed use of the Atlas launch vehicle, a proven and powerful rocket, demonstrated a pragmatic approach to leveraging existing technology for new frontiers.

Another strategic decision was to select astronauts from military volunteers. This ensured that the first spacefarers would have the necessary discipline, training, and experience to handle the rigors of space travel. These individuals were well-prepared to meet space missions' physical and psychological demands, bringing expertise and resilience to the program.

On June 22, discussions with Bureau of Budget officials focused on the proposed space agency's budget, particularly the manned satellite project. The meticulous planning and detailed budget considerations underscored the scale and ambition of the nation's space aspirations. These discussions were critical in securing the necessary funding and support for the ambitious projects ahead.

On June 26, representatives from NACA, AVCO, and Lockheed convened to discuss materials for the thermal protection of reentry vehicles. This collaboration emphasized the critical importance of heat management in ensuring the spacecraft's structural integrity and the astronaut's safety during reentry. The meetings showcased the collaborative spirit and shared goals of the various organizations involved, highlighting their commitment to overcoming the technical challenges of space travel.

Chapter 3 - The Mercury Spacecraft

Throughout June, personnel at Langley Aeronautical Laboratory worked diligently on drafting preliminary specifications for the first manned satellite vehicle. Under the supervision of Maxime Faget and Charles W. Mathews, these specifications evolved through numerous revisions, eventually forming the foundation for the Project Mercury spacecraft contract with McDonnell Aircraft Corporation. Establishing a working group between Langley and the Lewis Flight Propulsion Laboratory facilitated detailed planning and coordination, ensuring that all aspects of the spacecraft's design and functionality were meticulously addressed.

The Mercury spacecraft, meticulously designed by Maxime Faget, was a marvel of engineering that played a pivotal role in the United States' first human spaceflight program. Faget, who had begun his research during the National Advisory Committee for Aeronautics (NACA) era, brought his vision to life through the Mercury project. The spacecraft was compact yet sophisticated, measuring 10.8 feet (3.3 meters) in length and 6.0 feet (1.8 meters) in width. With the launch escape system attached, its overall length extended to 25.9 feet (7.9 meters).

Faget's design was revolutionary for its time. The Mercury spacecraft was a conical capsule optimized for reentry from space. It was built to be light yet strong, capable of withstanding the extreme conditions of space travel, including the intense heat generated during reentry into Earth's atmosphere. The outer shell was made of an ablative material designed to burn away and dissipate heat, protecting the astronaut inside.

Inside, the spacecraft was equipped with a life support system that provided oxygen and removed carbon dioxide, ensuring a breathable atmosphere. The cramped interior was just big enough for a single astronaut seated in a custom-fitted couch to absorb the forces of launch and reentry. The control panel, though limited compared to modern spacecraft, was advanced for its time, giving the astronaut some control over its orientation and systems.

The launch escape system was a critical safety feature designed to pull the capsule away from the rocket in the event of a malfunction during launch. This system consisted of a solid-fuel rocket attached to a tower on top of the capsule. If activated, it would rapidly separate the spacecraft from the booster, allowing for a safe descent by parachute.

In the Mercury and subsequent early human spaceflight programs, including Apollo, Gemini, and Soyuz, a launch escape system (LES) was designed to ensure the safety of the astronauts in the event of an emergency during launch or ascent.

During a typical launch escape system operation, the abort Trigger: Sensors would detect the anomaly and trigger an abort sequence if a critical issue arose with the launch vehicle or spacecraft before reaching a safe altitude.

Firing of LES: Upon detection of an abort condition, the launch escape system would activate. This involved firing the LES main rockets, typically solid rocket motors, to rapidly pull the spacecraft away from the booster rocket.

Separation: Once the LES rockets fired, the spacecraft would separate from the malfunctioning booster. This maneuver was crucial to ensure that the astronauts were safely away from any potential explosion or catastrophic failure of the launch vehicle.

Descent by Parachute: After separation, the spacecraft would continue its ascent trajectory for a short distance before beginning its descent back towards Earth. The spacecraft would deploy parachutes to slow its descent and ensure a safe landing for the astronauts.

The launch escape system was a critical safety feature designed to provide a reliable means of aborting a launch in emergencies such as engine failures, structural issues, or other serious malfunctions. Its successful deployment relied on rapid and precise response to ensure the astronauts' survival and recovery, demonstrating the importance of safety in human spaceflight operations.

The Mercury spacecraft's development involved rigorous testing and innovation. Engineers used wind tunnels and rocket sleds to simulate and refine its performance. The spacecraft was subjected to countless tests to ensure it could withstand the harsh environment of space and the stresses of launch and reentry. Each component was meticulously checked and rechecked to guarantee astronaut safety.

Faget's design proved highly successful, providing the United States with a reliable and safe means of human spaceflight. The Mercury program's achievements were a testament to the ingenuity and determination of the engineers and scientists who worked tirelessly to make it a reality. Their efforts not only paved the way for future space exploration but also demonstrated the incredible potential of human ingenuity in overcoming the challenges of space travel.

Design and Specifications

Inside the Mercury capsule, the habitable volume was a mere 100 cubic feet (2.8 cubic meters), just enough to accommodate a single astronaut. This tight, instrument-filled environment featured 120 controls: 55 electrical switches, 30 fuses, and 35 mechanical levers, all meticulously arranged to optimize functionality within the limited space. The heaviest model, Mercury-Atlas 9, weighed 3,000 pounds (1,400 kilograms) when fully loaded.

The interior of the Mercury capsule was a marvel of compact design and efficiency. Every inch of space was utilized, with instruments and controls positioned within easy reach of the astronaut, who would be seated in a custom-fitted couch designed to absorb the forces experienced during launch and reentry. The astronaut's suit, while adding to the cramped conditions, was essential for providing life support and protection against the harsh environment of space.

The outer skin of the Mercury spacecraft was crafted from René 41, a nickel alloy known for its ability to withstand extreme temperatures. This material was crucial for the spacecraft's reentry phase, where temperatures could soar to thousands of degrees Fahrenheit. The René 41 skin, combined with an ablative heat shield, ensured the capsule and its occupant would survive the intense heat of atmospheric reentry.

The spacecraft's life support system was designed to maintain a stable internal environment, providing oxygen and removing carbon dioxide. This system ensured the astronaut could breathe and remain relatively comfortable during the mission. In addition to life support, the Mercury capsule was equipped with navigation and communication systems that allowed for constant contact with ground control. The Manned Space Flight Network facilitated this, providing guidance and support throughout each mission.

The launch escape system, an essential safety feature, was designed to separate the capsule from the launch vehicle in case of a malfunction. This system consisted of a solid-fuel rocket mounted on a tower atop the capsule. If activated, it would pull the spacecraft away from the rocket, allowing for a safe descent via parachute.

The development and testing of the Mercury capsule involved extensive simulations and real-world trials. Engineers used wind tunnels, rocket sleds, and high-altitude balloon drops to refine the spacecraft's design and ensure its reliability. Each component underwent rigorous testing to meet the stringent safety standards required for human spaceflight.

Maxime Faget's visionary design and the relentless efforts of the engineers and scientists behind Project Mercury resulted in a robust and functional spacecraft. The Mercury capsule's successful missions demonstrated the feasibility of manned spaceflight and laid the foundation for future programs like Gemini and Apollo. These achievements propelled the United States to the forefront of space exploration and showcased the extraordinary potential of human ingenuity and perseverance in overcoming the challenges of venturing into space.

Structure and Components

The Mercury spacecraft had a distinctive conical shape with a convex base, meticulously designed to optimize aerodynamics during ascent and to dissipate heat during reentry. This shape reduced aerodynamic drag and helped stabilize the spacecraft's orientation in the upper atmosphere.

At the base of the capsule was the heat shield, a critical component responsible for protecting the spacecraft and its occupant during the intense heat of atmospheric reentry. This heat shield was composed of an aluminum honeycomb structure covered with multiple layers of fiberglass. The innovative design allowed the shield to absorb and dissipate the extreme temperatures encountered during reentry, which could exceed 10,000 degrees Fahrenheit.

The retropack was attached to the heat shield, vital in the spacecraft's reentry process. The retropack consisted of three main rockets designed to slow the spacecraft for reentry and three smaller rockets used for separation from the launch vehicle at orbital insertion. This system ensured the spacecraft could safely transition from orbit to a controlled descent.

The retropack's three main rockets were positioned around the base of the heat shield. They were crucial for reducing the spacecraft's velocity, allowing it to reenter the Earth's atmosphere at the correct angle and speed. The three smaller rockets provided the necessary thrust to separate the Mercury capsule from the launch vehicle once it had reached orbit, ensuring a smooth and precise orbital insertion.

The combination of the heat shield and retropack exemplified the sophisticated engineering and attention to detail that characterized the Mercury program. The heat shield's ablative material effectively managed the reentry heat, gradually burning away to dissipate the thermal energy, while the retropack's rockets provided precise control over the spacecraft's descent trajectory.

Inside the capsule, the astronaut was surrounded by a highly efficient and compact arrangement of instruments and controls. With only 100 cubic feet (2.8 cubic meters) of habitable volume, the interior was a tight, instrument-filled environment featuring 120 controls: 55 electrical switches, 30 fuses, and 35 mechanical levers. This arrangement allowed the astronaut to manage all necessary functions within arm's reach.

Crew Compartment

The pressurized crew compartment of the Mercury spacecraft was strategically positioned just above the heat shield, ensuring the astronaut was well-protected during the critical phases of launch and reentry. Inside this compartment, the astronaut was secured to a custom-fitted seat designed to absorb the intense forces experienced during these phases. The seat was tailored to fit each astronaut's body, providing both comfort and support in the confined space.

Various instruments and controls, carefully arranged for easy access and optimal functionality within the limited interior space, surrounded the astronaut. The capsule featured 120 controls, including 55 electrical switches, 30 fuses, and 35 mechanical levers, all essential for managing the spacecraft's operations and systems.

Beneath the seat was the environmental control system, a critical component for maintaining a habitable environment inside the capsule. This system provided a continuous oxygen supply and regulated the cabin temperature, ensuring the astronaut remained comfortable and safe throughout the mission. It also filtered out carbon dioxide, vapor, and odors, maintaining air quality. Additionally, the system managed waste on orbital flights, ensuring that the confined space remained hygienic and livable.

The environmental control system's complexity and reliability were vital for the success of the Mercury missions. It ensured that the astronaut could breathe and function effectively, even in the harsh space environment. This system's careful design and placement beneath the seat maximized available space within the compact capsule.

Recovery System

At the narrow end of the Mercury spacecraft, the recovery compartment housed three essential parachutes designed to ensure a safe landing. This compartment contained a drogue chute deployed first to stabilize the spacecraft's descent. Once the drogue chute had done its job, the two main parachutes came into play: a primary chute and a reserve chute, providing redundancy to guarantee a secure landing even if one failed.

As the spacecraft descended through the Earth's atmosphere, the drogue chute slowed and stabilized the capsule, preventing it from tumbling and ensuring a controlled descent. Following this, the primary main chute deployed, dramatically reducing the descent speed to a safe level. The reserve chute, though rarely needed, was a critical backup in case of any issues with the primary chute.

A landing skirt was deployed just before landing by lowering the heat shield. This landing skirt cushioned the final impact, absorbing some of the landing force to protect the astronaut inside. The combination of the parachute system and the landing skirt ensured that the spacecraft could safely land in the ocean, where recovery teams would be ready to retrieve the astronaut and the capsule.

Antenna and Communication

The antenna section, located on top of the recovery compartment of the Mercury spacecraft, played a crucial role in communication and orientation. This section housed the communication antennas, vital for maintaining contact between the spacecraft and ground control throughout the mission. Additionally, it contained orientation scanners that helped the spacecraft maintain the correct position in space.

One of the key features of the antenna section was a small flap, often referred to as the "spoiler" or "drogue stabilization flap." This flap ensured that the spacecraft faced heat shield-first during reentry, providing additional stability and protection. By helping to orient the spacecraft correctly, the flap ensured that the heat shield could effectively dissipate the intense heat generated during reentry, safeguarding the astronaut and the integrity of the spacecraft.

The correct orientation was critical during reentry, as the heat shield needed to absorb and deflect the extreme temperatures encountered when the spacecraft re-entered the Earth's atmosphere. The small flap worked in tandem with the spacecraft's overall aerodynamic design to maintain this orientation, contributing to the mission's success and the astronaut's safety.

Launch Escape System

The launch escape system (LES), mounted at the narrow end of the Mercury spacecraft, was a critical safety feature designed to protect the astronaut in the event of a launch failure. This system comprised three small solid-fueled rockets. These rockets were designed to quickly and effectively separate the capsule from the booster, ensuring the astronaut's safety during the most dangerous phases of the launch.

In the event of a launch failure, the solid-fueled rockets of the launch escape system would ignite, generating enough thrust to pull the Mercury capsule away from the malfunctioning booster. This rapid separation was essential to prevent the capsule from being caught in an explosion or other catastrophic failure of the launch vehicle.

Once the capsule was safely away from the booster, the launch escape system would deploy the parachutes to ensure a controlled descent. The system was designed to deploy a drogue chute to stabilize the capsule, followed by the main chutes to slow the descent further. This process ensured that the capsule could achieve a safe splashdown in a nearby body of water, where recovery teams would be ready to retrieve the astronaut and the spacecraft.

The launch escape system was a marvel of engineering, providing a reliable and quick response in an emergency. Its design ensured that the astronaut had a high chance of survival even in a launch failure, demonstrating the priority placed on safety within the Mercury program.

Ground-Controlled Reentry

Unlike modern spacecraft, the Mercury capsule lacked an onboard computer and relied instead on ground-based computing resources. During each mission, powerful computers at NASA facilities performed all critical reentry calculations. These ground-based computers calculated essential parameters such as retrofire times and firing attitudes with precision.

These calculations were transmitted to the Mercury spacecraft via radio throughout the flight. The ground control team constantly monitored the spacecraft's status and provided real-time updates and instructions to the astronaut. This approach ensured that the complex and critical computations required for a safe reentry were handled by the best available technology at the time.

The absence of an onboard computer meant that the astronaut had to execute the commands received from ground control manually. The astronaut's ability to accurately interpret and implement these instructions was vital to the mission's success. Training for the Mercury astronauts included extensive simulations and practice following ground-based guidance, preparing them to manage the spacecraft effectively.

Retropack: Retrorockets with red posigrade rockets

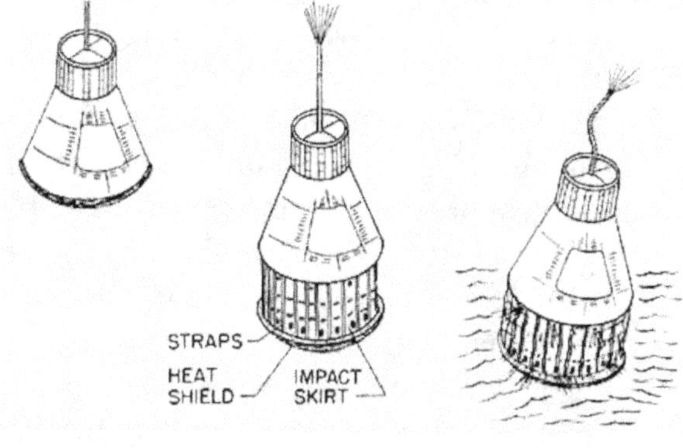

Landing skirt (or bag) deployment: skirt is inflated; on impact the air is pressed out (like an airbag)

Ground-based computing and radio transmission methods underscored the Mercury missions' collaborative nature. The success of each flight depended on seamless communication and coordination between the astronaut and the ground control team. The engineers and scientists on the ground played as crucial a role as the astronaut in ensuring the mission's success.

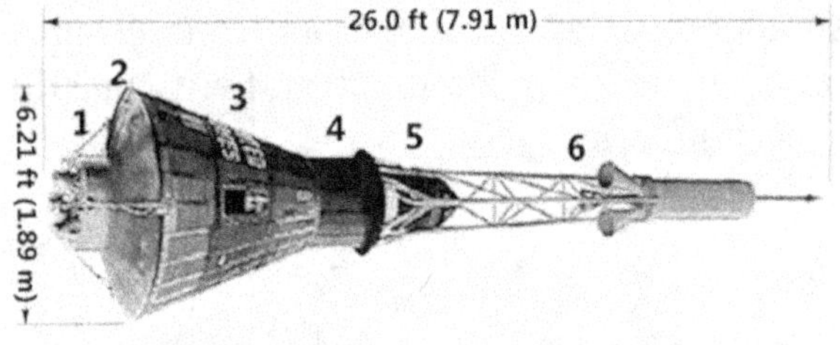

1. Retropack. 2. Heatshield. 3. Crew compartment. 4. Recovery compartment. 5. Antenna section. 6. Launch escape system.

The reliance on ground-based computers also highlighted the technological limitations and innovations of the era. While today's spacecraft are equipped with advanced onboard computers capable of performing complex calculations autonomously, the Mercury program showcased the ingenuity required to achieve spaceflight with the resources available in the late 1950s and early 1960s.

Sensory Experience

One experienced a blend of sensations and emotions as an astronaut inside the Mercury spacecraft. The tight quarters amplified every sound—the hum of the environmental control system, the clicking of switches, and the faint creaking of the metal structure. The small portholes provided glimpses of the outside world, transitioning from the deep blue of Earth's atmosphere to the inky blackness of space dotted with stars.

The vibrations and forces during launch were intense, pressing the astronaut firmly into their seat. The roar of the rocket engines grew louder, and the g-forces increased, making even the simplest movements a challenge. As the spacecraft ascended, the violent shaking subsided, replaced by the calmness of weightlessness. Floating in space, the astronaut could see the curvature of Earth, its vast oceans, and swirling clouds from a perspective few had ever witnessed.

In orbit, the view outside the portholes was both awe-inspiring and humbling. The stark contrast between the bright daylight side of Earth and the pitch-black expanse of space was a reminder of the unique position the astronaut held. The sight of distant stars, undistorted by the atmosphere, added to the sense of being on the edge of a vast, unexplored frontier.

The reentry phase brought a new set of sensations. The spacecraft began to heat up as it encountered the friction of Earth's atmosphere, and the heat shield's ablative material started to burn away, protecting the capsule. The vibrations returned with a vengeance, pressing the astronaut back into their seat as the g-forces built up again. The capsule shook and rattled; the air outside roaring as it decelerated.

As the parachutes deployed, the descent became more controlled, leading to a final splashdown in the ocean. The astronaut washed over with relief from a successful mission and pride from contributing to a pioneering program. The sight of recovery teams approaching and the feel of the capsule being hoisted out of the water marked the end of a remarkable journey.

The anticipation before launch, the awe of orbiting Earth, and the relief of successful reentry and splashdown were all part of the Mercury astronaut's journey. This pioneering program demonstrated the feasibility of human spaceflight and laid the foundation for future missions, inspiring a generation to look to the stars with hope and determination.

Launch Vehicles

Project Mercury utilized a variety of launch vehicles for its different flight profiles, from uncrewed tests to suborbital and orbital missions. Each type of launch vehicle played a crucial role in the program's overall success, tailored to meet specific mission requirements.

For orbital flights, the Mercury-Atlas rocket was employed. This powerful rocket, developed by Convair, could propel the Mercury spacecraft into Earth orbit. The Mercury Atlas was an adaptation of the Atlas D intercontinental ballistic missile (ICBM), modified to carry the Mercury capsule. Its significant thrust and reliability made it the ideal choice for missions that required reaching and maintaining orbit. John Glenn's historic flight on February 20, 1962, aboard Friendship 7, was launched using a Mercury-Atlas rocket, marking a major milestone in the space race.

John Glenn honored by the President. February 1962

Suborbital flights, which were shorter and did not require the same level of propulsion as orbital missions, used the Mercury-Redstone rocket. Developed by the Army Ballistic Missile Agency and designed by a team led by Wernher von Braun, the Mercury-Redstone was a single-stage rocket adapted from the Redstone ballistic missile. It provided sufficient power to carry the Mercury spacecraft on a suborbital trajectory, reaching the edge of space before descending back to Earth. Alan Shepard's pioneering flight on May 5, 1961, aboard Freedom 7, utilized the Mercury-Redstone rocket, making him the first American in space.

John Glenn wearing his Mercury space suit

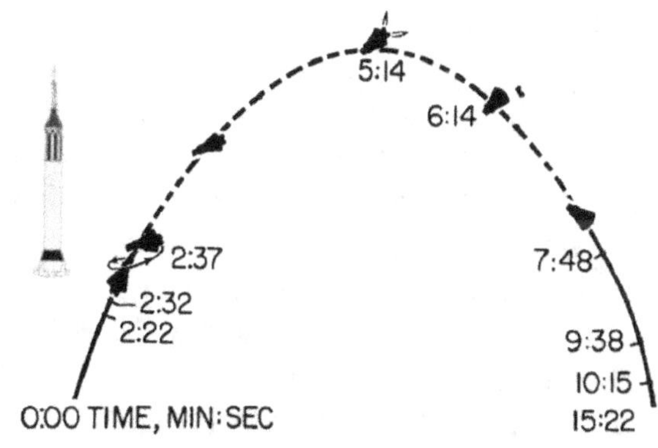

Suborbital Profile. Dashed line: region of weightlessness.

Little Joe Rocket

The name "Little Joe" for the rockets used in the Mercury program was indeed derived from their distinctive four-rocket configuration, which resembled the "double deuce" dice throw in a game of craps. This arrangement was notable for its compact and efficient design, fitting for its role in testing critical components such as the launch escape system.

These Little Joe rockets played a pivotal role in the early stages of Project Mercury. They were specifically designed to simulate various launch scenarios and emergency procedures, including testing the escape systems crucial for astronaut safety during a launch abort or emergency. By subjecting these systems to rigorous testing under controlled conditions, the Little Joe rockets contributed significantly to refining and validating the technologies that would ensure the success of manned spaceflight missions in the challenging space environment.

The Little Joe rocket was employed for uncrewed tests of the launch escape system. This smaller, less complex launch vehicle was designed to test the Mercury capsule's abort and recovery systems. The Little Joe rocket allowed engineers to conduct high-stress tests of the launch escape system under various conditions, ensuring it would function correctly in an emergency. These tests were critical for validating the safety mechanisms to protect the astronaut during a launch failure.

Each launch vehicle—Mercury-Atlas, Mercury-Redstone, and Little Joe—played a distinct and vital role in the Project Mercury program. The combination of these rockets enabled a comprehensive approach to testing and executing missions, addressing different phases of spaceflight and ensuring the safety and success of the astronauts.

Launch Escape System Testing

To ensure the safety of the astronauts, Project Mercury employed a 55-foot-long (17-meter) launch vehicle known as Little Joe for uncrewed tests of the launch escape system. This system, which included a Mercury capsule equipped with an escape tower, was rigorously tested under maximum aerodynamic pressure, or max q. At this critical point, aerodynamic forces against the spacecraft were at their peak, making separating from the launch vehicle the most challenging. It was also when astronauts would experience the heaviest vibrations, posing a significant challenge to the integrity and function of the escape system.

The Little Joe rocket was powered by solid-fuel propellant and originally designed by the National Advisory Committee for Aeronautics (NACA) in 1958 for suborbital crewed flights. However, it was adapted for Project Mercury to simulate the conditions of an Atlas-D launch. Built by North American Aviation, the Little Joe's flight path was determined solely by its launch angle, as it could not change direction mid-flight. When fully loaded, Little Joe could reach a maximum altitude of 100 miles (160 kilometers).

The Little Joe tests were crucial for validating the launch escape system, ensuring that the Mercury capsule could be safely separated from the launch vehicle in an emergency during launch. These tests involved subjecting the capsule to the intense forces experienced during max q, providing engineers with valuable data on how the escape system would perform under the most demanding conditions.

In addition to the Little Joe tests, a Scout launch vehicle was used once to evaluate the tracking network's performance. Unfortunately, this mission ended in failure when the vehicle was destroyed shortly after launch. Despite this setback, the lessons learned from the Scout mission contributed to the overall knowledge and improvement of the tracking systems used in subsequent missions.

The rigorous testing conducted with the Little Joe rocket and other launch vehicles demonstrated the thoroughness and attention to detail that characterized Project Mercury. By meticulously testing and validating every aspect of the launch and escape systems, the engineers and scientists involved in the program prioritized the safety of the astronauts above all else.

On July 9, General Electric Company personnel briefed NACA headquarters on their studies related to manned space flight. Although NACA did not officially comment, the briefing showcased the widespread interest and diverse approaches being explored across the industry. These months leading up to the formal start of Project Mercury were characterized by intense activity, collaboration, and innovation. From early discussions and technical evaluations to high-stakes conferences and detailed program outlines, each step brought the United States closer to achieving manned spaceflight. The dedication and ingenuity of those involved laid a solid foundation for the ambitious goals of Project Mercury, setting the stage for America's historic leap into space.

On July 15, the offices of McDonnell Aircraft Corporation were abuzz with activity. Engineers and designers huddled over blueprints and models, their discussions punctuated by occasional exclamations of discovery or frustration. On this day, Cook Electric Company submitted a proposal to McDonnell as part of a preliminary study for a manned satellite. McDonnell, keen on securing the Mercury prime development contract, had dedicated 11 months and significant company resources to developing a manned orbital spacecraft concept. The proposal from Cook Electric was a crucial piece of the puzzle, reflecting the intense collaborative efforts required to advance the project.

On July 16, the halls of Congress were filled with anticipation and determination. Lawmakers passed the National Aeronautics and Space Act of 1958, a landmark decision that would pave the way for the creation of NASA. This legislative milestone marked a significant step toward organizing the United States' efforts in space exploration, centralizing various space activities under one cohesive agency.

On July 18, in a meticulously worded memorandum to Dr. James R. Killian, Jr., Special Assistant to the President for Science and Technology, Dr. Hugh L. Dryden, Director of NACA, outlined the wealth of experience NACA would bring to NASA. Dryden highlighted years of research in vehicle stabilization, control systems, high-temperature structures, and reentry challenges. The X-15 program had provided invaluable data on human factors in spaceflight, making NACA a natural fit to lead the manned satellite program envisioned by the Space Act of 1958.

On July 29, President Eisenhower signed the National Aeronautics and Space Act into law, formally establishing NASA. The ceremony, attended by key figures in aerospace and government, was a momentous occasion that signaled the United States' commitment to leading the charge in space exploration. This formal establishment of NASA marked the beginning of a new era in American scientific and exploratory endeavors.

On July 30, at the centrifuge in Johnsville, Pennsylvania, the air was filled with the powerful machine's whirring sound as Carter C. Collins prepared for a critical test. Using the Mercury contour couch development model designed by Maxime A. Faget and his team, Collins endured a grueling 20g load. The successful test proved that astronauts could withstand the extreme accelerations experienced during reentry, a significant milestone for Project Mercury and a testament to the meticulous engineering and testing underpinning the program.

On July 31, representatives from Republic Aviation briefed NACA Headquarters personnel on their man-in-space studies. The room was filled with engineers and scientists eager to hear about the Republic's vision of a four-stage solid launch vehicle system. The proposed vehicle, a triangular-shaped lifting reentry vehicle termed a "sled," would feature aerodynamic and reaction controls for the pilot. The detailed briefing covered launch sequences, reentry procedures, and recovery plans, sparking intense discussions among the attendees and highlighting the diverse array of innovative ideas being considered.

Throughout July, Maxime A. Faget suggested using a tractor rocket as an escape device, an idea that would evolve into the Mercury escape rocket. This concept was a critical safety feature, ensuring that astronauts could be safely ejected from the spacecraft in an emergency. The development of the escape rocket underscored the comprehensive approach to address every aspect of astronaut safety and mission success.

On August 1, 1958, Dr. Hugh L. Dryden presented a comprehensive program on manned space flight vehicle technology to the Select Committees of Congress on Astronautics and Space Exploration. The presentation highlighted the technological advancements and strategic plans necessary for successful manned space missions, emphasizing the critical role of NACA's research and expertise. The committee members listened intently as Dryden detailed cutting-edge developments in vehicle stabilization, control systems, and reentry technology, all essential components of the emerging space program.

A week later, on August 8, a significant memorandum from the Secretary of the Army to the Secretary of Defense recommended Project Adam, a manned space flight program proposing a ballistic suborbital flight using existing Redstone hardware. The memo requested $9 million for fiscal year 1959 and $2.5 million for 1960, aiming to demonstrate the nation's capabilities and bolster its political and psychological standing in the Cold War context. The proposal reflected the Army's eagerness to contribute to the burgeoning field of manned space exploration, leveraging its existing missile technology for new horizons.

Throughout August, President Eisenhower underscored the importance of consolidating space exploration efforts under a single civilian agency by assigning the development and execution of the manned space flight program to NASA. Although NASA would not become operational until October 1, this decision significantly shifted towards a more coordinated and focused national space effort. The assignment emphasized the necessity of centralizing expertise and resources to achieve the ambitious goals for manned space missions.

These pivotal developments in August 1958 were instrumental in shaping the future of America's space exploration endeavors. Dr. Dryden's presentation to Congress, the Army's ambitious Project Adam proposal, and President Eisenhower's strategic consolidation of space efforts under NASA all contributed to a unified and purposeful approach to manned space flight. This intense planning and collaboration period set the stage for the historic achievements that would soon follow, propelling the United States into a new scientific discovery and exploration era.

In September 1958, the USAF transferred responsibility for MISS to NASA. On November 5, 1958, NASA established the Space Task Group (STG) at the Langley Research Center, with Robert R. Gilruth as its director. Project Mercury's objective was to launch a man into Earth orbit, return him safely, and evaluate his capabilities in space. Organizational Efforts

Under the leadership of T. Keith Glennan and Hugh L. Dryden, NASA organized the Space Task Group to oversee Project Mercury. Their goals were clear: achieve orbit, study human capabilities in space, and ensure safe return. Using existing technology where possible, the team focused on simplicity and reliability.

The establishment of NASA and Project Mercury marked a pivotal moment in American space exploration. The goals were ambitious: to demonstrate that humans could survive and operate in the harsh space environment. The project called upon the best minds in science and engineering, requiring innovative solutions to unprecedented challenges. The rapid progression and success of Project Mercury set the stage for the subsequent achievements of Gemini and Apollo, ensuring that the legacy of human space exploration would continue to expand the boundaries of what was possible.

During a pivotal Army Advanced Research Projects Agency (ARPA) conference on September 11, officials received significant news regarding Project Adam's future. The consensus was clear: Project Adam was unlikely to receive the green light for further development. This decision marked a strategic pivot for ARPA, signaling a shift in focus towards projects with greater potential and more promising outcomes.

Moving away from Project Adam was not merely a dismissal of one initiative but a reflection of a broader realignment within the agency. ARPA increasingly prioritized collaborations and ventures in harmony with the nation's long-term space policy goals. This strategic reorientation aimed to harness cutting-edge technologies and foster innovative partnerships that could propel the United States to the forefront of space exploration and defense.

The decision underscored a deliberate preference for projects that offered clear, tangible benefits and aligned with the overarching mission to enhance national security and technological supremacy. By redirecting resources and attention towards more viable and strategically beneficial initiatives, ARPA demonstrated its commitment to advancing the nation's space capabilities in a rapidly evolving global landscape.

By September 17, a significant step was taken with the formation of a joint NASA/ARPA Manned Satellite Panel. This panel brought together a diverse group of experts from the Langley and Lewis Research Centers alongside representatives from various military services. The primary mandate of this newly formed panel was to draft comprehensive plans for a research program aimed at achieving manned space flight.

This collaborative effort was a crucial foundation for what would later become known as Project Mercury. The panel's work involved integrating the extensive expertise and resources from both NASA and the military, ensuring a unified approach to tackling the formidable challenges of manned space flight.

The contributions from the Langley and Lewis Research Centers were particularly vital, as they provided advanced aeronautical research and engineering capabilities. The military's involvement brought strategic planning and operational experience, essential for the program's practical implementation.

The formation of the Manned Satellite Panel marked a pivotal moment in the history of space exploration. It exemplified the power of collaboration between civilian and military institutions, combining their strengths to achieve a common goal.

On September 25, NASA Administrator Dr. T. Keith Glennan announced that NASA would officially commence operations on October 1. This declaration was met with widespread enthusiasm and a renewed sense of purpose within the scientific and engineering communities. The formation of NASA was hailed as a bold and decisive step forward, galvanizing the nation's efforts in space exploration and solidifying its commitment to pushing the boundaries of human knowledge and capability.

Throughout September, extensive studies were initiated on the tracking and ground instrumentation networks essential for the manned satellite project. These studies were critical to ensuring reliable communication and control during space missions. The detailed planning and analysis underscored the complexity and technical challenges of maintaining contact with spacecraft in orbit.

The establishment of these networks required sophisticated technology and meticulous coordination. Engineers and scientists focused on developing systems that could provide continuous, precise tracking of satellites, ensuring that data could be transmitted and received seamlessly. This work was foundational to the success of future missions, as effective communication would be paramount to the safety and success of manned space flights.

From September 24 to October 1, the esteemed Robert R. Gilruth chaired a series of intensive meetings in Washington, D.C. These gatherings brought together some of the most influential minds in the field of space exploration, including S. B. Batdorf, A. J. Eggers, Maxime A. Faget, George Lowg, Warren North, Walter C. Williams, and Robert C. Youngquist. Their mission was to draft a comprehensive manned satellite program that required both visionary thinking and meticulous planning.

The meetings were characterized by rigorous debate and dynamic collaboration. Participants brought their unique expertise, contributing to the rich tapestry of ideas and strategies. Maxime A. Faget's innovative engineering concepts, George Low's managerial acumen, and Walter C. Williams's operational insights were just a few examples of the diverse strengths that fueled the discussions.

Despite the intense pressure and the high stakes, the team worked tirelessly to iron out the details of their ambitious objectives. They meticulously examined every aspect of the proposed program, from the technological specifications of the spacecraft to the logistics of mission control and astronaut training.

The outcome of these meetings was a detailed and robust plan that would guide the United States in its quest to achieve manned spaceflight. This plan laid the groundwork for Project Mercury, the first American manned spaceflight program, which aimed to send astronauts into orbit and return them safely to Earth.

On October 1, 1958, a significant transition in American space exploration history occurred: NASA was officially activated, taking over the nonmilitary space projects previously managed by the Advanced Research Projects Agency (ARPA). This momentous day also saw the inactivation of the National Advisory Committee for Aeronautics (NACA), which had spent 43 years pioneering aeronautical research. NACA's facilities and personnel were seamlessly integrated into NASA, bringing a wealth of knowledge and experience to the newly formed agency.

The activation of NASA marked the beginning of a new era in space exploration for the United States. This shift underscored a strategic commitment to harnessing the collective expertise and resources of the nation's top scientists and engineers under one unified agency. The consolidation of efforts aimed to streamline the country's space research and development, ensuring a focused and coordinated approach to achieving ambitious goals in space.

NASA's formation was met with great enthusiasm and anticipation, as it signified the nation's resolve to lead in the emerging frontier of space exploration. With the combined strengths of NACA's rich legacy in aeronautics and ARPA's advancements in space technology, NASA was poised to spearhead groundbreaking missions and innovations. The agency's mandate included reaching new milestones in space travel and fostering scientific discoveries that could benefit all of humanity.

This organizational transformation laid the groundwork for a series of historic achievements, from the early manned spaceflights of Project Mercury to the monumental Apollo moon landings. By uniting the brightest minds and cutting-edge facilities, NASA set the stage for America to make its mark on the cosmos, embarking on a journey that would inspire generations and shape the future of space exploration.

As these milestones unfolded, the groundwork for Project Mercury was laid meticulously. The collective efforts of engineers, scientists, and policymakers coalesced into a unified vision. From bustling conference rooms and laboratories to the tense moments of rocket launches, the period from January to October 1958 was marked by dynamic progress and unwavering determination. The air was thick with the scent of rocket fuel and the hum of innovation, heralding the dawn of America's journey into manned spaceflight.

Fervent discussions and strategic planning sessions occurred in conference rooms nationwide. Engineers sketched out blueprints and debated design specifications while scientists presented their latest research findings. Policymakers worked tirelessly to secure funding and navigate the bureaucratic challenges of such an ambitious national endeavor. Each meeting was a step forward, bringing clarity and direction to the multifaceted challenges of sending a human into space.

Laboratories buzzed with activity, filled with the whirr of machinery and the clatter of tools. Technicians and researchers conducted experiments, testing materials and components to ensure they could withstand the harsh conditions of space. The development of life support systems, reentry techniques, and propulsion technologies was a priority, each breakthrough bringing Project Mercury closer to reality.

Rocket launch sites, such as Cape Canaveral, became the focal points of this extraordinary period. The thunderous roar of test launches echoed across the landscape, each one a testament to the progress being made. These launches were not without their tense moments—every successful lift-off was met with cheers and applause, while any setbacks only strengthened the resolve of those involved.

The vision for Project Mercury was clear: to send a man into space and safely return him to Earth. This goal required an unprecedented level of collaboration and innovation. The team had to develop a spacecraft capable of supporting human life and mastering the complexities of launch, orbit, and reentry. Each component had to be meticulously engineered and rigorously tested.

During these months, the dedication of those involved was unwavering. Long hours and sleepless nights were the norm as the team worked to overcome the myriad technical challenges. The spirit of innovation was palpable, driving them to push the boundaries of what was possible. They were not just building a spacecraft but forging a path to the stars.

By October 1958, significant strides had been made. The framework for Project Mercury was solidly in place, setting the stage for the United States to embark on its manned spaceflight program. The project symbolized the culmination of intense effort and the beginning of a new era in space exploration.

The progress during this period was not merely technical but also symbolic. It demonstrated America's commitment to leading the space race and exploring the unknown. The nation watched in anticipation, inspired by the possibility of human spaceflight.

In summary, the period from January to October 1958 was marked by dynamic progress and unwavering determination. The collective efforts of engineers, scientists, and policymakers coalesced into a unified vision, laying the meticulous groundwork for Project Mercury. This era was characterized by the scent of rocket fuel, the hum of innovation, and the dawn of America's journey into manned spaceflight. The achievements of these months set the stage for the following historic missions, propelling humanity into the cosmos and cementing America's place at the forefront of space exploration.

Throughout 1958, the Navy's ambitious space proposal, Project Mer, introduced a novel concept for manned spaceflight. The plan envisioned launching an astronaut into orbit in a collapsible pneumatic glider. Once in orbit, the glider would be inflated, allowing it to be flown back down for a water landing. Although Project Mer was ultimately not pursued, it exemplified the diverse and imaginative approaches being considered for manned spaceflight during this transformative period. Such innovative concepts reflected the spirit of ingenuity and determination that characterized the early days of the American space program.

The events of 1958 were pivotal in setting the stage for Project Mercury and the broader objectives of the United States' space program. This period saw innovative thinking, rigorous testing, and strategic planning, laying the foundation for the historic achievements that would soon follow. The inception of Project Mercury began with a flurry of intense activity and meticulous preparation.

In 1958, America found itself in a heated race to put a man in space, fueled by the geopolitical tensions of the Cold War. The newly formed NASA was determined to achieve a monumental leap for mankind. The urgency of the space race galvanized the nation's top scientists and engineers, driving them to push the boundaries of technology and human endurance.

Project Mercury aimed to send the first American astronauts into space and return them safely to Earth. This ambitious goal required overcoming numerous technical and logistical challenges. The project's success depended on the collaborative efforts of NASA's newly integrated team, drawing on the rich legacy of NACA's aeronautical research and the innovative spirit of ARPA's space projects.

As the year progressed, NASA's teams worked tirelessly to develop the spacecraft, train astronauts, and establish the necessary support infrastructure. Each step was marked by careful planning and relentless testing, ensuring that every aspect of the mission was meticulously prepared. The determination and ingenuity of the people behind Project Mercury exemplified the pioneering spirit that would define America's endeavors in space.

From October 3 to 7, the initial discussions regarding a manned satellite project were filled with both excitement and trepidation. The atmosphere was charged with anticipation as detailed plans were presented to the Advanced Research Projects Agency (ARPA) and NASA Administrator Dr. T. Keith Glennan. Each meeting buzzed with the energy of possibility and the weight of the enormous challenges ahead.

On October 6, 1958, personnel from Langley Research Center visited the Army Ballistic Missile Agency (ABMA) with a palpable sense of urgency and purpose. Their mission was clear: to negotiate the procurement of Redstone and Jupiter launch vehicles, which were crucial for the impending manned space missions. The meetings, held in the utilitarian offices of the ABMA, were filled with technical jargon and strategic discussions.

Engineers and officials from both sides delved into the specifics of rocket design, propulsion systems, and delivery timelines. These detailed negotiations were critical to securing the necessary hardware to propel a man into space. The atmosphere was charged with a mix of tension and excitement as each participant understood the gravity of their task and the historical implications of their work.

The discussions covered every aspect of the launch vehicles, from the reliability of their engines to the intricacies of their guidance systems. The Redstone rocket, a direct descendant of the German V-2, had been successfully used in ballistic missile tests and was seen as a reliable candidate for launching

America's first astronauts. With its greater thrust capacity, the Jupiter rocket was also considered for future missions requiring more powerful launch vehicles.

Securing these rockets was a critical step forward in Project Mercury. The successful procurement of the Redstone and Jupiter launch vehicles would provide NASA with the tools needed to carry out its ambitious goal of manned spaceflight. This procurement marked a significant milestone in the United States' quest to lead in space exploration, reflecting the nation's commitment to innovation and technological advancement.

On October 7 and 8, 1958, members of the Space Task Group embarked on a crucial journey to the Air Force Wright Air Development Center in Dayton, Ohio. Their task was to study reentry methods critical for astronauts' safe return. This mission ensured that astronauts could survive the intense heat and forces experienced during reentry into Earth's atmosphere.

The group, joined by Air Force Ballistic Missile Division experts, then traveled to the Chicago Midway Laboratories to explore various ablation techniques. These techniques involved materials designed to absorb and dissipate the extreme heat generated during reentry. The laboratories buzzed with activity as scientists and engineers scrutinized high-temperature test results, their faces illuminated by the glow of experimental apparatuses. The air was thick with the smell of heated metal and chemicals, underscoring the intense nature of their work.

Scientists methodically reviewed data from countless tests, each result bringing them closer to finding the optimal materials and designs for the spacecraft's heat shield. The Space Task Group and the Air Force Ballistic Missile Division collaboration was pivotal, combining expertise and resources to tackle one of the most significant challenges in manned spaceflight. The insights gained from these studies would prove invaluable,

contributing to developing a reliable reentry system that would ensure the safety of America's first astronauts.

On October 7, the anticipation peaked when Dr. Glennan approved with a resounding directive: "Let's get on with it." This momentous decision marked the official green light for Project Mercury, setting the stage for one of the most ambitious endeavors in human history.

The approval of Project Mercury was more than just an administrative milestone; it was a bold declaration of intent. The project aimed to send the first American astronauts into space, achieving national prestige and significant scientific milestones. This directive rallied the entire team, infusing them with purpose and urgency.

Dr. Glennan's approval catalyzed an intense period of activity and innovation. Engineers, scientists, and administrators across NASA and ARPA began working in unison, tackling the myriad technical challenges ahead. Their goal was clear: safely sending a man into space and returning him to Earth, proving the feasibility of human spaceflight, and laying the groundwork for future exploration.

Contractors and Facilities

Numerous contractors and facilities were involved in Project Mercury. McDonnell Aircraft built the spacecraft, while North American Aviation developed the Little Joe rockets for testing the escape system. Western Electric Company established the communication network and launched vehicles manufactured by Chrysler and Convair. The Cape Canaveral Air Force Station became the launch site, with additional facilities across the US supporting various aspects of the program.

Project Mercury officially began on October 7, 1958. In an impressively short span of three years and two weeks, NASA, in collaboration with McDonnell Aircraft Corporation of St. Louis, Missouri, launched

John Glenn aboard the Mercury spacecraft "Friendship 7." This rapid progression prompted many to question how such a feat was swiftly accomplished.

Creation and Development

Officially approved on October 7, 1958, Project Mercury was initially known as Project Astronaut. President Dwight Eisenhower, seeking to emphasize the broader scope of the mission, chose the name Mercury from classical mythology. The project absorbed various military initiatives, consolidating efforts to put a man in space. The Cold War context added urgency and intensity to these efforts as the US and the Soviet Union sought to demonstrate their technological superiority.

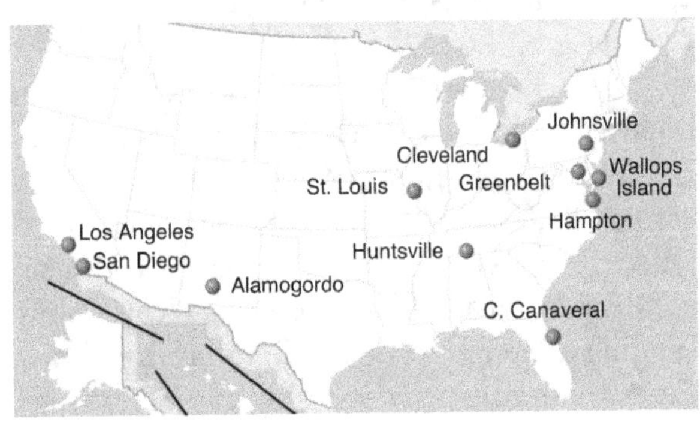

Project Mercury's Facilities

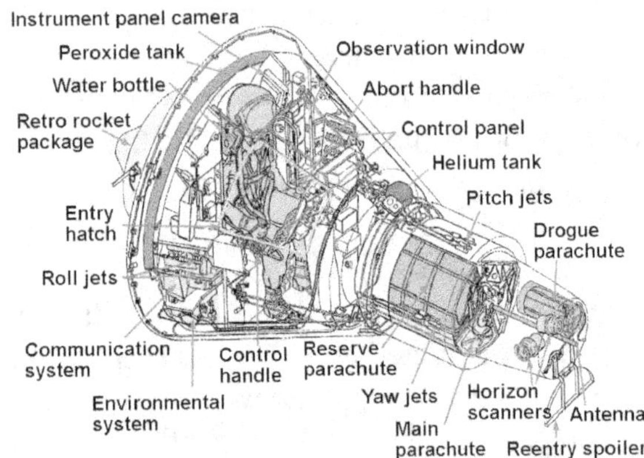

Project Mercury's formal program approval date was October 7, 1958. Just three years and two weeks after awarding the development and production contract to the McDonnell Aircraft Corporation of St. Louis, Missouri, John Glenn's orbital flight aboard the Mercury spacecraft "Friendship 7" transpired. This brief time scale is extraordinary compared to other major programs of national importance and urgency. How was man committed so soon to orbital flight?

The key to this phenomenal success was the concurrency of effort. All facets of the program leading to manned spaceflight were guided along a simultaneous route rather than by the concept of qualifying each phase before development work began on another. From the outset, work was being accomplished on all spacecraft components, adapting the launch vehicles, readying the worldwide tracking network, selecting and training astronauts, and developing ground support equipment for systems checkout and astronaut training. No detail was too small for the attention of scientists and engineers making the critical decisions that would commit a man to orbital flight. Every organization with technical proficiency or capability in a relevant field was visited, and arrangements were made for assistance, facilities, or the use of equipment. The test and reliability program to which Mercury hardware was subjected was exhaustive. This close attention refutes the "crash program" connotation often cited. The term "accelerated" more aptly describes the effort. The managers resisted the urge for a crash program, even with the American public anxiously awaiting the advent of manned spaceflight.

Several catalysts created the conditions leading to the approval of the Mercury project, contributing directly to the goal of attaining manned spaceflight. Shortly after World War II, experimental missile tests were conducted in the White Sands, New Mexico area to altitudes beyond the sensible atmosphere. During the same period, rocket aircraft research aimed at breaking the sound barrier was initiated. From the early to mid-1950s, the National Advisory Committee for Aeronautics (NACA) and industry scientists and engineers tackled the thermal barrier to resolve the reentry problem for ballistic missiles. These research endeavors naturally progressed towards addressing the challenges of manned spaceflight. Another factor contributing to the national space program's growing interest was the planning and research devoted to the artificial Earth satellite program for the International Geophysical Year. The launch of Sputnik I in 1957 provided the necessary impetus for public support of a manned spaceflight project. At that time, the Atlas launch vehicle had reached a point where it could be seriously considered for manned spaceflight. It was the only American launch vehicle capable of lifting a payload for the manned orbital requirements.

On October 9, 1958, back at Langley, an airdrop program was initiated to develop full-scale parachute and landing systems for Project Mercury. The test field was a hive of activity, with engineers and technicians meticulously preparing parachutes and deployment systems. The air was filled with the hum of aircraft engines and the crisp rustle of fabric as parachutes were carefully packed and readied for deployment.

The sight of parachutes billowing open against the sky became a regular spectacle as tests progressed. Each successful deployment marked a critical step closer to ensuring safe landings for future astronauts. Engineers watched intently, noting the parachutes' performance and making adjustments as necessary. The thud of capsules hitting the ground and the applause following each successful test echoed across the field, a testament to the relentless pursuit of perfection.

Technicians on the ground hurried to retrieve and inspect the deployed parachutes, analyzing their condition and performance. Data collected from these tests were meticulously recorded and studied, providing invaluable insights into the design and functionality of the landing systems. The dedication and precision of the Langley team

were evident in every aspect of the program, each test bringing them closer to achieving a reliable and safe return system for America's first astronauts.

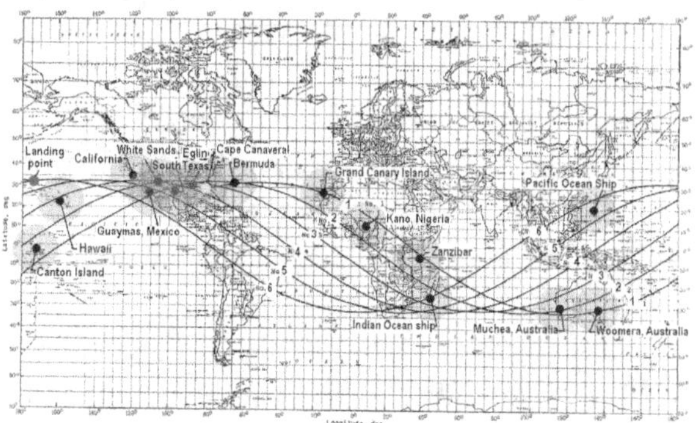

Mercury Tracking Network

On October 14, 1958, Project Mercury's administrative wheels turned swiftly. The Assistant Secretary of Defense for Supply and Logistics invited NASA to submit nominations for procurement urgency, commonly known as the DX priority rating. This rating would ensure that materials and components for the project were given the highest priority, underscoring the mission's national importance.

The DX priority rating was a critical designation, reflecting the urgent need to expedite the acquisition of essential equipment and resources. By granting this priority, the government acknowledged Project Mercury's paramount significance in national security and technological advancement during the Cold War era.

In the following days, NASA administrators and project managers worked diligently to compile and submit the necessary documentation. Their offices buzzed with activity as lists of required materials and components were meticulously prepared and reviewed. Telephones rang incessantly, and meetings were held to coordinate the efforts of various departments and contractors involved in the project.

The DX priority rating facilitated the rapid procurement of resources and galvanized the entire team, imbuing their work with a heightened sense of urgency and purpose.

Negotiations continued on October 17 and 18 as Langley personnel visited the Air Force Ballistic Missile Division in Inglewood, California. They sought to secure Atlas launch vehicles, essential for the manned satellite project. The discussions were intense, characterized by the strategic back-and-forth typical of high-stakes negotiations, each side keenly aware of the project's critical importance.

On October 21, a bidders' briefing for the Little Joe launch vehicle was held, drawing significant attention from numerous aerospace manufacturers. The Little Joe, a towering 48-foot-tall rocket, became a focal development point. The briefing was a critical step in selecting the right contractors to bring this innovative project to fruition.

Engineers and representatives from leading aerospace companies gathered to scrutinize Little Joe's specifications and consider its technical challenges. The rocket was designed to serve as a test vehicle for the Mercury spacecraft, capable of simulating various phases of a manned flight, including launch, reentry, and recovery operations.

The atmosphere at the briefing was charged with excitement and anticipation. Industry leaders and engineers marveled at the rocket's potential, discussing its design and the innovative approaches needed to overcome the challenges ahead. The Little Joe was not just another rocket; it was crucial to the nation's efforts to achieve manned spaceflight. Its successful development would provide invaluable data and experience, paving the way for the more complex missions.

Participants weighed in on the ambitious project as the briefing proceeded, assessing its feasibility and the resources required. The discussions were marked by a sense of urgency and purpose, as everyone present understood the importance of their contributions to Project Mercury's overall success.

The Little Joe briefing underscored the American aerospace industry's collaborative spirit and technical prowess. It highlighted the collective effort needed to tackle the daunting challenges of space exploration. By bringing together the best minds and companies, NASA aimed to ensure that the United States remained at the forefront of technological innovation and space achievement. The excitement and determination displayed at this briefing indicated the broader commitment to pushing the boundaries of human capability and making the dream of manned spaceflight a reality.

On October 23, NASA took a significant step forward by issuing preliminary specifications for a manned spacecraft to industry leaders. These detailed documents outlined the program's scope and proposed methods for analysis and construction. The release of these specifications was a crucial milestone, marking the transition from conceptual designs to tangible engineering plans. This move set the stage for the meticulous development processes, ensuring that every aspect of the spacecraft would meet the rigorous demands of manned spaceflight.

A few days later, on October 27, a Special Committee on Life Sciences was established at Langley. The committee's mandate was to

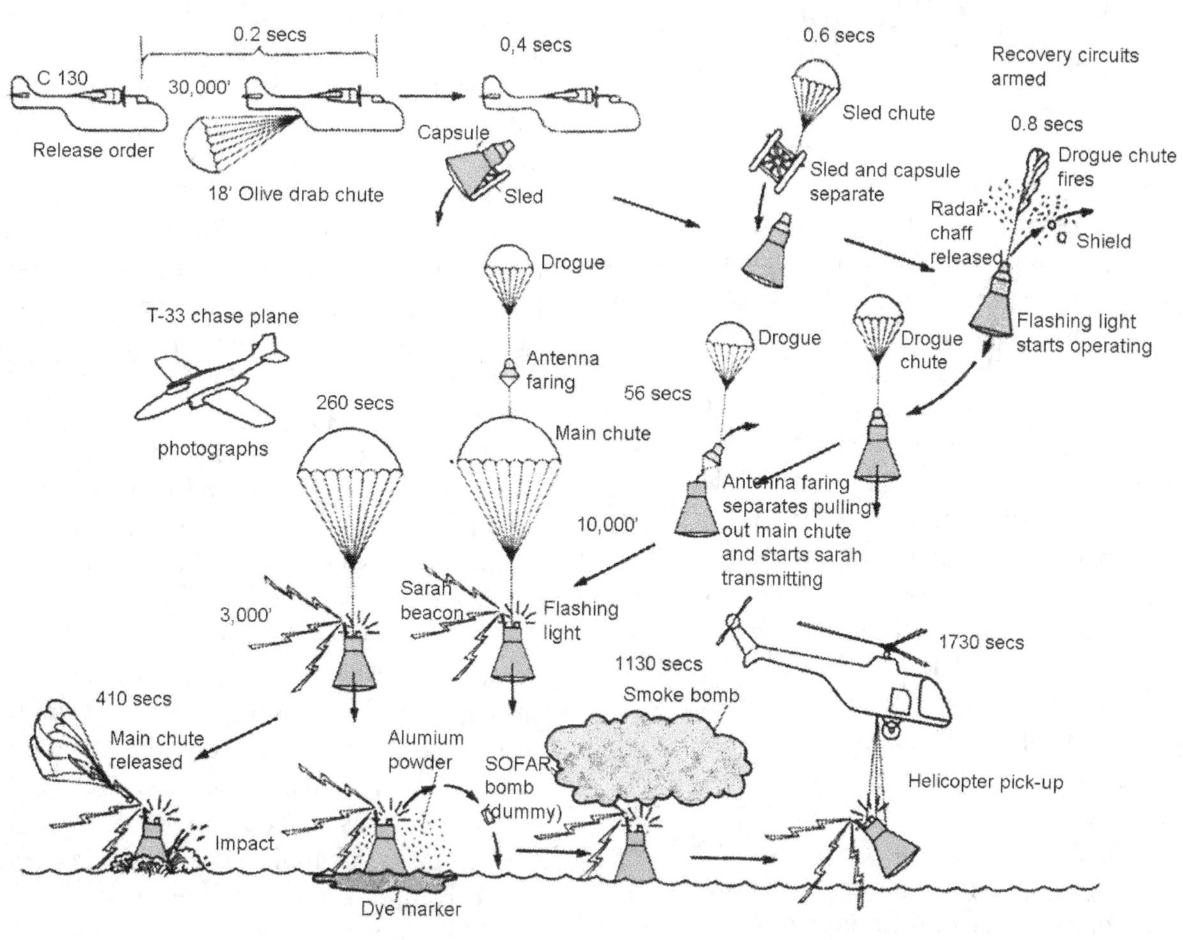

define the qualifications and attributes required for America's first astronauts and to provide guidance on the program's human aspects. This committee's work was pivotal in shaping the selection criteria and training programs for the pioneers of space travel. Their efforts ensured that the human element of the space program was meticulously planned and executed, addressing the physiological and psychological demands of spaceflight.

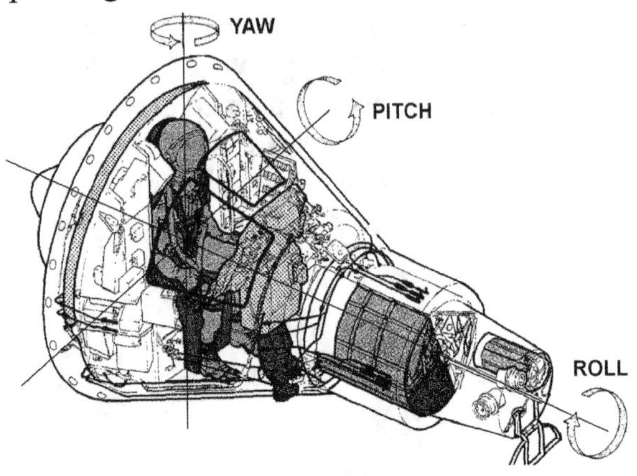

Throughout October, drop tests were conducted using full-scale capsules and parachute systems. These tests were essential for perfecting the deployment and stability of the parachutes, a critical component for the safe return of astronauts. Initially, concrete-filled drums were used to simulate the weight and dynamics of the capsules. As the tests progressed, actual spacecraft models were employed, providing more accurate data.

On November 3, 1958, military aeromedical personnel reported for duty, beginning an intensive research and planning focused on human factors, crew selection, and training plans. These specialists brought a wealth of knowledge and experience in aeromedical science, essential for addressing space travel's physiological and psychological challenges.

The collaboration between NASA and military experts underscored the interdisciplinary effort required for successful space exploration. These professionals collaborated to develop comprehensive selection criteria, ensuring that only the most capable individuals would be chosen as astronauts. Their studies delved into every aspect of human performance and well-being in the extreme conditions of space, from the effects of prolonged weightlessness to the stresses of reentry and landing.

Integrating military aeromedical expertise into NASA's program was a strategic move that leveraged the strengths of both organizations. The rigorous standards and protocols established by the military complemented NASA's innovative approach to space exploration, creating a robust framework for astronaut training and mission planning.

Training and Testing

Astronaut training was rigorous, taking place at Langley Research Center, Lewis Flight Propulsion Laboratory, and other locations. Wind tunnels, rocket sled tracks, and aircraft were utilized to study and perfect the spacecraft's systems. Recovery operations involved Navy ships and helicopters, ensuring that each mission could end safely.

On November 5, 1958, the Space Task Group was officially established at Langley Field, Virginia, although the groundwork for its formation had begun in early October. Under the leadership of Robert R. Gilruth and Charles J. Donlan, this group quickly began solidifying its workforce. The initial team comprised 35 dedicated individuals from the Langley Research Center, supplemented by additional personnel from the Lewis Research Center.

This core team was often called the "embryo workforce" of Project Mercury. Their primary mission was to lay the foundational groundwork necessary for the success of America's first manned spaceflight program. The Space Task Group focused on various critical tasks, including designing and testing spacecraft, developing life-support systems, and establishing safety protocols and procedures.

Under Gilruth and Donlan's guidance, the team brought together some of the country's most talented engineers, scientists, and technicians. They worked tirelessly, often pushing the boundaries of contemporary technology and knowledge. The collaborative environment fostered at Langley Field was instrumental in driving innovation and problem-solving, ensuring that each aspect of the project was meticulously planned and executed.

The formation of the Space Task Group marked a significant milestone in the history of NASA and Project Mercury. This small but highly skilled team laid the critical foundation for one of the most significant achievements in human space exploration. Their efforts were crucial in turning the ambitious goal of sending a man into space and returning him safely to Earth into a reality. Through their dedication and ingenuity, the Space Task Group set the stage for the following historic missions, helping to establish the United States as a leader in the space race.

On November 7, 1958, a pivotal contractor briefing was held at Langley Field, drawing the attention of 40 prospective bidders for the development of the manned spacecraft. The briefing was a key event in the timeline of Project Mercury, as it laid out the detailed specifications and high expectations for the ambitious program.

The atmosphere at the briefing was one of intense focus and anticipation. Engineers and representatives from leading aerospace companies gathered to absorb NASA's comprehensive technical requirements. These specifications were meticulously prepared, reflecting the project's technical rigor and the high standards set by NASA for the spacecraft that would eventually carry America's first astronauts into space.

The briefing highlighted various aspects of the spacecraft design, including structural integrity, life-support systems, propulsion, navigation, and safety features. Each component was scrutinized to ensure that the final product would meet the stringent demands of manned spaceflight. The prospective bidders were acutely aware of the challenges involved and the historic opportunity to contribute to a landmark project in human space exploration.

The detailed presentations underscored NASA's commitment to pushing the boundaries of technology and innovation. The specifications not only set the technical parameters but also conveyed the strategic importance of the project in the context of the Cold War and the space race. This was more than just a technological challenge; it was a national priority with profound implications for the United States' position on the global stage.

On November 14, 1958, the meticulous specifications for the first manned spacecraft were finalized and distributed to twenty leading aerospace firms. This dissemination of detailed requirements was more than a procedural step; it marked the transition from theoretical planning to concrete action.

The comprehensive specifications outlined every aspect of the spacecraft, from structural design and material selection to propulsion systems and life support mechanisms. These documents represented the culmination of countless hours of research and collaboration among scientists, engineers, and visionaries dedicated to making human spaceflight a reality.

The distribution of these specifications ignited a flurry of activity within the aerospace industry. Each of the twenty firms began to evaluate the requirements, assemble their best teams, and draft proposals meeting the stringent criteria. This project phase was crucial, as it would determine which companies would ultimately contribute to constructing the spacecraft that would carry humans into space.

This milestone was not just a technical achievement; it symbolized the space program's collective ambition and forward momentum. It underscored the United States' commitment to advancing human space exploration and set the stage for the following groundbreaking missions.

On November 20, 1958, the Space Task Group took a significant step to enhance its collaborative efforts by inviting the three branches of the U.S. military to assign liaisons to the project. This strategic move aimed to ensure seamless coordination and tap into the military's wealth of expertise.

The invitation extended to the Army, Navy, and Air Force was met with prompt action. By January 1959, each branch had selected and assigned their representatives to the Space Task Group. These liaisons brought invaluable experience in engineering, aeronautics, and logistics, further bolstering the project's multidisciplinary approach.

Integrating military expertise into the Space Task Group was more than a formality; it was a deliberate effort to unify the nation's best resources in pursuit of a common goal. The military's experience in advanced technologies, project management, and operational planning complemented the civilian scientific community's innovative drive and theoretical knowledge.

This collaboration ensured that the project benefited from a holistic perspective, blending the rigor and discipline of military precision with the creativity and ingenuity of civilian research. Military liaisons helped streamline communication, foster mutual understanding, and create a cohesive strategy for overcoming the myriad challenges of human spaceflight.

By the beginning of 1959, the Space Task Group was not just a collection of scientists and engineers; it had evolved into a well-rounded team poised to tackle the ambitious goal of manned space exploration. This multidisciplinary approach set the stage for the groundbreaking achievements that would soon follow, marking another critical milestone in the journey toward space.

On November 24, 1958, NASA took a pivotal step in the evolution of its space program by placing an order for an Atlas launch vehicle. This decision marked a significant milestone in the Mercury program, which aimed to propel the United States to the forefront of space exploration. The Atlas, a powerful intercontinental ballistic missile (ICBM) repurposed for human spaceflight, was chosen for its capability to deliver the necessary thrust to break free of Earth's gravitational pull and reach orbit.

This order was part of a broader preliminary research program designed to test and refine the technologies and methodologies essential for human spaceflight. Engineers and scientists worked tirelessly to adapt the Atlas for manned missions, modifying its structure to accommodate the Mercury capsule, which would house the astronaut.

Chapter 4 - The Technology Behind Mercury

The Mercury capsule, crafted by McDonnell Aircraft, was a marvel of engineering. It was designed to house enough water, food, and oxygen for about a day in a pressurized cabin. Each launch from Cape Canaveral Air Force Station was a spectacle, with the rockets roaring to life and piercing the sky with a trail of fire and smoke.

Spacecraft production in clean room at McDonnell Aircraft, St. Louis, 1960

Safety was paramount in every aspect of the project. The capsule featured a launch escape rocket, designed to propel it away from the main vehicle in case of a failure, ensuring the astronaut's survival. Ground control, via the Manned Space Flight Network, maintained constant communication, providing guidance and support throughout each mission. Retrorockets and an ablative heat shield were crucial for safe reentry, culminating in a parachute-assisted water landing. Recovery teams, often stationed on US Navy ships, swiftly moved in to retrieve the astronaut and the capsule from the ocean.

The first launch of this modified Atlas was scheduled for May 1959, setting the stage for the United States' entry into the era of manned space exploration. Although unmanned, this initial launch was critical for testing the rocket's performance and ensuring the safety and reliability of future missions.

Integrating the Atlas launch vehicle into the Mercury program was a testament to NASA's innovative spirit and commitment to overcoming the immense technical challenges of space travel. This rapid development and testing laid the groundwork for the Mercury mission's eventual success, culminating in Alan Shepard's historic flight on May 5, 1961, making him the first American to travel into space.

Shepard's flight watched on TV in the White House. May 1961.

On November 26, 1958, NASA officially named its manned satellite program Project Mercury. This designation was carefully chosen to symbolize NASA's unwavering commitment to advancing human spaceflight. The name "Mercury" evoked images of the swift and powerful Roman god, reflecting the program's ambitious goals and the agility required to achieve them.

USS Kearsarge with crew spelling Mercury-9. May 1963.

Project Mercury represented NASA's first significant effort to send humans into space and safely return them to Earth. It was an era marked by intense scientific inquiry and engineering innovation. The program aimed to investigate man's ability to function in space, understand the effects of space travel on the human body and develop the necessary technology to support these endeavors.

The choice of the name Mercury was apt. Just as the Roman god Mercury was known for his speed and ability to traverse great distances, NASA's Project Mercury sought to achieve rapid advancements in space technology and human space travel. The program's objectives included orbiting a manned spacecraft around Earth, investigating human capabilities in space, and safely recovering the astronaut and the spacecraft.

This period saw NASA engineers and scientists pushing the boundaries of what was possible, developing new materials, propulsion systems, and life-support mechanisms. The project laid the foundation for future space exploration missions and was a critical learning platform.

The successes and lessons of Project Mercury would pave the way for subsequent programs, such as Gemini and Apollo, ultimately leading to the historic moon landings and beyond. Through Project Mercury, NASA demonstrated America's technological prowess, inspiring a nation to look to the stars and dream of exploring the vast unknown.

On November 28, 1958, NASA successfully achieved a significant milestone by launching an Atlas launch vehicle. The rocket soared 6,300 miles downrange from Cape Canaveral, demonstrating its remarkable capability and setting a crucial precedent for future manned missions.

This successful launch was more than just a technological achievement; it demonstrated the Atlas rocket's reliability and power. The Atlas, initially designed as an intercontinental ballistic missile, had undergone significant modifications to serve the needs of space exploration. Its successful flight validated these modifications and provided the confidence to integrate the Atlas into NASA's burgeoning manned spaceflight program, Project Mercury.

The test flight's success was a testament to the rigorous testing and relentless efforts of the engineers and scientists involved. It marked a turning point, proving that the Atlas could be trusted to deliver a spacecraft into orbit, a critical capability for the Mercury missions that would soon follow.

Today's Atlas launch underscored the United States' growing rocketry and space technology capabilities. It was a clear signal to the world that NASA was making rapid progress in the space race, laying the groundwork for historic achievements in the following years. This launch was a technical success and a morale booster for the nation, fueling public interest and support for America's space endeavors.

November-December 1958 marked a period of intense activity and progress for Project Mercury. During these months, a scale model of the Mercury spacecraft was subjected to rigorous testing at the Arnold Engineering Development Center. The transonic test tunnel roared as the model, oriented for reentry, underwent comprehensive aerodynamic assessments. These tests were crucial for understanding the behavior of the spacecraft during its return to Earth, ensuring the safety and success of future manned missions.

Concurrently, studies on spacecraft recovery operations were initiated. These studies addressed the complexities of retrieving spacecraft after their missions, focusing on potential solutions for safe and efficient recovery. The challenges were immense, requiring innovative approaches and meticulous planning to guarantee that astronauts and their vehicles could safely recover from the ocean after splashdown.

As 1958 drew to a close, the foundation of Project Mercury was firmly established. The year had been marked by relentless effort, collaboration, and groundbreaking advancements. Scientists, engineers, and military personnel worked together with a shared vision, transforming the project from a bold concept into a tangible reality. Their collective determination and innovative spirit propelled the project forward, setting the stage for America's first steps into space.

December 1, 1958, marked a bustling day at the Space Task Group headquarters, where the chilly air was alive with the hum of activity and anticipation. Engineers and scientists were deeply engrossed in their work, huddling over blueprints and models of the Big Joe spacecraft. This spacecraft was slated for a critical reentry test, and its success was paramount to the progress of Project Mercury.

An Atlas launch vehicle was to boost the Big Joe spacecraft, a cornerstone of NASA's efforts to achieve manned spaceflight. The design and construction of Big Joe were collaborative efforts, with responsibilities divided between two key NASA research centers. The Langley Research Center focused on the spacecraft's aerodynamic design, while the Lewis Research Center took on the challenge of developing the instrument package.

The Lewis team, understanding the importance of their mission, worked tirelessly on the intricate instrument package that would provide crucial data during the test. Their efforts were instrumental in shaping the core of the Flight Operations Division at Cape Canaveral, which would oversee the launch and recovery of spacecraft.

This period was marked by an intense drive to solve the myriad technical challenges of human spaceflight. The collaborative efforts and relentless dedication of the engineers and scientists at both research centers were vital in ensuring that every component of the Big Joe spacecraft was meticulously designed and tested.

As 1958 came to a close, the groundwork laid during these months was pivotal for the future success of Project Mercury. The progress made on the Big Joe spacecraft and its forthcoming tests demonstrated the ingenuity and determination of NASA's teams. This foundational work was critical in advancing the United States 'capabilities in space exploration and setting the stage for the historic manned missions that would soon follow.

On December 8, 1958, a pivotal moment in the history of American space exploration occurred. The Space Task Group, responsible for the nation's burgeoning space program, ordered nine Atlas launch vehicles from the Air Force Ballistic Missile Division. This order was not just a procurement of rockets but a significant logistical advancement in the United States' efforts to lead in space exploration.

The Atlas launch vehicles were integral to the success of Project Mercury, America's first human spaceflight program. Project Mercury aimed to send an American astronaut into space and ensure their safe return, an ambitious goal that required reliable and powerful launch vehicles. The decision to order these Atlas rockets underscored the Space Task Group's commitment to overcoming the technical challenges of spaceflight.

The Atlas rockets were originally developed as intercontinental ballistic missiles (ICBMs) but were soon recognized for their potential in launching spacecraft. Their selection for Project Mercury highlighted a critical shift in their application—from instruments of war to tools of exploration and scientific discovery.

The impact of this order extended beyond Project Mercury's immediate goals. It marked a collaboration between military technology and space exploration, setting a precedent for future programs. The successful use of Atlas rockets in manned and unmanned missions provided valuable data and experience in shaping subsequent projects, such as Gemini and Apollo.

Moreover, this milestone played a crucial role in the broader context of the Cold War space race. The Atlas launch vehicles helped propel the United States forward, contributing to its prestige and demonstrating its technological prowess. This boosted national morale and fostered international recognition of America's capabilities in space exploration.

In summary, the December 8, 1958, order for nine Atlas launch vehicles was a defining moment in America's journey to space. It facilitated the achievements of Project Mercury, laid the groundwork for future missions, and symbolized the nation's commitment to exploring the final frontier.

On December 9, 1958, within the confines of a small, cramped office, an aeromedical selection team meticulously drafted a tentative astronaut selection procedure. This detailed plan represented a crucial step in the process of identifying the individuals who would become America's first astronauts. According to the proposal, 150 men were to be nominated by January 21, 1959. From this pool, 36 candidates would be chosen to undergo further testing.

The procedure outlined was rigorous and designed to ensure that only the most qualified candidates would advance. After an intensive nine-month training period, the ultimate goal was to narrow this group down to a core team of six astronauts. The selection criteria were stringent, reflecting the high physical, mental, and emotional standards required for the unprecedented challenges of space travel.

Despite the thoroughness and precision of the plan, it faced significant scrutiny. By the end of December, the selection procedure was ultimately rejected. This setback highlighted the complexities and evolving nature of the astronaut selection process, which required balancing technical expertise, physical endurance, and the ability to perform under extreme conditions.

The rejection of this initial plan underscored the iterative nature of developing a successful astronaut program. It prompted further refinements and adjustments, eventually adopting a robust selection process to yield the first group of NASA astronauts, the Mercury Seven. These astronauts became national heroes and laid the foundation for America's future achievements in space exploration.

This early effort, though unsuccessful, was an essential learning experience. It reflected the determination and adaptability of the Space Task Group as they navigated the uncharted territory of human spaceflight, contributing to the United States' progress in space and its enduring legacy in exploration.

On December 10, 1958, the Space Task Group took a pivotal step in advancing the United States' efforts in space exploration by appointing a Technical Assessment Committee. This committee, chaired by the esteemed aeronautical engineer Charles H. Zimmerman, was tasked with evaluating contractor proposals for the manned space capsule. Zimmerman's expertise, particularly his innovative work on vertical takeoff and landing aircraft, made him an ideal leader for this committee.

The committee comprised a team of highly skilled technical specialists from various fields, including aerodynamics, materials science, and systems engineering. Their mission was to meticulously scrutinize every component of the proposed designs to ensure they met the highest safety, reliability, and performance standards. This rigorous evaluation process was essential, as the success of the United States' manned space program hinged on the integrity and capability of the space capsule.

Establishing the Technical Assessment Committee marked a significant milestone in the nation's progress toward space exploration. The careful selection of contractors and the emphasis on technical excellence underscored the United States' commitment to achieving manned spaceflight. This initiative laid the groundwork for the development of the Mercury capsule, which would eventually carry the first American astronauts into space, symbolizing a leap forward in the space race and solidifying the United States' position as a leader in space technology.

As the committee's work progressed, it influenced broader technological advancements and innovations within the space program and in various industrial and scientific applications. The meticulous standards set by the committee ensured that the technologies developed during this period would have far-reaching impacts, contributing to the growth of aerospace engineering and the overall progress of the nation's space endeavors.

On December 11, 1958, the Lewis Research Center (now known as the NASA Glenn Research Center) made a pivotal presentation outlining its funding requirements for the Big Joe flight test spacecraft. This spacecraft was critical for testing the Mercury program's heat shield and reentry systems. The presentation detailed the budget for developing the attitude control and instrumentation systems, which were vital for ensuring the spacecraft's stability and precise navigation during flight.

The proposal was meticulously crafted, including comprehensive technical specifications and detailed financial projections. These projections were essential for demonstrating the project's feasibility and importance to the Space Task Group and other decision-makers. The Lewis Research Center's thorough presentation played a crucial role in illustrating the need for adequate funding to achieve the ambitious goals of the Mercury program.

By February 1959, after rigorous reviews and discussions, the Space Task Group confirmed the agreements and allocated the required funds to the Lewis Research Center. This financial transfer marked a significant step forward, enabling the team at Lewis to proceed with the development and testing of the Big Joe spacecraft's systems. The successful funding and subsequent advancements in the Big Joe project underscored the collaborative efforts and strategic planning necessary for the United States to achieve its space exploration objectives.

This milestone advanced the Mercury program and contributed to the broader progress of the nation's space capabilities. The development of reliable attitude control and instrumentation systems set a foundation for future spacecraft, enhancing the technological prowess that would propel the United States into a leading position in the space race. The Big Joe flight test ultimately provided critical data and confidence needed for the subsequent manned missions, highlighting the importance of early investments and detailed planning in the success of space exploration endeavors.

On December 12, 1958, Robert R. Gilruth, the esteemed Mercury Project Manager, sought crucial technical support from the Lewis Flight Research Branch. Recognizing the vital role of precise control and monitoring in the upcoming Big Joe tests, Gilruth's request underscored the importance of leveraging Lewis' advanced instrumentation facilities. These facilities were renowned for their state-of-the-art technology, capable of providing the high accuracy and reliability necessary for such a critical phase of the Mercury program.

Simultaneously, the Space Task Group intensified its efforts in the technical assessment of proposals for the manned spacecraft. Under the leadership of Charles H. Zimmerman, the team meticulously evaluated each submission, ensuring that only the most robust and innovative designs would be selected for further development. Zimmerman's expertise and the rigorous scrutiny applied by the committee were crucial in identifying the best solutions to meet the program's stringent requirements.

The involvement of the Lewis Flight Research Branch and the thorough technical assessment of spacecraft proposals marked significant steps in the preparation for manned spaceflight. These efforts highlighted the collaborative nature of the space program, where various branches and experts came together to address complex challenges. Lewis's precise control and monitoring capabilities were instrumental in the Big Joe tests, which aimed to validate the Mercury capsule's heat shield and reentry systems.

As these assessments and collaborations progressed, they not only paved the way for the successful execution of the Big Joe tests but also laid the groundwork for the broader objectives of the Mercury program. The detailed planning and integration of advanced technologies ensured that the United States was on a solid path toward achieving its goal of manned spaceflight. These foundational efforts were critical in propelling the nation's space program forward, ultimately contributing to the following historic achievements, including the successful orbital flights of American astronauts.

On December 13, 1958, excitement and tension filled the air as Gordo, a Navy-trained squirrel monkey, was prepared for a groundbreaking mission. Launched into space aboard an Army Jupiter missile nose cone, Gordo's journey was a pivotal moment in the early stages of space exploration. Although the mission ultimately faced a setback due to a malfunctioning float mechanism that prevented the recovery of the nose cone, it provided invaluable biomedical data that significantly advanced our understanding of space travel's physiological impacts on living organisms.

During his brief flight, Gordo experienced 8.3 minutes of weightlessness and extreme takeoff and reentry pressures. These conditions offered critical insights into how space travel affects the body, helping scientists understand the challenges humans would later face in similar environments. The data collected from Gordo's mission were meticulously analyzed, contributing to developing safety protocols and life support systems for future manned spaceflights.

Despite its imperfect outcome, Gordo's mission underscored the importance of animal testing in the early phases of space exploration. The findings from this flight informed subsequent missions and played a crucial role in shaping the biomedical research strategies that would ensure the safety and well-being of astronauts. This pioneering effort marked a significant step forward in the United States' quest to send humans into space, demonstrating the country's commitment to understanding and overcoming the challenges of extraterrestrial travel.

The excitement surrounding Gordo's launch reflected the broader national enthusiasm for space exploration during this era. Each mission, whether deemed a success or a failure, provided essential knowledge that fueled the progress of the space program. Gordo's flight, in particular, highlighted the complexities and risks of space travel while also showcasing the resilience and dedication of the scientists and engineers working tirelessly to push the boundaries of human knowledge and capability.

In a defining policy speech on December 17, 1958, Dr. T. Keith Glennan, the first Administrator of NASA, officially introduced the manned satellite project by its new name: Project Mercury. This designation was not merely a label but a symbol of the mission's profound significance, representing a new era in the United States' space exploration endeavors.

Dr. Glennan's announcement of Project Mercury set a clear and ambitious direction for NASA, encapsulating the nation's aspirations to send humans into space and safely return them to Earth. The name "Mercury," inspired by the Roman god known for his speed and mobility, aptly reflected the project's goal of achieving swift advancements in manned spaceflight. It also underscored the urgency and competitive spirit of the era, driven by the space race with the Soviet Union.

Project Mercury became the cornerstone of NASA's human spaceflight program. It aimed to orbit a manned spacecraft around Earth, investigate human capabilities in space, and safely recover the astronaut and spacecraft. The project involved rigorous planning, technological innovation, and extensive testing, setting the stage for the following historic missions.

Dr. Glennan's speech highlighted the collaborative efforts and dedication of countless scientists, engineers, and astronauts working towards this monumental goal. It inspired a sense of national pride and commitment to achieving what once seemed impossible. The introduction of Project Mercury marked the beginning of a series of groundbreaking achievements in space exploration, culminating in the successful orbital flights of astronauts like John Glenn and paving the way for future endeavors such as the Apollo moon landings.

On December 29, 1958, North American Aviation was awarded a pivotal contract for designing and constructing the Little Joe airframe. This contract represented a crucial step in developing the launch vehicle required to test and refine the Mercury spacecraft's capabilities.

The Little Joe rocket was conceived as a low-cost, reliable test vehicle that could simulate spaceflight conditions. Its primary purpose was to conduct high-altitude abort tests, validating the Mercury spacecraft's escape system and other critical components. These tests were essential to ensure the safety and reliability of the manned missions, which were the ultimate goal of Project Mercury.

North American Aviation, renowned for its expertise in aerospace engineering, was tasked with creating a robust and efficient airframe that could withstand the rigorous demands of these tests. The Little Joe rocket would carry the Mercury capsule to suborbital altitudes, allowing engineers to assess the spacecraft's performance under various conditions.

The award of this contract underscored NASA's methodical and progressive approach to human spaceflight. By ensuring that each component and system of the Mercury spacecraft was thoroughly tested and refined, NASA aimed to minimize risks and enhance the chances of success for its astronauts.

The development of the Little Joe airframe was a critical milestone in the Mercury program. It provided the necessary platform for conducting a series of crucial tests that would ultimately contribute to the success of America's first manned space missions. These efforts advanced the United States' technical capabilities and demonstrated the nation's commitment to leading the way in space exploration.

The success of the Little Joe tests helped build confidence in the Mercury spacecraft's design and performance, paving the way for the historic manned missions that would follow. This contract with North American

Aviation was thus a key element in the broader narrative of America's journey to space, highlighting the importance of careful planning, rigorous testing, and collaborative effort in achieving groundbreaking accomplishments.

As the year drew to a close on December 30, 1958, the Space Task Group's technical assessment teams reached a significant milestone by completing their evaluation of industry proposals for manned spacecraft. This comprehensive review process involved meticulous scrutiny of each proposal, assessing the technical merits, feasibility, and potential of the designs submitted by various aerospace contractors.

The findings from these evaluations were compiled into detailed reports and forwarded to the Source Selection Board at NASA Headquarters. This board decided which contractor would be awarded the project to develop the Mercury spacecraft. The Source Selection Board's task was to carefully consider the technical assessments alongside other critical factors such as cost, schedule, and contractor capability.

Completing the technical assessments marked a crucial phase in the Project Mercury timeline. It demonstrated the dedication and expertise of the Space Task Group's teams, who worked tirelessly to ensure that the selected design would meet the rigorous demands of human spaceflight. This step was essential in laying a solid foundation for the next development and testing stages.

The forwarding of the assessment findings to NASA Headquarters set the stage for the final selection process, a decision that would shape the future of America's space program. The chosen contractor would play a pivotal role in designing and building the spacecraft that would eventually carry the first American astronauts into space, a monumental achievement in the space race.

On December 31, 1958, the Space Task Group solidified its plans for the Mercury program by placing a letter of intent with North American Aviation to fabricate the Little Joe Test vehicle airframe. This step was crucial for maintaining the project's momentum and ensuring that the ambitious goals of Project Mercury remained within reach.

The aggressive schedule detailed in the letter of intent called for the delivery of airframes every three weeks starting in June 1959. This rapid production timeline was essential to keeping pace with the Mercury program's testing and development needs. The Little Joe rockets were specifically designed to conduct high-altitude abort tests, which were critical for validating the Mercury spacecraft's escape system and other vital components.

With all major rocket motors already ordered and scheduled for early delivery, the program was well-positioned to meet its flight test schedule. The Space Task Group's careful coordination and forward planning underscored its commitment to methodically advancing the United States' capabilities in human spaceflight.

The meticulous planning and coordination extended beyond just hardware; it involved synchronizing efforts across multiple teams and contractors to ensure that all components of the Mercury program were progressing in unison. This included detailed logistical planning, rigorous timelines, and a focus on quality and reliability at every stage.

By the end of 1958, the Space Task Group had established a robust foundation for Project Mercury. The letter of intent with North American Aviation and the ensured timely delivery of rocket motors highlighted the group's strategic foresight and dedication. These preparations were crucial for the upcoming tests and eventual manned missions, reflecting a concerted effort to push the boundaries of space exploration.

In January 1959, the Mercury project faced a significant 22-month delay from its inception to its first orbital mission. This ambitious endeavor involved a complex network of contributors, including a dozen prime contractors, 75 major subcontractors, and approximately 7,200 third-tier subcontractors.

By 1969, NASA's estimated cost for Project Mercury amounted to $392.6 million. This budget was meticulously allocated across various essential components: $135.3 million for spacecraft development, $82.9 million for launch vehicles, $49.3 million for operations, $71.9 million for tracking operations and equipment, and $53.2 million for necessary facilities.

Originally conceptualized as "Man in Space Soonest," the precursor program to Project Mercury aimed for an audacious goal—a manned Moon landing by 1965. This ambitious initiative was envisioned to utilize a "Super Titan" rocket and was projected to cost $1.5 billion at the time (equivalent to approximately $15.7 billion in today's terms). While this specific timeline and approach evolved, the foundational work and technological advancements from this early program laid crucial groundwork for NASA's subsequent missions, notably the Gemini and Apollo programs, which ultimately realized the dream of landing humans on the Moon.

Selecting the Mercury Astronauts

In early February 1959, the quest to find America's first astronauts intensified under NASA's Mercury program. Detailed scrutiny of 508 service records identified potential pilot candidates, each evaluated against stringent criteria that mirrored the exacting demands of pioneering space missions. From this exhaustive review, approximately 110 pilots emerged as initial contenders, distinguished by their blend of skills, experience, and robust physical fitness.

The selection process was rigorous and methodical. Initial briefings and interviews further winnowed the field, ensuring that only those with unwavering determination and commitment to exploration advanced. By mid-February, 53 pilots had volunteered for the Mercury program, demonstrating their readiness to venture into the unknown for scientific advancement and national pride.

From this group of eager volunteers, 32 candidates were chosen to undergo the next evaluation phase. This phase rigorously tested their flying proficiency and evaluated their physical and mental endurance, preparing them for the daunting challenges that lay ahead.

Scheduled for comprehensive physical examinations at the esteemed Lovelace Clinic in Albuquerque, New Mexico, these candidates faced a battery of tests designed to assess their fitness for space travel. The Lovelace Clinic, renowned for its expertise in aerospace medicine, conducted thorough evaluations encompassing cardiovascular health, neurological stability, and overall physical resilience.

Beyond selecting the Mercury astronauts, these evaluations held profound significance. They laid the groundwork for understanding how the human body responds to the extreme spaceflight conditions, informing future missions and shaping the ongoing assessment and training protocols for astronauts.

As the 32 candidates underwent their assessments at the Lovelace Clinic, they stood on the precipice of history. Their journey through the selection process underscored the meticulous preparation and unwavering dedication with which the United States approached manned spaceflight. These men were not merely pilots but trailblazers destined to lead the nation into a new era of exploration and discovery.

Establishing these rigorous qualifications in early 1959 marked a pivotal milestone for Project Mercury. By meticulously selecting individuals who met exacting technical and physical standards while embodying exceptional courage and adaptability, NASA laid the foundation for successfully executing its inaugural space missions. The meticulous preparation of the Mercury Seven underscored NASA's commitment to safety, excellence, and the limitless possibilities awaiting humanity beyond Earth's atmosphere.

On January 6, 1959, a pivotal meeting occurred at NASA Headquarters to debate the methods for spacecraft heat protection, a critical component for ensuring the safety and success of the Mercury missions. The discussion centered on two primary approaches: the beryllium heat sink and ablation methods.

The beryllium heat sink method used a thick layer of beryllium to absorb and dissipate the heat generated during reentry. While this method was reliable, it added significant weight to the spacecraft, potentially limiting its performance and payload capacity.

In contrast, the ablation method used a heat shield material that gradually eroded and burned away during re-entry, carrying the heat away. Though more complex, this method offered the advantage of being lighter and more efficient in managing the intense heat of reentry.

After thorough deliberation, the team decided to modify the spacecraft structure to accommodate both methods interchangeably. This flexibility ensured that NASA could adapt to the latest advancements and choose the most effective heat protection method as the project progressed.

The team selected fiberglass bonded with a modified phenolic resin for the ablation heat shields. This material was chosen due to its favorable structural properties after reentry heating, providing reliable protection while maintaining the integrity of the spacecraft. The phenolic resin offered excellent heat resistance and mechanical strength, making it an ideal choice for withstanding the extreme temperatures encountered during reentry.

This decision marked a significant milestone in the development of the Mercury spacecraft, reflecting NASA's commitment to innovation and safety. By preparing the spacecraft to support both heat protection methods, NASA ensured that they could leverage the best available technology, enhancing the likelihood of mission success.

On January 9, 1959, the Source Selection Board, composed of top NASA officials, concluded their meticulous evaluation of industry proposals for the Mercury spacecraft. The board's comprehensive report was presented to Dr. T. Keith Glennan, the NASA Administrator. After thorough consideration, the board selected McDonnell Aircraft Corporation as the prime contractor to develop and produce the Mercury spacecraft.

This decision was a pivotal moment in the history of American space exploration. McDonnell Aircraft Corporation, known for its expertise and innovation in aerospace engineering, was tasked with designing and building the spacecraft that would carry the first Americans into space. The company's proven track record and technical capabilities made it the ideal choice to tackle the formidable challenges of human spaceflight.

Selecting McDonnell Aircraft Corporation set the course for the Mercury program, ensuring that the spacecraft would be developed with the highest safety, reliability, and performance standards. The decision also began a close and productive collaboration between NASA and McDonnell, characterized by a shared commitment to pioneering space exploration.

The McDonnell team quickly set to work, leveraging their expertise to address the unique challenges of the Mercury project. Their efforts included developing advanced life support systems, perfecting the spacecraft's heat shield technology, and ensuring the capsule could withstand the harsh conditions of space travel and reentry.

This partnership and the subsequent development of the Mercury spacecraft laid the groundwork for America's success in the space race. The Mercury program's achievements, culminating in the historic flights of astronauts like Alan Shepard and John Glenn, showcased the capabilities and resilience of both NASA and McDonnell Aircraft Corporation.

The selection of McDonnell as the prime contractor was more than a business decision; it was a defining moment that shaped the future of American manned spaceflight. It reflected NASA's strategic vision and commitment to excellence, setting a precedent for the collaborative efforts that would drive the nation's achievements in space exploration for decades.

On January 14, 1959, preliminary negotiations with McDonnell Aircraft Corporation commenced, marking the beginning of a critical phase in the Mercury spacecraft's research and development program. These initial discussions focused on the project's intricate technical and legal aspects essential for setting clear expectations and guidelines for the work ahead.

In conference rooms filled with blueprints, technical documents, and detailed project plans, representatives from NASA and McDonnell engaged in in-depth discussions. Engineers, scientists, and legal experts collaborated to address various technical specifications, design requirements, and contractual obligations. The atmosphere was intense focus and determination, as both parties understood the significance of their mission.

Key negotiation topics included the spacecraft's design features, such as its structural integrity, life support systems, and heat shield technology. The teams also delved into the testing and validation protocols required to ensure the spacecraft's safety and reliability under the extreme conditions of spaceflight and reentry.

Additionally, the legal teams worked to finalize the terms of the contract, covering aspects such as timelines, deliverables, and intellectual property rights. Ensuring clear communication and mutual understanding between NASA and McDonnell was paramount to the project's success.

These preliminary negotiations laid the essential groundwork for the Mercury spacecraft's design and construction. NASA and McDonnell established a solid technical and legal agreement foundation, setting the stage for a collaborative and efficient development process. This phase was crucial in aligning the efforts of both organizations toward the common goal of achieving manned spaceflight.

On January 16, 1959, NASA formally requested the Army Ordnance Missile Command to construct and launch eight Redstone and two Jupiter launch vehicles to support Project Mercury. This request highlighted the project's dependence on military technology and expertise to achieve its ambitious manned spaceflight goals.

The Redstone and Jupiter rockets were chosen for their proven reliability and performance. The Redstone rocket, originally developed as a ballistic missile, had already demonstrated its capabilities in previous tests, making it a suitable choice for launching the Mercury spacecraft on suborbital missions. With its greater thrust and payload capacity, the Jupiter rocket was selected for orbital missions and other critical tests.

NASA's reliance on military technology was a testament to the collaborative spirit that characterized the early days of the American space program. The expertise and resources of the Army Ordnance Missile Command were invaluable in meeting the technical and logistical challenges of launching manned spacecraft. This partnership allowed NASA to leverage existing technology and infrastructure, accelerating the timeline for achieving its objectives.

The construction and launch of these vehicles involved a complex and coordinated effort. Engineers and technicians from NASA and the military worked closely together to adapt the Redstone and Jupiter rockets for their new roles in the Mercury program. This included modifications to accommodate the Mercury capsule, ensure astronaut safety, and integrate sophisticated tracking and telemetry systems.

This request to the Army Ordnance Missile Command underscored the critical role of military support in Project Mercury's success. NASA's collaboration with the military facilitated the development of reliable launch vehicles and contributed to the advancement of space technology.

As these rockets were built and prepared for launch, they would undergo rigorous testing and evaluation to ensure their readiness for manned missions. The successful deployment of the Redstone and Jupiter launch vehicles would pave the way for the historic flights of the Mercury astronauts, marking significant milestones in America's journey to space.

This strategic partnership between NASA and the military exemplified the nation's unified effort to achieve space exploration goals, demonstrating how combined resources and expertise could overcome the formidable challenges of launching humans into space.

On January 21, 1959, NASA initiated the meticulous process of screening records for prospective astronauts, a crucial step in selecting the individuals who would become America's first space explorers. This comprehensive review involved sifting through detailed records to identify the most qualified candidates for the historic missions of Project Mercury.

The screening process was rigorous and exacting, reflecting the high standards and significant responsibilities that would be placed on the astronauts. Candidates were evaluated based on stringent criteria, including their physical and mental fitness, professional qualifications, and flight experience.

Medical records were scrutinized to ensure that candidates were in peak physical condition and free from any medical issues that could compromise their performance or safety during spaceflight. Psychological evaluations were also critical in assessing the candidates' ability to handle the intense stress and isolation of space missions.

In addition to health and psychological fitness, educational and professional backgrounds were thoroughly reviewed. Candidates were required to have a strong foundation in engineering or a related field and extensive flight experience. Military test pilots, with their rigorous training and familiarity with high-performance aircraft, were among the primary candidates considered for the program.

Flight records were analyzed to verify that candidates had the necessary experience and expertise in handling advanced aircraft, particularly those with experience in jet piloting. This ensured that the selected astronauts possessed the skills and judgment needed to manage the complexities of spaceflight and respond effectively to any emergencies.

This detailed and methodical screening process ensured that only the best and most capable individuals were chosen to become the first American astronauts. It was a crucial step in preparing for the manned missions that would soon captivate the nation and the world.

The selected candidates would eventually undergo further rigorous training and testing, culminating in the formation of the "Mercury Seven," the first group of astronauts who would carry the hopes and dreams of a nation into space. The meticulous screening of records on January 21, 1959, marked the beginning of this historic journey, laying the groundwork for America's achievements in human spaceflight.

The Little Joe Program

On January 23, 1959, NASA allocated an additional $1,556,200 to Langley Research Center to develop the Little Joe program. Combined with previous allocations, this substantial funding boost ensured that the Little Joe program had the necessary financial support to proceed without delays, maintaining its critical role in the broader Project Mercury timeline.

The Little Joe program was integral to Project Mercury. It was designed to provide a low-cost, reliable test vehicle for the Mercury spacecraft. Its primary purpose was to conduct high-altitude abort tests to validate the spacecraft's escape system and other crucial components. These tests would simulate emergency conditions, ensuring that the Mercury capsule and its systems could protect astronauts in the event of a launch failure.

The Little Joe program played a significant role in testing the Mercury spacecraft. It utilized seven airframes for eight flights, achieving mixed results, with three successful flights. These tests were critical in evaluating the launch escape system, spacecraft aerodynamics, and structural integrity under various conditions. Production spacecraft and boilerplates were employed in these test flights, providing valuable data for the program.

Heat Shield Testing: The uncrewed flights tested the spacecraft's heat shield, ensuring it could withstand the intense heat generated during reentry into Earth's atmosphere. This was vital for protecting future astronauts.

Aerodynamics and Structural Integrity: The flights assessed the spacecraft's aerodynamics and structural integrity, verifying that it could maintain stability and withstand the stresses of launch and reentry.

Launch Escape System: The launch escape system was tested to ensure it could safely propel the spacecraft away from the launch vehicle in case of an emergency during ascent. This system was crucial for astronaut safety.

The additional funds allocated to Langley Research Center expedited the construction, testing, and refinement of the Little Joe rockets. This financial support enabled the procurement of materials, the assembly of test vehicles, and the execution of numerous test flights, all of which were vital to the success of Project Mercury.

Langley Research Center, renowned for its aeronautical research and engineering capabilities, was pivotal in developing the Little Joe rockets. The center's expertise ensured that the rockets were designed and tested to meet the rigorous demands of spaceflight. The additional funding reinforced Langley's capacity to address technical challenges and advance the program according to schedule.

The timely infusion of funds on January 23, 1959, underscored NASA's commitment to maintaining momentum in the Mercury program. By securing the financial resources needed for the Little Joe development, NASA ensured that this crucial testing phase could proceed smoothly, providing valuable data and insights to enhance the safety and reliability of the manned missions.

On January 25, 1959, McDonnell Aircraft Corporation delivered a crucial piece of equipment to NASA: the pilot egress trainer. This trainer was essential for preparing astronauts to handle emergencies, specifically the challenges they might encounter during a splashdown in the ocean following reentry.

The Space Task Group personnel wasted no time in putting the trainer to use. They immediately began conducting rough water evaluations, simulating the harsh and unpredictable conditions astronauts could face upon landing in the ocean. These tests ensured the astronauts could efficiently and safely exit the spacecraft in various emergencies, including strong waves, high winds, and other maritime hazards.

The pilot egress trainer was designed to replicate the Mercury spacecraft's cabin, providing astronauts with a realistic environment to practice egress procedures. These procedures included releasing harnesses, opening hatches, and deploying flotation devices, all while dealing with the disorienting effects of being in the water.

The rough water evaluations involved placing the trainer in water tanks and subjecting it to wave machines and other equipment miming the ocean's tumultuous conditions. Astronauts practiced getting out of the spacecraft quickly and safely, often under timed conditions to simulate the urgency they would face during an actual splashdown.

These rigorous and demanding training sessions were crucial for building the astronauts' confidence and competence. By practicing in controlled yet challenging environments, the astronauts could better prepare for the real-life scenarios they might encounter during their missions.

The arrival and immediate utilization of the pilot egress trainer underscored NASA's commitment to comprehensive preparation and safety. It also highlighted the collaborative efforts between NASA and McDonnell Aircraft Corporation, as they worked together to equip the astronauts with the skills and knowledge needed for their historic journeys.

These training exercises significantly ensured that the Mercury missions would reach space and bring their astronauts back safely.

On January 26, 1959, NASA completed contract negotiations with McDonnell Aircraft Corporation to design and develop the Mercury spacecraft. This milestone marked the formal agreement that would propel the construction of the spacecraft tasked with carrying America's first astronauts into space.

During these negotiations, McDonnell estimated that the first three Mercury spacecraft could be delivered within ten months. This ambitious timeline underscored McDonnell's confidence in its engineering capabilities and commitment to the Mercury program. However, as the project progressed, it became clear that some refinements and adjustments were necessary to ensure the spacecraft met the rigorous standards required for manned spaceflight. Despite these enhancements, the timeline was extended by only two months, demonstrating efficient project management and a steadfast dedication to meeting deadlines.

The contract outlined the critical design and development phases, including the construction of the spacecraft, integration of systems, and extensive testing to ensure safety and reliability. McDonnell's engineers and technicians would work closely with NASA's Space Task Group to address technical challenges and incorporate technological advancements.

Completing these negotiations was a significant step forward for Project Mercury. It provided the financial and logistical framework necessary to move from planning to tangible development. It also reinforced the collaborative spirit between NASA and McDonnell, setting the stage for a productive partnership to achieve the historic goal of manned spaceflight.

With the contract, McDonnell could now focus on transforming conceptual designs into operational spacecraft. Though slightly adjusted, the timeline remained aggressive and reflected the urgency of the space race. Both organizations' efficient coordination and mutual dedication were pivotal in maintaining momentum and ensuring that the Mercury spacecraft would be ready for its critical missions.

This agreement marked the beginning of an intensive period of development and testing, ultimately leading to the successful launch of the first American astronauts. The contractual foundation laid on January 26, 1959, was instrumental in advancing the United States' efforts in space exploration, paving the way for the remarkable achievements that would soon follow.

On January 29, 1959, NASA completed the draft plan for the Little Joe flight test program, an essential component of the

Mercury project. With scheduled updates on April 14, 1959, this comprehensive test program outlined several primary objectives to ensure the Mercury spacecraft's safety and functionality for future manned missions.

The primary objectives of the Little Joe flight test program included:

Investigating Flight Dynamics: Understanding the Mercury spacecraft's flight dynamics was critical for mission success. The tests aimed to gather data on the spacecraft's stability, control, and behavior during different phases of flight, including launch, ascent, and re-entry. This information would help engineers refine the spacecraft's design and flight procedures.

Checking Drogue Parachute Operations: The drogue parachute system was a vital safety feature designed to decelerate the spacecraft before deploying the main parachutes. The Little Joe tests were intended to verify the deployment sequence and functionality of the drogue parachutes under various conditions, ensuring they could reliably perform during reentry and splashdown.

Determining Physiological Effects of Acceleration on a Small Primate: To simulate the conditions that human astronauts would experience, small primates were to be flown aboard the Little Joe rockets. These tests aimed to study the physiological impacts of rapid acceleration and deceleration on living organisms, providing crucial data to ensure the health and safety of future astronauts.

Assessing Spacecraft Aerodynamic Characteristics: Evaluating the Mercury spacecraft's aerodynamic performance was essential for understanding how it would behave under the stresses of launch, flight, and re-entry. The tests would measure aerodynamic forces and moments, helping engineers optimize the spacecraft's shape and structural integrity.

The draft plan for the Little Joe flight test program demonstrated NASA's thorough and methodical approach to developing the Mercury spacecraft. The program aimed to identify and mitigate potential risks by addressing these key objectives, thereby enhancing the reliability and safety of manned space missions.

The planned April 14, 1959, updates reflected NASA's commitment to continuous improvement and iterative development. This approach allowed the integration of new findings and technological advancements, ensuring the Mercury program remained at the forefront of space exploration.

The Little Joe flight test program was a cornerstone of Project Mercury, providing essential validation and data that would pave the way for the successful launch of America's first astronauts.

On January 30, 1959, Admiral Arleigh Burke, the Chief of Naval Operations, informed Dr. T. Keith Glennan, NASA Administrator, that Navy candidates had officially begun the selection process for Project Mercury. This marked a significant step in the interdisciplinary collaboration critical for the success of America's first manned spaceflight program.

The involvement of Navy personnel underscored the importance of leveraging the expertise and skills of military aviators, who were highly trained in handling advanced aircraft and operating under extreme conditions. The rigorous selection process focused on identifying individuals who met the stringent criteria established by NASA, including age, height, physical fitness, education, and flight experience.

Navy candidates brought a wealth of experience, particularly in high-performance jet piloting and advanced navigation techniques. Their rigorous training and familiarity with high-stress environments made them ideal candidates for the challenges of spaceflight. The collaboration between NASA and the Navy exemplified the broader national effort to pool resources and expertise from various disciplines to achieve the ambitious goals of Project Mercury.

This integration of military and civilian efforts highlighted the interdisciplinary nature of the space program. It demonstrated how different U.S. government and military branches could work together towards a common goal. The Navy's participation in the astronaut selection process was crucial in building a team capable of pioneering human space exploration.

The selection of Navy candidates, alongside those from other military branches and civilian test pilots, ensured a diverse pool of highly qualified individuals from which NASA could choose its first astronauts. This collaborative approach was instrumental in assembling the group that would eventually become known as the "Mercury Seven," the first American astronauts who would go on to make history with their groundbreaking missions.

Throughout January 1959, significant progress was made in developing the Mercury spacecraft, as McDonnell Aircraft Corporation selected key subcontractors to contribute to various critical systems. These selections and subsequent contract awards underscored the collaborative effort required to achieve Project Mercury's ambitious goals.

McDonnell chose Minneapolis-Honeywell as the subcontractor for the Mercury stabilization system. Minneapolis-Honeywell, known for its expertise in control systems, was tasked with developing the technology necessary to maintain the spacecraft's orientation and stability during flight. This system was crucial for ensuring the safe and precise control of the spacecraft throughout its mission.

In addition to Minneapolis-Honeywell, several other subcontractors were considered for vital components of the Mercury spacecraft. Bell Aircraft Rockets Division was selected for the reaction control system, responsible for providing the small thrusts needed to maneuver the spacecraft in space. General Electric, Barnes Instruments, and Detroit Controls were evaluated for the horizon scanner, a key instrument for determining the spacecraft's orientation relative to Earth. Ultimately, contracts were awarded to Bell Aircraft Rockets Division for the reaction control system and Barnes Instruments for the horizon scanner, recognizing their capabilities to deliver reliable and innovative solutions.

The extensive subcontracting process reflected the complex and multidisciplinary nature of the Mercury program. By leveraging the specialized expertise of these companies, NASA and McDonnell ensured that each critical system of the spacecraft was developed to the highest standards of performance and reliability.

Furthermore, study contracts were awarded to Aeronutronics, Space Electronics, and MIT Lincoln Laboratory. These organizations assisted in developing tracking and ground instrumentation plans for Project Mercury. Their work was essential for establishing a robust and effective network to monitor and communicate with the spacecraft throughout its mission.

Aeronutronics focused on advanced electronics and instrumentation, while Space Electronics contributed its knowledge of space systems. MIT Lincoln Laboratory, renowned for its cutting-edge research, provided expertise in developing sophisticated tracking technologies. Together, these organizations played a pivotal role in creating the infrastructure needed to support the Mercury missions, ensuring precise tracking and reliable communication with the spacecraft.

The careful selection of subcontractors and the awarding of study contracts in January 1959 demonstrated the meticulous planning and strategic coordination behind Project Mercury. These efforts laid the groundwork for successfully developing and testing the spacecraft, ultimately leading to the historic achievements of America's first manned space missions. This collaborative approach highlighted the importance of integrating diverse expertise and resources to overcome the challenges of human space exploration.

Chapter 5 - Astronaut Selection

On February 7, 1959, the selection process for America's first astronauts reached a pivotal stage with the commencement of medical evaluations at the Lovelace Clinic in Albuquerque, New Mexico. This marked a crucial phase in identifying the individuals who would venture into space as part of Project Mercury, the United States' inaugural human spaceflight program.

The Lovelace Clinic, renowned for its specialization in aerospace medicine, was the ideal facility for conducting these rigorous examinations. Prospective astronauts underwent a comprehensive battery of tests meticulously designed to assess their physical endurance and psychological resilience. These evaluations were crucial in ensuring that only the most qualified candidates, capable of withstanding the demanding conditions of space travel, were selected.

The medical tests encompassed a range of assessments, including cardiovascular evaluations to gauge heart health under stress, endurance trials to simulate the physical strains of spaceflight, and sensory assessments to evaluate balance and coordination. Each test aimed at identifying individuals who could perform effectively in the challenging space environment.

The selection criteria were stringent, reflecting the mission's pioneering nature and the high stakes involved. Beyond physical fitness, candidates were also evaluated for their psychological fortitude and ability to handle the psychological stresses inherent in space exploration.

The rigorous evaluations at the Lovelace Clinic underscored NASA's commitment to ensuring the safety and success of its astronauts. The selected individuals would not only make history by becoming America's first space travelers. Still, they would also pave the way for future missions, eventually leading to lunar landings and beyond.

Overall, the medical tests conducted at the Lovelace Clinic on February 7, 1959, marked a significant milestone in the early days of America's space program. They represented a crucial step toward fulfilling the nation's aspirations of achieving human spaceflight and set the stage for the historic achievements that would follow in the years to come.

In addition to physical assessments, psychological evaluations were conducted to gauge the candidates' mental resilience and stability. Space travel posed unprecedented challenges, not only to the body but also to the mind. The missions' isolation, confinement, and high-stakes nature required individuals who could maintain composure and clarity under extreme conditions.

The medical screening at the Lovelace Clinic was a pivotal step in the astronaut selection process for Project Mercury. It ensured that those chosen to represent the United States in space would be capable of withstanding the physical rigors and psychological pressures of spaceflight. This comprehensive approach to astronaut selection underscored the meticulous planning and attention to detail that characterized the early days of America's space program.

The Mercury astronauts underwent a series of medical tests at the Lovelace Clinic in New Mexico as part of their selection and preparation for spaceflight. Some of the specific tests included:

Physical Examinations: Comprehensive medical evaluations to assess overall health and fitness.

Cardiovascular Testing: Evaluation of heart function and cardiovascular fitness.

Psychological Testing: Assessment of psychological traits and stress tolerance.

Endocrinological Tests: Hormonal assessments to ensure metabolic health.

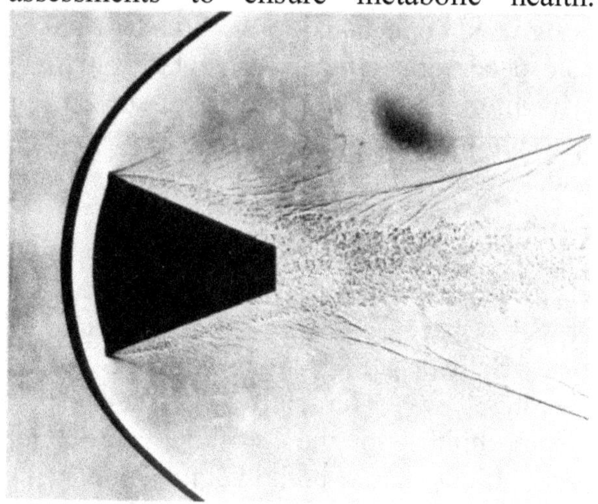

Shadowgraph of the reentry shock wave simulated in a wind tunnel, 1957

Radiological Examinations: X-rays and other imaging techniques to examine bone structure and overall health.

Nutritional Assessments: Dietary evaluations to optimize nutrition for spaceflight.

Environmental Stress Tests: Simulation of various environmental stressors to gauge astronaut response.

These tests were crucial in selecting astronauts who were physically and psychologically fit for the challenges of space travel during the Mercury program.

The selection of astronauts who passed these rigorous tests led to the formation of the "Mercury Seven," the first group of American astronauts. Their successful missions marked a significant achievement in space exploration. They paved the way for future endeavors, including the Apollo moon landings and establishing a long-term human presence in space. The Lovelace Clinic's role in this process highlighted the critical intersection of medicine and space technology, a partnership that continues to evolve and support human space exploration.

Wind tunnel tests were essential for understanding the aerodynamic properties of the spacecraft models. Engineers and scientists conducted these tests at several prominent facilities, including the Arnold Engineering Development Center in Tennessee and NASA's Langley, Ames, and Lewis Research Centers. Each facility brought unique capabilities and expertise to the testing process.

Experiment with boilerplate spacecraft, 1959

On February 10, 1959, the testing phase of Project Mercury reached a new milestone as wind tunnel tests of various spacecraft configuration models commenced. These tests were crucial in validating the designs that would ultimately carry America's first astronauts into space.

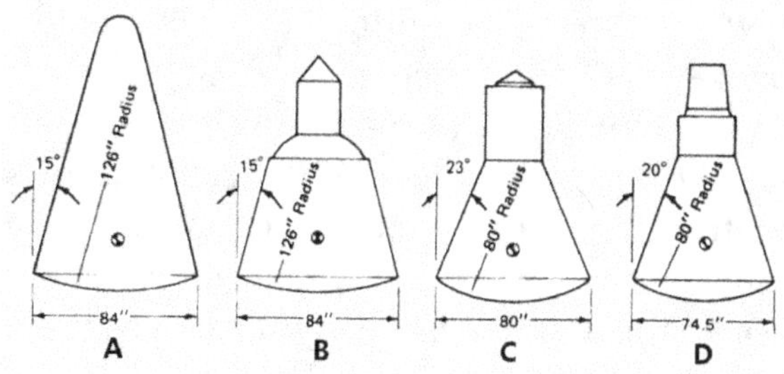

Evolution of capsule design, 1958–59

Throughout the year, over 70 different models of the Mercury spacecraft were subjected to rigorous wind tunnel testing. These tests simulated the aerodynamic forces and conditions the spacecraft would encounter during launch, flight, and re-entry. Engineers analyzed data on drag, lift, stability, and control to refine the spacecraft's design and ensure its safe and reliable performance.

At the Arnold Engineering Development Center, advanced wind tunnel facilities allowed for high-speed testing, crucial for understanding the conditions during the spacecraft's rapid ascent and reentry. NASA's Langley Research Center, known for its pioneering work in aeronautics, provided insights into the spacecraft's stability and control characteristics. Ames Research Center contributed its expertise in high-speed aerodynamics and thermal protection, vital for ensuring the spacecraft could withstand

Launch vehicles: 1. Mercury-Atlas (orbital flights). 2. Mercury-Redstone (suborbital flights). 3. Little Joe (uncrewed tests)

the intense heat of reentry. Meanwhile, Lewis Research Center focused on propulsion and power systems, ensuring the spacecraft's engines and systems could perform under the extreme conditions of space travel.

The wind tunnel tests of Project Mercury's configuration models were a testament to the collaborative effort and technical ingenuity that defined America's early space program. The data gathered from these tests led to significant refinements in the spacecraft's design, enhancing its safety and performance. These meticulous preparations were crucial for Project Mercury's success, ultimately enabling the United States to achieve its goal of putting a man in orbit and setting the stage for future milestones in space exploration.

The comprehensive wind tunnel testing program advanced the Mercury spacecraft's design and contributed to the broader field of aerospace engineering. The insights from these tests informed the development of subsequent spacecraft and aviation technologies, underscoring Project Mercury's enduring impact on the nation's progress in space.

On February 11, 1959, a pivotal meeting occurred in Huntsville, Alabama, between NASA's Space Task Group personnel and the Army Ballistic Missile Agency (ABMA). This meeting was crucial in delineating responsibilities for the Redstone and Jupiter rocket flight phases, both integral components of Project Mercury.

Huntsville, home to the ABMA and the visionary rocket engineer Wernher von Braun, was a hub of rocket development and innovation. During this meeting, representatives from both organizations engaged in detailed discussions about the technical aspects of the Redstone and Jupiter rockets, which were to serve as the launch vehicles for early Project Mercury missions.

The Redstone rocket, a direct descendant of the German V-2, had already proven its reliability and performance in various military applications. It was selected to launch the suborbital flights for Project Mercury, providing the initial boost to propel the spacecraft into space. Meanwhile, the Jupiter rocket, a more powerful vehicle, was considered for its potential to support longer, more complex missions.

During the Huntsville meeting, several critical technical details were addressed. These included propulsion systems, guidance and control mechanisms, structural integrity, and integration processes. The discussions ensured that the Redstone and Jupiter rockets would meet the stringent requirements for human spaceflight, emphasizing safety and reliability.

The meeting established several working panels to facilitate ongoing collaboration and streamline the integration of efforts. These panels brought together NASA and ABMA experts to focus on specific technical challenges and ensure seamless coordination between the two organizations. The working panels covered propulsion, guidance and control, aerodynamics, and launch operations.

On February 12, 1959, NASA officials convened with representatives from the U.S. Navy to discuss the critical aspect of search and recovery operations for Project Mercury. Recognizing the importance of safely retrieving astronauts and spacecraft after their missions, this meeting was a key step in ensuring comprehensive support from the Navy.

The discussions led to a dedicated NASA-Navy Committee tasked with developing a detailed plan for search and recovery efforts. This committee addressed the logistical and operational challenges in locating and recovering the Mercury capsules and their occupants from the ocean, where most of the landings were expected to occur.

Establishing the NASA-Navy Committee underscored the necessity of meticulous planning and coordination between the two organizations. With its extensive maritime capabilities and experience in search and rescue operations, the Navy was uniquely positioned to provide the necessary support. The committee's responsibilities included outlining the procedures for tracking the spacecraft, deploying recovery forces, and ensuring the swift and safe retrieval of astronauts and equipment.

The initial meeting of the NASA-Navy Committee was scheduled for February 17, 1959. This meeting would set the groundwork for the detailed planning and coordination required to execute successful recovery operations. The committee aimed to address various technical and logistical aspects, including the deployment of ships and aircraft, communication protocols, and contingency plans for different recovery scenarios.

This collaborative effort between NASA and the Navy was vital for Project Mercury's success. The detailed planning ensured that every mission phase, from launch to recovery, was thoroughly covered, minimizing risks and enhancing mission safety. The Navy provided a robust framework for recovering astronauts and capsules, ensuring the program's overall success.

The formation of the NASA-Navy Committee exemplified the interagency cooperation that became a hallmark of America's early space endeavors. The Navy's successful recovery operations during the Mercury missions set a standard for future programs, contributing to astronauts' safe return from space and solidifying the United States' leadership in human space exploration.

On February 12-13, 1959, crucial discussions occurred between NASA and the Air Force Ballistic Missile Division at Langley Field. These talks focused on integrating Atlas launch vehicles into Project Mercury, a key component for achieving America's goal of manned spaceflight.

The Atlas rocket, originally designed as an intercontinental ballistic missile, was chosen for its powerful thrust and proven reliability. During the meetings, participants delved into the Atlas's technical specifics, particularly its use in Project Mercury's first major test flight, known as "Big Joe." This mission aimed to validate the heat shield technology and overall spacecraft design in preparation for future manned missions.

The "Big Joe" test flight demonstrated the spacecraft's ability to withstand the intense heat and pressure of reentry into the Earth's atmosphere. Success in this test would provide confidence in the spacecraft's structural integrity and the effectiveness of its heat shield, both essential for the safety of future astronauts.

In addition to the technical details of the Big Joe mission, discussions also covered abort-sensing capabilities for later flights.

Abort systems were designed to detect anomalies during launch and safely jettison the spacecraft away from the rocket in case of a malfunction. These systems were crucial for protecting the lives of astronauts, ensuring they could be safely recovered even in the event of a launch failure.

The collaboration at Langley Field marked a significant step in integrating military and civilian efforts to achieve human spaceflight. The expertise and resources of the Air Force Ballistic Missile Division were invaluable in adapting the Atlas rocket to Project Mercury's needs.

The success of the Big Joe test flight, conducted later in September 1959, validated the spacecraft's heat shield and reentry capabilities, providing a critical milestone in Project Mercury. The advancements in abort sensing technologies further enhanced the safety protocols for manned missions, paving the way for the successful orbital flights that followed.

These discussions and the subsequent technical achievements highlighted the importance of interagency cooperation in advancing the United States' space program. Integrating the Atlas launch vehicle into Project Mercury demonstrated the feasibility of manned spaceflight and laid the foundation for future missions, including the historic Apollo moon landings. The collaborative efforts at Langley Field were a testament to the nation's commitment to exploration and innovation, driving progress in space exploration and technology.

On February 15, 1959, the final phase of medical examinations for selecting the Mercury astronauts commenced at the Wright Air Development Center in Ohio. This crucial step ensured that only the most physically and mentally fit individuals would be chosen for America's first manned space missions.

The Wright Air Development Center, a premier facility for aeromedical research, provided an ideal setting for these rigorous examinations. Candidates underwent comprehensive tests, including cardiovascular assessments, neurological evaluations, and extensive endurance trials. These tests were meticulously designed to simulate the extreme conditions of space travel and to evaluate the candidates' ability to withstand them.

In addition to physical assessments, the candidates faced a battery of psychological evaluations. These tests measured their mental resilience, stress tolerance, and cognitive abilities. Spaceflight required individuals who could remain calm, focused, and efficient under the pressures of isolation, confinement, and high-stakes situations. The psychological evaluations were essential in identifying individuals who possessed the mental fortitude necessary for the challenges of space travel.

The selection process at the Wright Air Development Center was about physical and psychological fitness and identifying individuals who could work effectively as a team. Interpersonal skills, the ability to communicate clearly, and the capacity to make quick, sound decisions were all critical traits evaluated during this phase.

Completing these final medical examinations culminated in an exhaustive selection process that began with hundreds of applicants. From this group, a select few would emerge as the Mercury Seven, the first astronauts of the United States. These individuals would become national heroes, blazing a trail for human space exploration and inspiring future generations.

The rigorous selection process at the Wright Air Development Center exemplified the meticulous attention to detail and commitment to excellence that characterized Project Mercury. The chosen astronauts were pioneers in space exploration and embodied human endurance and determination. Their selection was a testament to the rigorous standards set by NASA and the critical role of medical science in ensuring the safety and success of human spaceflight.

On February 17, 1959, the NASA-Navy Committee on Project Mercury search and recovery operations held its inaugural formal meeting. This meeting began a comprehensive collaboration to ensure the safe retrieval of astronauts and spacecraft from the ocean following their missions.

The committee, comprising experts from both NASA and the U.S. Navy, convened to initiate joint recovery exercises and meticulously plan the detailed operations necessary for effective search and recovery efforts. The focus was on the Atlantic Missile Range, a key area for Project Mercury's launch and recovery operations.

During this first meeting, the committee outlined the scope of their collaboration, identifying key responsibilities and establishing protocols for coordinated efforts. Discussions covered various aspects of recovery operations, including tracking and locating the spacecraft, deploying recovery forces, and ensuring astronauts' rapid and safe retrieval.

The committee planned a series of joint recovery exercises to achieve these objectives. These exercises simulated real mission scenarios, allowing both NASA and Navy personnel to practice and refine their procedures. The exercises would involve deploying ships, aircraft, and specialized recovery teams to locate and retrieve mock spacecraft from the ocean, ensuring that all team members were well-prepared for actual recovery missions.

The committee also addressed the logistical challenges of coordinating recovery efforts along the Atlantic Missile Range. This included establishing communication protocols, defining the roles of various naval assets, and ensuring that all recovery operations could be conducted efficiently and safely.

The formation of the and its initial meeting were critical steps in the success of Project Mercury. The detailed planning and joint exercises provided a robust framework for recovery operations, significantly enhancing the safety and reliability of the missions. The committee's work ensured that every aspect of search and recovery was meticulously planned, reducing risks and ensuring that astronauts could be swiftly and safely retrieved after their missions.

On February 19, 1959, Dr. T. Keith Glennan, the first Administrator of NASA, delivered a pivotal speech in which he provided a candid estimate of Project Mercury's cost. He projected that the ambitious program would exceed $200 million, a substantial investment reflecting the monumental technological challenges and the extensive infrastructure required to achieve human spaceflight.

Dr. Glennan emphasized the unprecedented nature of the technological hurdles faced by Project Mercury. The program aimed to safely send humans into space and return them to Earth, a feat that demanded advancements in rocket technology, spacecraft design, and human endurance studies. Each mission component—from the powerful Atlas and Redstone launch vehicles to the intricate Mercury capsule—required cutting-edge engineering and rigorous testing.

One of the most significant challenges highlighted by Dr. Glennan was the establishment of a comprehensive tracking and communication network. This network was essential for monitoring the spacecraft's journey, ensuring constant contact with the astronauts, and providing real-time data to mission control. The tracking network spanned the globe, involving ground stations, ships, and aircraft strategically positioned to maintain continuous communication with the spacecraft.

The extensive investment in Project Mercury was not just about overcoming technical challenges; it was also a testament to the United States' commitment to space exploration and its determination to lead in the new frontier of space. Dr. Glennan's speech underscored the broader implications of the program, framing it as a critical step in advancing scientific knowledge, enhancing national security, and inspiring the public.

Dr. Glennan's projection of the program's cost also highlighted the collaborative effort required to make Project Mercury a reality. It involved partnerships between NASA, the military, private industry, and international allies. This cooperative approach was essential in pooling resources, expertise, and technology to achieve the ambitious goals set by the program.

The financial commitment to Project Mercury laid the foundation for subsequent achievements in space exploration. The successful manned missions of the Mercury program validated the investments and technological advancements, paving the way for the Gemini and Apollo programs. These later programs built on the knowledge and infrastructure developed during Project Mercury, ultimately leading to the historic Apollo moon landings.

Dr. Glennan's speech clearly acknowledged the formidable challenges ahead and a rallying call for the nation. It encapsulated the spirit of innovation and perseverance that would drive the United States to achieve its space exploration milestones, establishing a legacy of scientific and technological progress that continues today.

On February 20, 1959, the Langley Research Center was formally assigned the responsibility for planning and contracting the tracking facilities for Project Mercury. This decision marked a critical step in ensuring the success of America's first human spaceflight program by providing a dedicated team to oversee the development of a comprehensive tracking network.

The Langley Research Center in Hampton, Virginia, was already a hub of aeronautical research and innovation. Its expertise made it the ideal choice for managing the complex requirements of Project Mercury's tracking facilities. The center's new role involved coordinating the design, construction, and implementation of a global network capable of monitoring the spacecraft throughout its mission.

Tracking the Mercury spacecraft was essential for communicating with the astronauts, ensuring their safety, and collecting vital data. The network needed to provide continuous coverage from launch to splashdown, which required an intricate arrangement of ground stations, tracking ships, and aircraft positioned around the globe.

The assignment of this responsibility to Langley Research Center underscored the importance of precise coordination and advanced technology in space missions. The center's tasks included selecting suitable sites for tracking stations, negotiating contracts with construction and technology providers, and integrating the various components into a seamless operation.

One of the primary challenges was ensuring that the tracking network could handle the Mercury spacecraft's high-speed, high-altitude flights. The data collected from these tracking facilities would be crucial for real-time decision-making during the missions, enabling mission control to respond promptly to any anomalies and ensure the astronauts' safe return.

The establishment of the tracking network also involved significant logistical planning. Langley Research Center had to coordinate with international partners to position tracking stations in remote and strategically important locations. This global effort highlighted the collaborative nature of the space program and the shared goal of advancing human space exploration.

By taking on the responsibility for Project Mercury's tracking facilities, Langley Research Center played a pivotal role in the program's success. The effective operation of the tracking network was demonstrated during the Mercury missions, where precise tracking and communication were maintained throughout the flights. This capability not only ensured the safety of the astronauts but also provided valuable data that informed future space missions.

The successful management of the tracking facilities by Langley Research Center set a standard for future space missions. It showcased the center's ability to handle complex logistical and technical challenges, contributing to the overall progress of the United States' space program. The legacy of this achievement continued to influence space exploration, providing a foundation for the sophisticated tracking and communication networks used in modern space missions.

On February 24, 1959, the Mercury-Redstone-Jupiter Study Panel convened at Redstone Arsenal in Huntsville, Alabama, to tackle some of the most critical technical aspects of the nascent Project Mercury. This meeting brought together leading experts in rocket science, aerodynamics, and engineering to discuss and refine the details of trajectory, aerodynamics, and flight loads for the Mercury missions.

The Redstone and Jupiter rockets were central to Project Mercury's goals. The Redstone rocket, a reliable and proven vehicle derived from the German V-2 missile, was designated for the suborbital flights that would precede orbital missions. The more powerful and capable Jupiter rocket was being evaluated for its potential use in longer missions. Ensuring that these rockets could safely and effectively launch the Mercury spacecraft was paramount.

During this initial meeting, the panel focused on key technical challenges. Trajectory analysis was critical for determining the flight path to safely carry the spacecraft into space and bring it back to Earth. Aerodynamics studies were essential to understand how the spacecraft and its booster would behave at various speeds and altitudes, especially during launch and reentry high-stress phases. Flight load assessments were necessary to ensure that the rocket and spacecraft could withstand the physical stresses encountered during these phases.

The detailed and intensive discussions addressed the complex interplay of forces that would act on the spacecraft. Experts analyzed data, ran simulations, and shared insights to develop robust solutions. This collaborative effort aimed to minimize risks and ensure the missions' safety and success.

The decision to reconvene the panel on March 13, 1959, underscored the importance of these technical discussions. This follow-up meeting allowed the experts to review progress, refine their analyses, and address any new challenges. The iterative nature of these meetings ensured that every aspect of the flight would be meticulously planned and tested.

The work of the Mercury-Redstone-Jupiter Study Panel was instrumental in Project Mercury's success. Their rigorous analysis and problem-solving laid the groundwork for safely and effectively using the Redstone and Jupiter rockets. The insights gained from these meetings informed the design and testing of the spacecraft, contributing to the overall reliability and safety of the missions.

As the research and development phase of Project Mercury progressed, the collective efforts of NASA, military personnel, and industry partners laid the essential groundwork for America's pioneering journey into space. The meticulous planning, rigorous testing, and interdisciplinary collaboration underscored the project's ambitious goals and the nation's determination to lead in space exploration.

The synergy between various organizations was evident in every aspect of Project Mercury. NASA's Space Task Group, responsible for managing the project, worked closely with the U.S. Navy, Air Force, and numerous industry contractors to address the myriad challenges of human spaceflight. The commitment of these diverse teams to a common goal facilitated remarkable advancements in aerospace technology and human endurance.

Key milestones, such as the rigorous medical examinations at the Lovelace Clinic and Wright Air Development Center, ensured that only the most physically and mentally fit individuals were selected as astronauts. The comprehensive wind tunnel tests conducted at the Arnold Engineering Development Center and NASA's research centers refined the spacecraft's aerodynamic properties, contributing to its safety and performance.

Crucial meetings, like those held at Redstone Arsenal, fostered in-depth discussions on trajectory, aerodynamics, and flight loads, leading to well-informed decisions and innovative solutions. The formation of the NASA-Navy Committee and their detailed planning of search and recovery operations highlighted the importance of collaboration in ensuring the astronauts' safe return.

Dr. T. Keith Glennan's public estimation of Project Mercury's cost at over $200 million underscored the project's scale and the nation's commitment to its success. This substantial investment reflected the unprecedented technological challenges and the extensive tracking network required to support the missions.

The formal assignment of tracking facility responsibilities to Langley Research Center marked another significant step in the project. This decision ensured that the necessary infrastructure was in place to monitor and communicate with the spacecraft throughout its journey.

Each of these efforts was a testament to the meticulous preparation and technical expertise that characterized Project Mercury. The collaborative spirit and shared vision of NASA, the military, and industry partners propelled the United States toward achieving its goal of human spaceflight. These foundational efforts not only led to the successful missions of Project Mercury but also paved the way for future endeavors, including the Gemini and Apollo programs, ultimately culminating in the historic Apollo moon landings.

On February 26, 1959, in the sterile, fluorescent-lit meeting room at Redstone Arsenal, Panel Number I, also known as the Design Subcommittee, convened for the first time. This gathering brought together engineers and scientists from NASA's Space Task Group and the Army Ballistic Missile Agency. Their primary focus was to discuss the integration requirements for the Mercury spacecraft with the Redstone and Jupiter launch vehicles. The meeting was a critical step in ensuring that the spacecraft and rockets would work seamlessly together, addressing the technical challenges of fitting the Mercury capsule onto these powerful boosters.

Concurrently, another Space Task Group and Langley Research Center team embarked on a mission to the Arnold Engineering Development Center (AEDC) in Tullahoma, Tennessee. Their visit aimed to evaluate the AEDC's capabilities for conducting tests on the Mercury spacecraft scale models. This assessment was vital to ensure the facilities could simulate the aerodynamic and thermal conditions the spacecraft would encounter during its missions.

The visit to AEDC was marked by meticulously examining the center's advanced testing infrastructure. The team inspected wind tunnels, thermal vacuum chambers, and other specialized equipment. Detailed discussions on logistics and scheduling were held to coordinate the testing phases and ensure that the Mercury program's stringent timelines were met.

These simultaneous efforts exemplified the interdisciplinary collaboration and meticulous planning that defined Project Mercury. At Redstone Arsenal, the integration discussions were pivotal in aligning the spacecraft's design with the launch vehicles' capabilities, ensuring that every component functioned as intended. At AEDC, the careful assessment and scheduling of tests ensured that the spacecraft would be thoroughly vetted under conditions replicating those of space.

The activities on February 26, 1959, highlighted the complex and multifaceted nature of Project Mercury. Integrating the spacecraft with its launch vehicles and validating its design through rigorous testing were critical steps toward the goal of human spaceflight. These efforts laid a solid foundation for the project's subsequent achievements, demonstrating the United States' resolve and capability to lead in space exploration. The collaboration among various agencies and the dedication to precision and safety set a high standard for all future space endeavors.

On February 27, 1959, the Space Task Group (STG) was abuzz with activity as personnel worked diligently to establish the design trajectory for the upcoming Big Joe flight test. This crucial test was intended to validate the heat shield technology and overall spacecraft design under realistic reentry conditions, and it required precise planning and coordination.

Amid the buzzing activity, the scent of freshly brewed coffee filled the air as engineers and scientists gathered for intense consultations with experts from Convair Astronautics and Space Technology Laboratories. These advisors brought a wealth of knowledge and experience, providing crucial insights into meeting the stringent trajectory requirements for the Big Joe mission.

The room was a hive of focused energy, with experts poring over diagrams, flight data, and technical specifications. The challenge was to align the spacecraft's trajectory with the mission's goals, ensuring that it would reenter the Earth's atmosphere at the correct angle and speed to test the heat shield's effectiveness. This required meticulous calculations and the integration of various data points to create a reliable flight plan.

Diagrams were scrutinized, equations were double-checked, and simulations were run to model the spacecraft's path. The trajectory design had to account for numerous factors, including the spacecraft's speed, altitude, and angle of descent. The team worked tirelessly to ensure that every detail was considered, knowing that the success of the Big Joe test was critical for the future of Project Mercury.

The collaborative efforts on that day exemplified the dedication and expertise that defined the early days of America's space program. The STG, alongside their partners from Convair Astronautics and Space Technology Laboratories, demonstrated an unwavering commitment to precision and excellence. Their work was about meeting technical specifications and paving the way for human spaceflight and ensuring that every mission component was thoroughly vetted and optimized.

By the end of the day, significant progress had been made. The design trajectory for the Big Joe flight test was taking shape, informed by the collective expertise and rigorous analysis of the assembled team. This milestone was a testament to the power of collaboration and the relentless pursuit of knowledge that would drive Project Mercury to its groundbreaking achievements.

The February 27, 1959, efforts were a crucial step in the journey to send Americans into space. The meticulous planning and detailed consultations ensured that the Big Joe test would provide valuable data and insights, helping to refine the Mercury spacecraft and bolster confidence in its design. This foundational work set the stage for the following successful missions, ultimately leading to the United States' historic achievements in human space exploration.

Throughout February 1959, a series of pivotal meetings and collaborative efforts occurred as Project Mercury's momentum continued to build. One of the key events was a significant meeting between personnel from the Space Task Group (STG) and the Air Force Ballistic Missile Division. This meeting was crucial in outlining the responsibilities for the first two Atlas rocket firings, which were integral to the success of the Mercury missions.

During the meeting, it was decided that the Space Technology Laboratories, under the direction of the Air Force Ballistic Missile Division, would be responsible for selecting design trajectories that would accurately simulate reentry conditions for the manned spacecraft. These trajectories had to be meticulously calculated to ensure that the spacecraft could safely reenter the Earth's atmosphere, providing vital data on heat shield performance and overall spacecraft integrity.

Detailed specifications were discussed to ensure the highest level of precision. This included collecting and analyzing impact dispersion data, which was essential for understanding where the spacecraft would land and ensuring the safety of populated areas. Range safety purposes were also a top priority, requiring stringent measures to protect the astronauts and the public. The reprogramming of guidance computers was another critical task, ensuring that the spacecraft could follow the designated trajectory with pinpoint accuracy.

In addition to these discussions, six specialized working panels were formed to address various critical aspects of the Mercury-Redstone program. These panels focused on:

Design Coordination: Ensuring that all components of the spacecraft and launch vehicles were seamlessly integrated and functioned as intended.

Pilot Safety: Developing and implementing safety measures to protect the astronauts during all mission phases.

Aerodynamics: Studying the aerodynamic properties of the spacecraft to optimize its performance during launch, flight, and re-entry.

Propulsion Systems: Ensuring the reliability and performance of the rocket engines and other propulsion components.

Guidance and Control: Refining the systems that would guide and control the spacecraft throughout its mission.

Recovery Operations: Plan and execute the safe retrieval of the spacecraft and astronauts after splashdown.

These working panels brought together experts from various fields, fostering an interdisciplinary approach to solving the complex challenges of human spaceflight. Each panel was crucial in addressing specific technical issues, ensuring that every aspect of the Mercury-Redstone program was thoroughly examined and optimized.

The efforts throughout February 1959 were marked by intense collaboration and a relentless pursuit of excellence. The meetings and panels demonstrated the high level of coordination and dedication required to achieve Project Mercury's ambitious goals. The meticulous planning and detailed technical discussions laid a solid foundation for the upcoming Atlas firings and the eventual manned missions, highlighting the nation's determination to lead in space exploration.

These foundational efforts propelled Project Mercury and set a standard for future space missions. The rigorous attention to detail, the interdisciplinary collaboration, and the commitment to safety and precision became hallmarks of NASA's approach to space exploration, guiding the United States to historic achievements in the years to come.

The Mercury Program at Arnold Engineering Development Center

In the early months of 1959, as the United States was fervently pushing the boundaries of space exploration, critical investigations were conducted at the Arnold Engineering Development Center (AEDC) in Tullahoma, Tennessee. This facility, renowned for its advanced aeronautical testing capabilities, became the proving ground for the Mercury spacecraft models, the vessels that would soon carry America's first astronauts into space.

From January to July, AEDC's cutting-edge wind tunnels roared to life, simulating the extreme conditions of space travel. Engineers and scientists subjected the Mercury spacecraft models to speeds of Mach 8, 16, and 20—velocities that replicated the intense conditions encountered during reentry into Earth's atmosphere. These tests were pivotal in ensuring the spacecraft's stability, heat transfer, and pressure distribution.

Stability was a paramount concern. At hypersonic speeds, even minor design flaws could result in catastrophic failures. The AEDC team meticulously analyzed the aerodynamic properties of the Mercury models, ensuring that they would remain steady and controllable during their perilous journey back to Earth. This stability was crucial not only for the safety of the astronauts but also for the precision required in reentry trajectories.

Heat transfer presented another formidable challenge. Reentry from space generates temperatures so extreme that they can melt most materials. The AEDC tests focused on how heat would be distributed across the spacecraft's surface, leading to the development of advanced heat shields. These shields, designed to absorb and dissipate the intense heat, became a cornerstone of spacecraft safety, protecting the vehicle and its occupants.

Pressure distribution tests were equally vital. The spacecraft would experience tremendous aerodynamic pressures, and understanding how these forces would impact the structure was essential for designing a robust and resilient vehicle. The data gathered from these tests informed crucial modifications to the Mercury spacecraft, enhancing its integrity and ensuring it could withstand the harsh realities of space travel.

These investigations at AEDC did more than advance the Mercury program; they laid the groundwork for the United States' burgeoning space ambitions. The knowledge and experience from these tests propelled the nation forward, culminating in the successful manned missions of the 1960s. The rigorous testing protocols developed at AEDC became standard practice in aerospace engineering, influencing subsequent space missions' design and safety protocols.

On March 6, 1959, officials from NASA's Space Task Group and McDonnell Aircraft Corporation convened in St. Louis, Missouri, for a critical meeting focused on Project Mercury's logistical aspects. The primary agenda was to discuss the requirements for spare parts and ground support equipment, essential components for ensuring the program's smooth operation and mission success.

The meeting was held at McDonnell's facilities, where engineers and project managers from both organizations engaged in thorough discussions. The goal was to identify and quantify the necessary spare parts to maintain the Mercury spacecraft and its associated systems. This included everything from electronic components to mechanical assemblies, ensuring that any potential issues could be quickly addressed without causing significant delays.

Ground support equipment was another vital topic of discussion. This encompassed the tools, test equipment, and infrastructure needed to prepare, launch, and recover the spacecraft. The teams meticulously reviewed the specifications and requirements for this equipment, considering factors such as reliability, availability, and ease of use. Ensuring that ground support operations were efficient and effective was crucial for the overall success of Project Mercury.

The productive collaborative discussions led McDonnell to swiftly develop and submit a preliminary plan outlining their proposed solutions. This plan detailed the spare parts inventory, logistics support, and the ground support equipment necessary for the program. The preliminary plan was then submitted for review by the Space Task Group and NASA Headquarters, allowing for further refinement and approval.

This meeting in St. Louis underscored the importance of meticulous planning and coordination in the early stages of Project Mercury. By addressing the logistical needs comprehensively, NASA and McDonnell Aircraft Corporation aimed to minimize risks and ensure that the spacecraft and its support systems were always ready for each mission phase.

The proactive approach taken in these discussions and the subsequent development of a detailed plan demonstrated the dedication and foresight of both organizations. This level of preparation was instrumental in achieving the reliability and success that characterized Project Mercury, ultimately leading to the historic manned missions that followed.

These efforts laid the groundwork for a robust support system sustaining the Mercury missions. The careful planning and coordination between NASA and its industry partners, like McDonnell Aircraft Corporation, set a precedent for future space endeavors, emphasizing the critical role of logistics and support in space exploration.

On March 8, 1959, the otherwise tranquil Wallops Island became a hub of intense activity as an abort test was conducted on a full-scale model of the Mercury spacecraft equipped with an escape tower. This critical test aimed to evaluate the effectiveness of the escape system designed to pull the spacecraft and its astronaut to safety in the event of a launch emergency.

As the test commenced, the roar of rocket engines echoed across the island, accompanied by the controlled chaos of the launch. Observers watched intently as the configuration took flight, exhibiting erratic motion that deviated from the expected path. Despite this, the test yielded invaluable data that would inform further refinements to the abort system.

The erratic motion observed during the test highlighted the complexities and challenges of designing a reliable escape system. Understanding these dynamics was crucial for ensuring that, in an actual emergency, the system would function correctly to safeguard the astronaut's life.

To delve deeper into the nuances of the abort system's behavior, the Langley Research Center was tasked with testing small-scale flight models. These models would be used to simulate various flight conditions and configurations, allowing researchers to analyze the abort system's motion and performance in greater detail. The data gathered from these smaller tests would be essential for refining the design and ensuring its reliability.

The atmosphere at Wallops Island during the test blended anticipation and technical rigor. Engineers and scientists worked diligently, surrounded by the sights and sounds of rocket engines and the bustling activity of a major test operation. The full-scale abort test and subsequent small-scale tests at Langley were part of a broader effort to perfect the Mercury spacecraft's systems and ensure the highest safety standards for the astronauts.

These tests were a testament to the iterative nature of spaceflight development. Regardless of its immediate success or challenges, each test provided critical insights that informed the next steps in the design and engineering process. The dedication to learning and improvement was a cornerstone of Project Mercury, driving the program toward its ultimate goal of successful human spaceflight.

The abort test on March 8, 1959, and the ongoing efforts at Langley Research Center underscored the meticulous planning and relentless pursuit of safety that characterized Project Mercury. These efforts laid the groundwork for the successful missions that followed, ensuring that the United States was prepared to meet the challenges of space exploration with innovation and resilience.

On March 9, 1959, exploratory noise transmission tests commenced at Langley Research Center, marking another critical step in the development of Project Mercury. These tests aimed to understand the levels of noise the Mercury spacecraft would experience during launch and flight, which was essential for ensuring the safety and comfort of the astronauts.

At Langley Research Center, engineers set up sophisticated equipment to measure and analyze how sound waves traveled through the spacecraft's structure. These exploratory tests were designed to simulate the intense noise environment generated by rocket engines during liftoff and ascent. The data gathered from these tests would inform the design of the spacecraft's interior, helping to mitigate the effects of high noise levels on the astronauts.

A comprehensive report on rocket engine noise was completed in tandem with these tests. This report detailed the expected noise levels the prototype Mercury spacecraft would endure based on previous tests and real-world data from rocket launches. Understanding these noise levels was crucial for developing effective noise reduction strategies and ensuring that communication systems within the spacecraft remained functional under high-noise conditions.

Meanwhile, parallel efforts were underway to test ablative materials, critical for the spacecraft's thermal protection during reentry. At both Langley Research Center and Wallops Island, engineers conducted rigorous tests on these materials under conditions simulating the extreme heat and pressure of reentry. Ablative materials work by gradually eroding and absorbing heat, protecting the spacecraft and its occupants from the intense temperatures encountered when returning to Earth's atmosphere.

These tests involved subjecting sample materials to high-temperature environments and measuring their performance and degradation. The goal was to identify the best materials that could withstand the harsh conditions of reentry, ensuring the safety and integrity of the Mercury spacecraft.

The combination of noise transmission tests, the comprehensive rocket engine noise report, and the ongoing tests of ablative materials highlighted NASA's multifaceted approach to addressing the various challenges of spaceflight. Each of these efforts was critical in ensuring that the Mercury spacecraft could withstand the harsh environments of launch, space, and reentry.

March 9, 1959, represented a day of intense scientific and engineering activity characterized by meticulous testing and analysis. The data and insights gained from these efforts would be pivotal in refining the Mercury spacecraft's design and ensuring the program's success. The dedication to thorough testing and problem-solving exemplified Project Mercury's rigorous standards, paving the way for the historic achievements in human space exploration that followed.

On March 10, 1959, the Space Task Group received troubling news from McDonnell Aircraft Corporation: several subcontractors encountered significant difficulties procuring the materials needed to fabricate critical components for Project Mercury. These delays were primarily due to lacking a DX priority procurement rating. This designation would expedite the acquisition of essential materials by prioritizing the project over other contracts.

The absence of this priority rating caused frustration and concern among the Project Mercury team. The DX rating, reserved for programs of the highest national importance, would have streamlined the procurement process, ensuring that materials were delivered promptly and reducing the risk of delays. Without it, subcontractors faced long lead times and supply chain bottlenecks, threatening the project's tight schedule.

McDonnell's notification highlighted the challenges faced by the burgeoning space program. The aerospace industry was experiencing unprecedented demand, and the competition for materials was fierce. This situation underscored the complexity of coordinating a project as ambitious as Project Mercury, which relied on a seamless supply chain to meet its ambitious goals.

The Space Task Group, recognizing the gravity of the situation, immediately began working to address the issue. They sought to secure the necessary DX priority rating for Project Mercury, leveraging their connections within the Department of Defense and other government agencies to advocate for the program's critical needs. The goal was to elevate the project's priority status, ensuring that the procurement of materials could proceed without further hindrance.

This challenge also emphasized the importance of effective project management and communication. The Space Task Group needed to maintain close coordination with McDonnell and its subcontractors, monitor the situation closely, and provide support where possible. It was a test of the project's resilience and the ability of its leaders to navigate the bureaucratic and logistical obstacles that inevitably arose.

Despite these setbacks, the Project Mercury team remained resolute. They understood that overcoming such hurdles was part of pioneering new frontiers in space exploration. Their determination to push forward, even in the face of procurement challenges, reflected the broader spirit of innovation and perseverance that defined NASA's early days.

The notification from McDonnell on March 10, 1959, reminded us of the many moving parts involved in bringing Project Mercury to fruition. It highlighted the critical role of supply chain management and the need for unwavering support at all levels of government and industry. As the team worked to resolve these issues, they continued to lay the groundwork for the successful missions that would eventually carry America into space, demonstrating the nation's commitment to leading the world in space exploration.

On March 11, 1959, the Langley Research Center's Pilotless Aircraft Research Division conducted a significant full-scale test at Wallops Island to simulate a pad-abort situation for Project Mercury. This test was crucial in assessing the spacecraft's abort system, designed to protect astronauts in the event of an emergency during launch.

The test involved a full-scale model of the Mercury spacecraft, matching the actual weight and dimensions of the final design. As the test commenced, the spacecraft launched from the pad and initially flew straight, demonstrating the initial thrust capabilities. However, soon after, a thrust misalignment caused the spacecraft to pitch unexpectedly, leading to an impact near the shore.

This outcome, while not entirely successful, provided invaluable data and insights. The test underscored the critical importance of precise engineering and thorough testing in developing reliable abort systems. The misalignment issue highlighted the need for rigorous examination of the thrust vector control mechanisms and the overall alignment of the propulsion system.

Engineers and scientists closely analyzed the test results, focusing on the root causes of the misalignment and the subsequent pitching motion. The data gathered from this test informed several key adjustments and improvements in the design and engineering of the abort system. This iterative process of testing, analyzing, and refining was essential to ensuring the safety and reliability of the Mercury spacecraft.

The test also reinforced the importance of conducting full-scale simulations under realistic conditions. By testing the spacecraft in a real-world environment, engineers could observe and address issues that might not have been apparent in smaller-scale or computer-simulated tests. This approach was critical to developing a robust and reliable abort system to safeguard astronauts' lives.

The efforts at Wallops Island on March 11, 1959, exemplified the rigorous testing regime that characterized Project Mercury. Regardless of its immediate success or failure, each test contributed to a deeper understanding of the spacecraft's behavior and performance. The commitment to learning from these tests and making necessary adjustments was a testament to the dedication and expertise of the engineers and scientists involved.

This pad-abort test was pivotal in perfecting the Mercury spacecraft's safety systems. The insights gained from this and subsequent tests helped build a solid foundation for the following successful missions. The relentless pursuit of excellence and the determination to overcome challenges were hallmarks of Project Mercury, driving the United States toward its goal of human space exploration

On March 16, 1959, the Space Task Group took a critical step forward in ensuring astronaut safety by requesting $125,000 from NASA Headquarters to procure five developmental pressure suits for Project Mercury. These suits were not merely protective gear; they represented a significant advancement in the project's progress, highlighting the meticulous attention to detail and commitment to astronaut safety.

The developmental pressure suits were designed to protect astronauts from the harsh conditions of space, including extreme temperatures, vacuum, and potential micrometeoroid impacts. The suits needed to be both durable and flexible, allowing astronauts to perform necessary tasks while providing life support and protection.

The procurement of these suits marked a pivotal moment in Project Mercury. As the first U.S. human spaceflight program, Mercury required innovations in all aspects of space travel, and the development of pressure suits was no exception. The suits had to be custom-fitted to each astronaut, ensuring both comfort and functionality.

The funding request underscored the project's priority and the importance of advancing human space exploration technology. The $125,000 investment was intended to cover the costs of designing, developing, and testing these pressure suits. This included materials, labor, and the integration of life-support systems. The suits needed rigorous testing to validate their performance under conditions simulating those in space.

The development and testing of pressure suits involved close collaboration with specialized contractors with aerospace and high-performance materials expertise. This collaboration was crucial for innovating suit designs that could meet the stringent requirements of space travel.

The Space Task Group's request also highlighted the importance of proactive planning and resource allocation in managing Project Mercury's complex logistics. Securing the necessary funds well in advance ensured that the pressure-suit development could proceed without delays, maintaining the project's overall timeline.

These developmental pressure suits would eventually evolve into the iconic Mercury suits worn by astronauts like Alan Shepard and John Glenn during their historic missions. The suits was vital in protecting the astronauts and ensuring their missions' success.

The March 16, 1959, request for funding was more than just a budgetary action; it represented a strategic investment in the safety and success of America's first human spaceflight program. This focus on astronaut safety and meticulous preparation paved the way for Project Mercury's achievements, setting the stage for future human space exploration missions. The successful development and deployment of these pressure suits were a testament to the ingenuity and dedication of the engineers and scientists who made Project Mercury a landmark in space exploration history.

On March 17, 1959, the Space Task Group requested funds to procure six main parachute canisters and twelve drogue parachute canisters from Goodyear Aircraft Corporation. These components were critical for the Little Joe and Big Joe phases of Project Mercury, designed to ensure the safe recovery of the spacecraft during critical test flights.

The main parachutes and drogue parachutes played essential roles in the reentry and landing phases of the missions. Drogue parachutes, deployed at high altitudes and speeds, were responsible for stabilizing the spacecraft and slowing it down sufficiently to allow the deployment of the main parachutes. The main parachutes then further decelerated the spacecraft, ensuring a safe and controlled descent into the ocean for recovery.

The Little Joe and Big Joe tests were integral to validating the Mercury spacecraft's design and its systems. Little Joe tests primarily focused on the spacecraft's launch escape system and high-altitude abort scenarios. Big Joe tests assessed the heat shield's performance and the overall reentry and recovery procedures. Both series of tests required reliable parachute systems to simulate real mission conditions and ensure the spacecraft could be safely recovered after each test.

Goodyear Aircraft Corporation, known for its expertise in aerospace components, was selected to provide these critical parachute systems. Procuring these canisters was about obtaining parts and ensuring they met the stringent specifications required for human spaceflight. Each parachute had to be rigorously tested to guarantee its performance under the demanding conditions of space reentry.

The funding request highlighted the meticulous planning and thorough testing that characterized Project Mercury. Securing the necessary parachutes well in advance allowed the team to proceed with scheduled tests without delays, maintaining the project timeline and ensuring that all systems were fully vetted before actual manned missions.

On March 17-18, 1959, a pivotal Mock-Up Inspection Board meeting convened at the McDonnell Aircraft Corporation's plant. The board gathered to thoroughly review the completed spacecraft mock-up, a crucial step in the early stages of America's burgeoning space program. This inspection was not merely a formality; it was an intense, critical evaluation designed to identify and rectify potential issues before actual construction and testing began.

The review team, composed of engineers, designers, and pilots, meticulously examined every aspect of the mock-up. Their goal was to ensure that the spacecraft met the stringent requirements necessary for manned spaceflight. The inspection highlighted several areas needing improvement, particularly concerning pilot safety and operational efficiency. Key directives for redesign emerged from the meeting, primarily focused on enhancing pilot egress and control accessibility. This included adjusting the cockpit layout to ensure that astronauts could quickly and safely exit the spacecraft in an emergency and reconfiguring control placements for optimal reach and usability.

These directives underscored the iterative nature of spacecraft development, where design is a continuous process of refinement and enhancement. Each review and subsequent redesign brought the spacecraft closer to operational readiness, reflecting the meticulous and cautious approach required for human spaceflight.

This meeting at McDonnell was more than just a technical review; it was a testament to the collaborative effort and rigorous scrutiny necessary to achieve success in space exploration. The insights gained, and the improvements mandated during these inspections were instrumental in advancing the Mercury program, NASA's first human spaceflight project. The iterative review and redesign process ensured that the spacecraft would not only meet but exceed the safety and functionality standards required for the astronauts who would pioneer America's journey into space.

On March 20, 1959, a significant step forward in the Mercury program occurred with the appointment of John H. Disher as the coordinator of the study panels. Disher's role was crucial; he was tasked with centralizing the program plans and proposals, ensuring that the various components of Project Mercury were aligned and integrated effectively. His coordination efforts were essential for streamlining the multitude of tasks and objectives that characterized America's first manned spaceflight program.

During this meeting, key discussions occurred regarding the test objectives for the Mercury-Redstone and Mercury-Jupiter rockets. It was decided that the initial flights would be unmanned, serving as critical tests of the launch systems and spacecraft. These unmanned missions would be followed by flights carrying primates, an important intermediate step designed to test life support systems and gather data on the biological effects of spaceflight. Finally, the program would culminate in manned flights, which would serve as training missions for the astronauts who would eventually venture into space.

In addition to these planning efforts, a detailed study titled "Recovery Operations for Project Mercury" was completed and forwarded to the Department of Defense. This study outlined the procedures and logistics necessary to recover the spacecraft and its occupants after reentry and splashdown. The collaboration with the Department of Defense highlighted the project's interdisciplinary nature, which required multiple agencies' expertise and resources to ensure the safety and success of the missions.

Disher's appointment and the decisions made at this meeting were instrumental in shaping the trajectory of Project Mercury. By establishing a clear sequence of test flights, the program could systematically address human spaceflight's technical and biological challenges. These early test missions were crucial for validating the spacecraft's design and systems, building confidence and knowledge to pave the way for the United States' first manned orbital missions.

On March 23, 1959, the McDonnell Aircraft Corporation faced significant supply chain challenges. It identified 32 critical items that required a DX priority procurement rating. This designation was essential for expediting the procurement process, ensuring that these crucial components were delivered promptly to keep the Mercury program on schedule.

The need for a DX priority rating underscored the complexities of sourcing specialized materials and parts necessary for the spacecraft. These items were routine supplies integral to the spacecraft's construction and functionality, each playing a vital role in the mission's success.

By addressing these supply chain issues head-on, McDonnell Aircraft Corporation demonstrated its commitment to overcoming obstacles and maintaining the momentum of Project Mercury. The corporation's ability to secure priority procurement status for these items was a testament to the collaborative efforts of various government agencies, all working together to achieve the ambitious goal of human spaceflight.

This proactive approach to managing supply chain difficulties indicated the broader challenges faced during the early days of America's space program. Ensuring the timely delivery of high-priority components was crucial for meeting the Mercury missions' tight deadlines and technical demands. The lessons learned in navigating these supply chain issues contributed to developing more efficient procurement processes, which would benefit future space endeavors and support the United States' continued progress in space exploration.

On March 26, 1959, Langley Research Center received the green light to conduct hypersonic flight tests for the Mercury spacecraft. This approval marked a significant milestone in the program, enabling crucial assessments of the spacecraft's performance under extreme conditions.

The hypersonic tests were designed to evaluate two critical aspects: heat transfer rates and dynamic behavior at high velocities. Understanding the heat transfer rates was essential for developing effective thermal protection systems as the spacecraft would encounter intense heat during reentry into Earth's atmosphere. These systems had to ensure that the spacecraft could withstand severe temperatures without compromising the safety of its occupants.

Additionally, the spacecraft's dynamic behavior at hypersonic speeds needed thorough examination. The tests aimed to analyze how the spacecraft would respond to the stresses and forces encountered at velocities exceeding five times the speed of sound. This data ensured that the spacecraft's structure and systems would remain stable and functional during ascent and reentry.

Langley Research Center's hypersonic flight tests represented a critical component of the iterative design and testing process that characterized the Mercury program. The insights gained from these tests would directly inform the design refinements necessary to enhance the spacecraft's durability and safety. By rigorously testing and validating the spacecraft's performance in extreme conditions, the team could mitigate risks and increase the likelihood of mission success.

These efforts at Langley were part of the broader, meticulous approach that defined the United States' early space exploration initiatives. The knowledge and experience gained from the hypersonic flight tests not only contributed to the success of Project Mercury but also laid the groundwork for future programs, such as Gemini and Apollo. The ability to ensure spacecraft durability and safety at hypersonic speeds was a crucial step in advancing America's capabilities in human spaceflight and solidifying its leadership in the space race.

On March 26, 1959, a collaborative meeting was held involving the Space Task Group, Langley Research Center, and Air Force School of Aviation Medicine personnel. The focus of this meeting was to plan bio-pack experiments for the upcoming Little Joe research and development test flights. These experiments were essential for studying the biological effects of spaceflight, ensuring the safety and well-being of astronauts who would eventually embark on manned missions.

Including bio-pack experiments in the Little Joe tests underscored NASA's comprehensive approach to testing. Addressing the technical aspects of space travel was not enough; the biological implications were equally critical. These bio-pack experiments aimed to gather vital data on how living organisms, including primates, responded to launch, microgravity, and reentry stresses. This information was crucial for developing life support systems and medical protocols for human spaceflight.

The Little Joe rockets, designed for suborbital test flights, provided a cost-effective and efficient means to conduct these experiments. NASA could identify and resolve potential issues by using Little Joe for preliminary testing before committing to more expensive and complex orbital missions.

As Project Mercury progressed, each meeting, test, and discussion brought NASA closer to achieving its goal of manned spaceflight. The dedication and meticulous planning of the Space Task Group and their partners at Langley Research Center and the Air Force School of Aviation Medicine were instrumental in overcoming the myriad challenges of early space exploration.

The collaborative efforts of these teams laid a robust foundation for the future of space travel. Their work addressed the immediate needs of Project Mercury and set the stage for the following Gemini and Apollo programs.

On March 27, 1959, a significant directive was issued from the office of Dr. T. Keith Glennan, the NASA Administrator. In a move aimed at fostering national pride and ensuring the visibility of America's space endeavors, Dr. Glennan decreed that "UNITED STATES" should be prominently painted in bold block letters on all launch vehicles, including the Mercury spacecraft. This decision was not merely about aesthetics; it was a symbolic gesture underscoring the national significance of the space program.

The markings served multiple purposes. They were a clear declaration of ownership and pride, showcasing the United States' advancements in space technology to domestic and international audiences. At a time when the space race with the Soviet Union was intensifying, such symbols were powerful tools for bolstering national morale and demonstrating technological prowess.

This directive from Dr. Glennan's office was part of a broader effort to ensure that NASA's missions were not just scientific and technological achievements but also symbols of American ingenuity and determination. By prominently marking the spacecraft, NASA aimed to instill a sense of collective accomplishment and national unity as the country ventured into the uncharted territory of space exploration.

Each decision and directive during this period was a step towards realizing Project Mercury's ambitious goals. The dedication of individuals at NASA, from administrative leaders to engineers and scientists, laid the groundwork for the program's success. Their efforts culminated in a series of milestones that would eventually see American astronauts orbit the Earth, paving the way for future exploration and solidifying the United States' position in the annals of space history.

On March 28, 1959, the Space Task Group convened to tackle a pivotal decision regarding the escape system for the Mercury spacecraft. Initially, McDonnell Aircraft Corporation had proposed an escape system involving eight small rockets mounted in a fin adapter. This design aimed to provide a reliable means of propelling the spacecraft and its astronaut to safety in an emergency during launch.

However, NASA had developed an alternative concept: a single-motor tripod system. This design was simpler and potentially more robust, but recent tests had cast doubt on its effectiveness. A troubling test result showed the spacecraft tumbling uncontrollably after launch, prompting serious concerns about the system's stability and reliability.

The meeting room was abuzz with tense discussions, with engineers and project managers debating the merits and drawbacks of each design. The atmosphere was charged, with the clinking of coffee cups providing a rhythmic backdrop to the heated exchanges. Despite the concerns raised by the recent test, the engineers ultimately concluded that the tripod concept, with its single-motor configuration, was more feasible and likely to be more reliable in the long run. Despite technical challenges, the decision to retain the tripod escape system underscored NASA's commitment to simplicity and reliability.

This critical choice was emblematic of the iterative process that characterized Project Mercury. Each test, failure, and subsequent redesign brought the team closer to achieving a safe and effective spacecraft. The engineers' ability to adapt and refine their designs in response to real-world data was crucial for the program's success.

The decision to stick with the tripod escape system also highlighted the collaborative nature of the project. The Space Task Group's ability to debate, evaluate, and ultimately reach a consensus was a testament to their dedication and expertise. This collective effort was essential for overcoming the many obstacles ahead, paving the way for the United States to achieve its ambitious goal of sending humans into space.

As the project moved forward, the lessons learned from each decision and test would continue to shape the development of the Mercury spacecraft. The team's perseverance and commitment to innovation were key factors in the program's eventual success, leading to historic milestones in space exploration and solidifying America's place in the space race.

On March 29, 1959, intense studies were conducted to determine the optimal altitude for separating the Little Joe spacecraft from its launch vehicle. The task was critical; identifying the perfect separation point was essential to ensuring the spacecraft's stability and safety during its suborbital test flights.

In the research labs, engineers, their brows furrowed in concentration, meticulously sifted through reams of data and conducted numerous simulations. The room buzzed with the quiet hum of machines and the murmur of focused discussions. Charts and graphs covered the walls, illustrating the complex interplay of forces that would act on the spacecraft during separation.

The Little Joe test flights were designed to simulate various phases of the Mercury missions, providing valuable data on the performance of the launch escape system and other critical components. Achieving the correct separation altitude was vital for these tests to yield meaningful and accurate results. If it is too low, the spacecraft might not experience the full spectrum of flight dynamics; if it is too high, it could encounter unexpected stresses or instability.

Through rigorous analysis and simulation, the engineers aimed to pinpoint the altitude allowing for a clean and controlled separation, ensuring that the spacecraft could transition smoothly from its booster to its independent flight. This level of precision was necessary to replicate the conditions that the Mercury spacecraft would face in actual missions, providing confidence in the system's reliability and safety.

The engineering team's dedication to this meticulous task reflected the broader commitment of everyone involved in Project Mercury. Each detail, no matter how small, was scrutinized and optimized to ensure the program's success. This careful, methodical approach was instrumental in overcoming the many challenges of early space exploration, ultimately leading to the United States' first manned spaceflights and setting the stage for future achievements in space.

On March 30, 1959, personnel from the Space Task Group made an important visit to the Atlantic Missile Range at the invitation of the Army Ballistic Missile Agency. This visit was a key opportunity to observe the intricate processes and technologies that would inform the development of the Mercury program.

The Space Task Group members witnessed the launch of a Jupiter rocket, gaining firsthand experience of the meticulous procedures and preparations that occurred in the days leading up to the launch. Observing the launch vehicle firing, they saw the complexities involved in managing such a powerful and sophisticated system, from the precise timing of engine ignitions to monitoring telemetry data.

Their tour included visiting the blockhouse, the nerve center of launch operations, where they received comprehensive briefings on the various recording devices and systems used to monitor and control the launch. These briefings were crucial, as the recorders would be instrumental in the centralized control facility for the upcoming Mercury-Redstone and Mercury-Jupiter flights.

The insights gained from this visit were invaluable. Understanding the detailed operation of the Jupiter launch vehicle and the associated control systems provided the Space Task Group with critical knowledge that could be applied to the Mercury program. The experience underscored the importance of precise control and data monitoring in ensuring the success of manned space missions.

This visit was part of the broader effort to integrate lessons learned from existing missile programs into the fledgling manned spaceflight initiatives. By leveraging the expertise and technologies developed for ballistic missile launches, the Space Task Group aimed to enhance the reliability and safety of the Mercury spacecraft.

On March 31, 1959, the range safety personnel at the Atlantic Missile Range were briefed on the functionality of the Mercury spacecraft during a typical flight on an Atlas launch vehicle. This meeting was crucial for ensuring the safety of the mission, as it provided a comprehensive overview of the spacecraft's operations and the procedures to be followed in case of an emergency.

During the briefing, members of the Space Task Group detailed the various phases of the Atlas-powered flight and explained how the Mercury spacecraft would perform under normal conditions. The presentation included intricate discussions about the spacecraft's systems, the sequence of operations, and the expected behavior during each launch phase.

A significant part of the briefing focused on potential abort scenarios. Space Task Group personnel proposed methods for initiating an abort during different stages of powered flight. They highlighted the importance of having reliable abort mechanisms to protect the astronaut in case of a malfunction or emergency. These discussions were technical and detailed, drawing on theoretical knowledge and practical experience from previous missile launches.

The exchange of ideas and expertise during this briefing marked the beginning of drafting a comprehensive range safety plan for Project Mercury. The plan aimed to ensure that all necessary precautions were in place to handle emergencies effectively. It outlined procedures for monitoring the flight, identifying potential hazards, and executing required aborts.

This collaborative effort between the Space Task Group and the range safety personnel was a critical step in enhancing the safety and reliability of the Mercury missions. By addressing potential risks and establishing clear safety protocols, the team worked to mitigate the dangers associated with human spaceflight.

Developing a thorough range safety plan indicated the meticulous and methodical approach that characterized Project Mercury. Each aspect of the mission was carefully considered and planned, reflecting the dedication and expertise of everyone involved. This commitment to safety and precision was essential for the program's success, ultimately paving the way for the United States to achieve its goal of sending astronauts into space and returning them safely to Earth.

On April 2, 1959, two significant milestones were achieved in the advancement of Project Mercury, marking a day filled with anticipation and strategic planning.

First, a preliminary briefing was held for prospective bidders on constructing a worldwide tracking range for Project Mercury. Representatives from 20 companies gathered to discuss the ambitious plan, which called for establishing an orbital mission tracking network of 14 sites spread across the globe. The room buzzed with the hum of strategic discussions as participants delved into the logistics, technical requirements, and global coordination needed to monitor spacecraft as they orbited the Earth.

This tracking network was essential for maintaining communication with the Mercury spacecraft, ensuring the collection of vital telemetry data, and providing continuous support throughout each mission. The project's scope underscored NASA's commitment to creating a robust and reliable infrastructure supporting the United States' entry into human spaceflight.

The April 2, 1959, events encapsulated the multifaceted efforts required to bring Project Mercury to fruition. Establishing the global tracking network and selecting the astronaut crew were both critical components in ensuring the success of America's first manned space missions. The dedication and strategic planning exhibited on this day reflected the overarching spirit of innovation and determination that defined the early years of the space race.

During the first half of April 1959, NASA and military services convened a series of critical meetings to finalize the Project Mercury animal payload program details. This comprehensive program was designed to include nine flights utilizing various launch vehicles such as Little Joe, Redstone, Jupiter, and Atlas. These flights were essential for gathering vital data on the biological impacts of spaceflight, a necessary precursor to sending humans into space.

The discussions during these meetings were extensive and detailed, covering the selection of animal subjects, the design of bio-pack containers, and the specific objectives of each flight. Including different launch vehicles allowed for thoroughly examining various flight profiles and conditions. This multi-faceted approach ensured that a wide range of data could be collected, providing a robust foundation for the subsequent manned missions.

Concurrently, the initial orientation for the seven newly selected astronauts was conducted. Alan Shepard, Gus Grissom, John Glenn, Scott Carpenter, Wally Schirra, Gordon Cooper, and Deke Slayton were introduced to Project Mercury's rigorous demands and high expectations. This orientation marked the beginning of their intensive training regimen, which would prepare them for the challenges of space travel.

The orientation sessions were designed to immerse the astronauts in the technical and operational aspects of the project. They received briefings on spacecraft systems, flight procedures, and emergency protocols. Additionally, the astronauts began physical conditioning and simulation exercises to build their endurance and familiarize themselves with the spacecraft's environment.

These two parallel efforts—the animal payload program's finalization and the astronaut crew's orientation—highlighted Project Mercury's comprehensive and systematic approach. By addressing the biological and human factors of spaceflight, NASA aimed to ensure the safety and success of its missions.

Simultaneously, another momentous event was taking place: the selection of the Mercury astronauts. The crew selection process, a rigorous and highly competitive undertaking, culminated in the selection of seven astronauts who would soon become household names. These pioneers—Alan Shepard, Gus Grissom, John Glenn, Scott Carpenter, Wally Schirra, Gordon Cooper, and Deke Slayton—were briefed on their roles and responsibilities and began their intensive training regimen.

The selection of these seven astronauts was a landmark moment in the history of space exploration. Each candidate was chosen for their exceptional skills, physical fitness, and mental fortitude, all necessary qualities for enduring the challenges of spaceflight. Their training would encompass various disciplines, from physical conditioning and simulations to classroom instruction and survival training, preparing them for the rigorous demands of their upcoming missions.

The Role of NASA's Space Task Group

The Space Task Group (STG) played a pivotal role within NASA. It oversaw the meticulous selection and rigorous training of the Mercury astronauts, laying the groundwork for America's early forays into space.

Comprising key figures such as Charles J. Donlan, Warren J. North, and Allen O. Gamble, the STG was instrumental in defining the stringent criteria that would shape the astronaut selection process. They identified essential skills, including monitoring the spacecraft's life support systems, operating reaction controls for orientation and trajectory adjustments, initiating descent procedures for safe reentry, and conducting vital scientific observations during missions. These capabilities were crucial, demanding astronauts who were not physically fit, technically adapted, or mentally resilient.

The term "astronaut" itself was carefully chosen to embody the courage, exploration, and pioneering spirit associated with those who journey beyond Earth's atmosphere. It symbolized the significant responsibilities and inherent risks that Mercury astronauts would shoulder.

Under the STG's meticulous planning and execution, every aspect of astronaut preparation was thoroughly addressed. Physical fitness was prioritized through rigorous training regimens to prepare astronauts for the physical demands of launch, orbit, and reentry. Psychological evaluations and stress tests ensured they possessed the mental fortitude to endure the isolation and pressures of space missions. Meanwhile, intensive training in spacecraft systems, navigation techniques, and emergency protocols equipped them with the technical proficiency needed to operate effectively in the unforgiving environment of space.

Through these comprehensive efforts, the STG not only selected and trained the Mercury astronauts and a template for future astronaut programs. Their dedication and foresight set a standard of excellence that would guide NASA through challenges and triumphs of subsequent missions, forging a path into the cosmos that continues to inspire generations of space explorers.

Setting the Stage for Future Achievements

Through this rigorous process, NASA successfully identified and prepared a team of pioneers who would lead the United States into the new frontier of human space exploration. The Mercury astronauts, equipped with unparalleled skills and training, set the stage for future achievements in the quest to reach the stars. Their successful missions demonstrated the effectiveness of the STG's selection and training programs and established a legacy of excellence that would continue to inspire generations of astronauts and space explorers.

NASA's Space Task Group played a pivotal role in shaping the early space program, ensuring that America's first astronauts were ready to meet the extraordinary demands of space travel. Their efforts laid the groundwork for the United States 'continued leadership in space exploration, marking the beginning of a journey that would take humanity to the Moon, Mars, and beyond.

On April 9, 1959, in a landmark moment for American space exploration, Dr. T. Keith Glennan, NASA Administrator, held a well-publicized press conference in Washington,

D.C., to announce the selection of the seven pilots chosen for the Mercury program. The event captured the nation's imagination and introduced the world to the men who would soon become symbols of courage and the spirit of discovery.

These seven individuals were meticulously chosen for their exceptional skills, physical fitness, and mental resilience, all critical attributes for the daunting challenges of spaceflight. As their names were announced, they quickly became synonymous with bravery and exploration, embodying the pioneering spirit that defined the early days of the space race.

NASA's press conference was a strategic move to inform the public and galvanize support for the Mercury program. It highlighted the human element of space exploration, making the astronauts household names and inspiring a generation to look to the stars. The media frenzy surrounding the event underscored the significance of their selection, as reporters and photographers eagerly captured every moment.

Public Communication

Communications between the Mercury spacecraft and ground control were occasionally broadcast live on television as the spacecraft passed over the United States during its missions. This broadcasting allowed the public to witness real-time interactions between the astronauts and mission control, providing a glimpse into the challenges and achievements of America's early manned spaceflights.

These live broadcasts were particularly significant during critical mission phases such as launch, orbital maneuvers, and reentry. They helped to engage and inform the public about the progress of the Mercury missions and the pioneering efforts of the astronauts involved. The visibility provided by these broadcasts also played a crucial role in building public support and enthusiasm for NASA's space exploration endeavors during the early 1960s.

Standing proudly before the cameras, the seven astronauts were now the face of America's ambitious quest to conquer space. Their training and missions would be followed closely by a captivated public, and their successes and challenges would be shared by millions. The announcement marked the beginning of a new era in space exploration, with these men at the forefront of NASA's efforts to achieve human spaceflight.

This press conference introduced the Mercury Seven to the world and set the stage for their rigorous training and the historic missions that would follow. The dedication and heroism of Shepard, Grissom, Glenn, Carpenter, Schirra, Slayton, and Cooper would leave an indelible mark on history, paving the way for future explorers and solidifying the United States' leadership in space exploration.

Public Perception and Legacy

NASA officially introduced the Mercury Seven in Washington, D.C., on April 9, 1959. They immediately captured the public's imagination and were hailed as national heroes. Time magazine compared the astronauts to historical explorers and pioneers, emphasizing the groundbreaking nature of their endeavors. The astronauts' introduction marked the beginning of a new era in American space exploration, showcasing the remarkable qualities of these men and setting the stage for America's achievements in space.

The Mercury Seven's educational achievements and extensive flying experience highlighted the critical role of higher education and specialized training in the success of space missions. Their contributions laid the groundwork for future astronauts and set a high standard for those who followed. Their legacy inspires and guides astronaut selection and training principles today, emphasizing the importance of academic excellence and practical experience in space exploration.

The Mercury Seven's journey from military test pilots to national icons was a testament to their character and dedication. They faced immense physical and mental challenges, yet their resilience and commitment to the mission never wavered. The Mercury Seven laid the foundation for future space exploration, and their legacy continues to inspire generations of astronauts and space enthusiasts.

On April 9, 1959, NASA introduced the world to the Mercury Seven, the first group of American astronauts selected for Project Mercury, the United States' inaugural human spaceflight program—these seven men—M. Scott Carpenter, L. Gordon Cooper, John H. Glenn Jr., Virgil I. "Gus" Grissom, Walter M. Schirra Jr., Alan B. Shepard Jr., and Donald K. "Deke" Slayton—became the faces of America's ambitious venture into space, embodying the nation's hopes and dreams during the early days of the space race.

Each member of the Mercury Seven brought a unique background and set of skills to the program. Yet, they shared a common bond of courage, determination, and an unyielding commitment to the mission. Selected from a pool of over 500 military test pilots, these men were chosen for their exceptional flying abilities, physical fitness, mental toughness, and ability to perform under extreme stress. They underwent rigorous training and preparation to ready themselves for the challenges of space travel.

Selection and Training Process

NASA sought individuals with the technical skills and experience as test pilots and the physical and mental fortitude to withstand the rigors of spaceflight. The candidates underwent a battery of tests, including medical evaluations, psychological assessments, and physical endurance trials designed to push them to their limits and identify those who could handle the extraordinary demands of space missions. This rigorous selection process ensured that only the most capable and resilient individuals were chosen to become America's first astronauts, setting a high standard for all future space programs.

Chapter 6 - The Mercury Seven

Ultimately, seven astronauts were chosen, known collectively as the "Mercury Seven." Each brought unique qualities and contributions to the program that would propel the United States into space.

Front to back: Alan Shepard, Gus Grissom, Gordon Cooper; Front: Wally Schirra, Deke Slayton, John Glenn, Scott Carpenter

The Mercury Seven set the stage for America's future space endeavors. Their courage and pioneering spirit achieved President Kennedy's goal of sending an American into space and laid the groundwork for the Gemini and Apollo programs. Each member of the Mercury Seven contributed to the foundation of human space exploration, their legacy continuing to inspire new generations of astronauts and space enthusiasts. Their collective efforts turned what once seemed an impossible dream into a reality, firmly establishing the United States as a formidable force in the realm of space exploration.

Left to right: Grissom, Shepard, Carpenter, Schirra, Slayton, Glenn and Cooper, 1962

The introduction of the Mercury Seven marked a new era in human exploration. Their missions, ranging from suborbital flights to multiple orbits around the Earth, tested the limits of human endurance and the capabilities of emerging technology. The lessons learned from Project Mercury laid the groundwork for the Gemini and Apollo programs, which would eventually land a man on the Moon.

The Mercury Seven in front of an F-106 Delta

These seven astronauts showcased American ingenuity and bravery, inspiring a nation to look to the stars. The Mercury missions were more than just technological feats; they were a testament to the human spirit and the relentless pursuit of knowledge. Each mission was meticulously planned yet fraught with risks. The successes and challenges faced by the Mercury Seven provided invaluable data on human spaceflight, influencing spacecraft design, life support systems, and reentry procedures.

(L to R) Cooper, Schirra (partially obscured), Shepard, Grissom, Glenn, Slayton, and Carpenter

Alan Shepard: Breaking the Boundaries

Standing at 5 feet 11 inches, Alan Shepard was the tallest astronaut in the group. Shepard's pioneering suborbital flight aboard Freedom 7 on May 5, 1961, made him the first American in space. His historic journey paved the way for future manned missions and marked a significant milestone in the Space Race.

On May 5, 1961, Alan Shepard etched his name into the annals of history as the first American in space, piloting the Freedom 7 capsule on a 15-minute and 28-second suborbital flight. The Mercury-Redstone 3 launch marked a pivotal moment in the Space Race, showcasing American prowess and determination.

The rocket beneath Shepard roared to life as the countdown reached zero, generating immense thrust that pressed him firmly into his seat. The forces he experienced were extraordinary, yet Shepard remained composed and focused. The capsule ascended rapidly, leaving behind a trail of smoke and fire, while Shepard's heart raced with anticipation and awe.

Alan Shepard's 1961 recovery seen from helicopter (Mercury-Redstone 3)

Shepard experienced intense g-forces during the ascent, a test of human endurance and resilience. His mission demonstrated that humans could withstand the severe physical stresses of launch and atmospheric reentry. Reaching the peak of his suborbital trajectory, Shepard experienced a brief period of weightlessness, a glimpse into the unique space environment. The capsule then descended, reentering Earth's atmosphere at high speeds. The heat shield protected Shepard from the searing temperatures generated by friction, and the parachutes deployed successfully, ensuring a safe splashdown in the Atlantic Ocean.

Shepard's successful mission paved the way for future manned spaceflights, proving that the United States was a formidable contender in space exploration. His achievement solidified his historical place and inspired a nation. The image of Shepard being hoisted from the ocean by a recovery helicopter became an iconic symbol of American ingenuity and courage.

Beyond his Mercury-Redstone 3 flight, Shepard's contributions to space exploration continued. He later commanded the Apollo 14 mission, becoming the fifth person to walk on the Moon. He famously hit two golf balls on the lunar surface during this mission, showcasing a lighter, more human side of space exploration. This moment, captured on film, became a memorable highlight of the Apollo program, symbolizing the technical achievements and the human spirit of adventure.

Alan Bartlett Shepard Jr.

Shepard's career was distinguished and decorated. Born in 1923, he served as a Lieutenant Commander in the US Navy (USN), bringing his military discipline and aviation expertise to NASA. His leadership and pioneering spirit left an indelible mark on the space program and the world. Shepard passed away in 1998, but his legacy endures, inspiring future generations to reach for the stars and explore the unknown.

Virgil I. "Gus" Grissom: The Tragic Hero

Gus Grissom, the shortest of the group at 5 feet 7 inches, flew the Liberty Bell 7 mission on July 21, 1961. Though marred by the premature blowing of the hatch upon splashdown, Grissom's flight demonstrated crucial improvements in spacecraft design. He later commanded the first Gemini and Apollo missions, underscoring his critical role in advancing human spaceflight.

Gus Grissom followed Alan Shepard into space on July 21, 1961, aboard the Liberty Bell 7. His flight on Mercury-Redstone 4 was another crucial step in the United States' burgeoning space program, reinforcing NASA's confidence in manned spaceflight. The mission lasted just over 15 minutes and demonstrated the growing capabilities of American space technology and the resilience of its astronauts.

During the ascent, Grissom experienced the same intense g-forces and vibrations that Shepard had faced, maintaining his composure throughout the launch. His flight was a testament to the rigorous training and preparation that NASA's astronauts underwent, ensuring they could handle the extreme conditions of space travel.

Recovery of Liberty Bell 7

The Liberty Bell 7 capsule, piloted by astronaut Gus Grissom during the second manned suborbital flight of the Mercury program, was recovered from the Atlantic Ocean, but not in 1999. It was recovered on July 20, 1999, nearly 38 years after its initial flight on July 21, 1961.

The recovery of Liberty Bell 7 was a significant event in space history. The capsule had sunk into the Atlantic shortly after splashdown due to a hatch prematurely blowing off. A team led by Curt Newport, an underwater salvage expert, used advanced sonar and robotic technology to locate and retrieve it.

The successful recovery of Liberty Bell 7 allowed historians and engineers to examine the spacecraft and its contents, shedding new light on the mission and the circumstances surrounding Grissom's dramatic ocean recovery. The capsule is now on display at the Kansas Cosmosphere and Space Center, where it continues to symbolize early human space exploration achievements.

Virgil Ivan (Gus) Grissom

However, Grissom's mission encountered a significant challenge upon splashdown. The side hatch of his capsule, Liberty Bell 7, blew prematurely, causing the spacecraft to fill with water and sink. Grissom found himself in a perilous situation, struggling to stay afloat in the choppy waters of the Atlantic Ocean. The swift actions of the recovery team were crucial; they managed to rescue him just in time, though the capsule was lost to the depths.

Despite this harrowing experience, Grissom's contributions to space exploration were invaluable. He demonstrated the ability to handle emergencies under extreme stress, a quality that underscored the importance of human presence in space missions. His performance bolstered NASA's understanding of spacecraft systems and the protocols necessary for astronaut safety.

Grissom's career flourished as he participated in the Gemini and Apollo programs. He served as the command pilot for Gemini 3, the first manned mission of the Gemini series, where he and John Young demonstrated critical maneuvers necessary for future lunar missions. His experience and leadership were instrumental in developing NASA's space exploration strategies.

Tragically, Grissom's career and life were cut short when he, along with astronauts Ed White and Roger B. Chaffee, died in the Apollo 1 fire on January 27, 1967, during a pre-launch test. The cabin fire was a devastating blow to NASA and a stark reminder of the inherent risks of space exploration. The accident significantly changed spacecraft design, safety protocols, and testing procedures, ultimately making future missions safer.

Grissom's dedication and sacrifice remain a somber yet powerful reminder of the dangers those who venture into space face. His legacy is a testament to the bravery and commitment of the early astronauts who paved the way for humanity's journey beyond Earth. Born in 1926, Captain Virgil I. "Gus" Grissom of the US Air Force (USAF) is remembered for his technical contributions and courage and perseverance in the face of unprecedented challenges. His memory continues to inspire new generations of explorers as they push the boundaries of what is possible.

John H. Glenn, Jr.: An Orbital Pioneer

The oldest of the Mercury Seven, at 37, John Glenn weighed a maximum of 180 pounds. On February 20, 1962, Glenn became the first American to orbit the Earth aboard Friendship 7. His successful three-orbit mission was a monumental achievement, solidifying America's presence in space and boosting national morale during the Cold War.

John Glenn's historic flight on February 20, 1962, aboard Friendship 7 marked a monumental achievement, making him the first American to orbit the Earth. During his mission on Mercury-Atlas 6, Glenn orbited the Earth thrice in a flight that lasted under five hours. This mission was crucial in demonstrating that the United States was a formidable competitor in the space race and could successfully send a human into orbit and bring them back safely.

John Herschel Glenn Jr.

The flight was not without its challenges. Midway through the mission, Glenn encountered issues with the spacecraft's automatic control system. The system, designed to maintain the spacecraft's proper orientation, began to malfunction. Demonstrating exceptional piloting skills and calm under pressure, Glenn manually adjusted the spacecraft's attitude, ensuring that it remained on the correct trajectory. His ability to take control and rectify the situation was a testament to his training, experience, and inherent skill as a pilot.

Following his successful mission, Glenn was celebrated as a national hero. His flight aboard Friendship 7 had advanced American space exploration and instilled a sense of pride and accomplishment across the nation. Despite this success, Glenn left NASA in 1964, believing he would not be selected for an Apollo mission. He transitioned to a new career in public service and was elected as a U.S. Senator from Ohio, serving from 1974 to 1999. During his tenure in the Senate, Glenn worked on issues ranging from science and technology to national security.

In a remarkable turn of events, John Glenn returned to space aboard the Space Shuttle Discovery on October 29, 1998 (STS-95). At the age of 77, he became the oldest person to fly in space. His participation in this mission was part of a study on the effects of spaceflight on the elderly, providing valuable data on how aging interacts with the stresses of space travel. This mission underscored Glenn's lifelong commitment to space exploration and scientific inquiry.

John Glenn's legacy is one of pioneering spirit, courage, and dedication. Born in 1921, Major John Glenn of the US Marine Corps (USMC) served his country both as a distinguished astronaut and a long-serving senator. His contributions to space exploration and his later public service made a lasting impact on the field of aerospace and the nation. Glenn passed away in 2016, leaving a legacy that inspires future generations to reach for the stars.

M. Scott Carpenter: The Second American in Orbit

Scott Carpenter orbited the Earth in Aurora 7 on May 24, 1962. Despite an off-target reentry, Carpenter's mission contributed significantly to space science, including experiments on human physiology and the effects of weightlessness. His thorough scientific approach provided valuable data for future missions.

Scott Carpenter's flight on Mercury-Atlas 7 on May 24, 1962, aimed to replicate the success of John Glenn's earlier mission. Carpenter's spacecraft, Aurora 7, followed a similar trajectory, orbiting the Earth thrice. This mission was designed to validate the Mercury program's capabilities further and gather more data on the effects of spaceflight on the human body.

Malcolm Scott Carpenter

Carpenter demonstrated his adept piloting skills during the mission and conducted several scientific experiments. However, the flight encountered challenges, particularly during reentry. A targeting error caused Aurora 7 to land 250 miles off-course in the Atlantic Ocean. This unexpected deviation led to a tense period as recovery teams worked to locate and retrieve Carpenter and his spacecraft. Despite the off-course landing, Carpenter remained calm and collected, utilizing his training to ensure his safety until rescue operations could reach him.

Carpenter's Overshoot

Scott Carpenter's overshoot of the intended landing site during his Aurora 7 mission was indeed attributed to a malfunction in the spacecraft's automatic stabilization system. This malfunction caused a slight misalignment during retrofire, altering the trajectory and resulting in a landing approximately 250 miles off-target in the Atlantic Ocean.

Following the splashdown, the U.S. Air Force offered to pick up Carpenter using a seaplane. However, the U.S. Navy, responsible for recovery operations under the Mercury program, declined this offer. This decision led to some controversy and, subsequently, a Senate hearing to investigate the circumstances surrounding Carpenter's recovery and the decision-making process.

Carpenter's flight on Mercury Atlas 7 was his only journey into space, but his contributions to exploration did not end there. After his spaceflight, he transitioned to the Navy's "Man in the Sea" program. This initiative aimed to explore and develop underwater habitats and technologies, pushing the boundaries of human exploration into the depths of the oceans. Carpenter became the only American to be both an astronaut and an aquanaut, showcasing his versatility and pioneering spirit.

In the "Man in the Sea" program, Carpenter participated in the SEALAB project, living and working for extended periods in underwater habitats. His work in this program advanced our understanding of human endurance and capabilities in extreme environments in space and under the sea.

Scott Carpenter's career reflected a remarkable dedication to exploration and innovation. Born in 1925, Lieutenant Carpenter of the US Navy (USN) was an accomplished astronaut and a pioneering aquanaut. His dual contributions to space and underwater exploration highlighted his unique historical place as a trailblazer in both domains. Carpenter passed away in 2013, leaving a legacy of courage, curiosity, and a relentless pursuit of knowledge. His achievements continue to inspire future generations to explore the unknown, whether it be in the vastness of space or the depths of the oceans.

Wally M. Schirra, Jr.: A Veteran of Three Programs

Wally Schirra lost weight to meet the Mercury program's stringent requirements but later set a U.S. duration record with his nine-hour flight on Sigma 7 on October 3, 1962. Schirra's precise piloting and engineering assessments during his flight ensured that the spacecraft's systems operated optimally, enhancing the reliability of future missions.

Walter Marty (Wally) Schirra Jr.

Wally Schirra had the unique distinction of flying in the Mercury, Gemini, and Apollo programs, making him the only astronaut to participate in all three pioneering American space endeavors. His first spaceflight, Sigma 7, launched on October 3, 1962, and was a nearly perfect mission that demonstrated the reliability of the Mercury spacecraft. Schirra's meticulous attention to detail and technical expertise contributed to the mission's success, which set a new U.S. duration record with his nine-hour flight.

During the Sigma 7 mission, Schirra conducted experiments and tests to evaluate the spacecraft's systems and performance. His precision in executing the flight plan ensured that every aspect of the mission proceeded smoothly, providing NASA with critical data and confidence in the Mercury program. His flight emphasized the importance of technical proficiency and careful planning in achieving mission objectives.

Schirra's next mission was aboard Gemini 6A, where he and his crewmate, Thomas Stafford, achieved the first space rendezvous. Launched on December 15, 1965, Gemini 6A successfully met and maneuvered with Gemini 7, which was already in orbit. This milestone in spaceflight demonstrated the feasibility of rendezvous and docking, essential techniques for future lunar missions. Schirra's skillful piloting and the mission's success were crucial in advancing the Gemini program and paving the way for Apollo.

In 1968, Schirra commanded Apollo 7, the first crewed mission of the Apollo program. This mission was critical for testing the redesigned Command and Service Module (CSM) following the tragic Apollo 1 fire. Launched on October 11, 1968, Apollo 7 was an 11-day mission that orbited the Earth, thoroughly testing the spacecraft's systems in preparation for lunar missions. Schirra's leadership and attention to detail ensured a smooth and successful mission, providing NASA with the confidence needed to proceed with Apollo 8 and subsequent missions that would ultimately land humans on the Moon.

Schirra's contributions to space exploration were characterized by his focus on technical evaluation and precision. His ability to conduct thorough and accurate assessments of spacecraft systems was invaluable in each of his missions. Born in 1923, Lieutenant Commander Schirra of the US Navy (USN) was a dedicated and skilled astronaut whose career spanned the most formative years of American spaceflight. He passed away in 2007, leaving a legacy of excellence and a standard for future astronauts to aspire to.

Gordon "Gordo" Cooper: The Last Mercury Flight

Gordon Cooper, the youngest at 32 and lightest at 150 pounds, completed the longest Mercury flight aboard Faith 7, lasting over 34 hours from May 15-16, 1963. Cooper's mission demonstrated the feasibility of extended spaceflights and provided critical data on the endurance of the spacecraft and the astronaut under prolonged space conditions.

L. Gordon Cooper's historic flight on May 15, 1963, piloting the Faith 7 mission on Mercury-Atlas 9, was a landmark in American space exploration. As the final mission of the Mercury program, Faith 7 completed 22 orbits around the Earth, with Cooper spending 34 hours and 19 minutes in space. This mission marked the last time an American was launched solo into orbit and demonstrated significant advancements in space travel and astronaut endurance.

Cooper's mission was notable for its duration and the wealth of scientific data collected. During the flight, Cooper conducted various experiments, observed Earth's weather patterns, and took photographs contributing to meteorological and geographical studies. The mission's extended duration provided valuable insights into the effects of prolonged spaceflight on the human body, helping to inform the development of future space missions.

Cooper's manual reentry control was one of the Faith 7 mission's most remarkable aspects. Toward the end of the mission, an electrical malfunction caused the automatic stabilization and control system to fail. Demonstrating exceptional skill and composure, Cooper manually controlled the reentry process, using the spacecraft's orientation thrusters to guide Faith 7 back to Earth. His successful manual reentry showcased astronauts' capability to handle emergencies and reinforced the importance of human intervention in spaceflight.

Leroy Gordon (Gordo) Cooper Jr.

Cooper's achievements did not end with the Mercury program. He later participated in Project Gemini, commanding Gemini 5 alongside Charles "Pete" Conrad. Launched on August 21, 1965, Gemini 5 set a new endurance record, with the astronauts spending eight days in space. This mission proved that humans could survive and work in space for the duration required to reach the Moon and return. The endurance and resilience demonstrated by Cooper and Conrad during Gemini 5 were instrumental in advancing NASA's goals for the Apollo program.

Born in 1927, Captain L. Gordon Cooper of the US Air Force (USAF) was known for his technical expertise, bravery, and pioneering spirit. His contributions to the Mercury and Gemini programs were pivotal in the United States' efforts to explore space and land humans on the Moon. Cooper passed away in 2004, leaving behind a legacy of innovation, courage, and dedication to space exploration.

Cooper's missions underscored the importance of human presence in space exploration for conducting scientific research and managing unforeseen challenges. His ability to manually control Faith 7's reentry and his endurance during Gemini 5 highlighted the critical role of astronauts in ensuring mission success.

Donald K. "Deke" Slayton: The Persistent Aviator

Deke Slayton, initially grounded due to a medical condition, eventually flew in the Apollo-Soyuz Test Project in 1975, a mission that united American and Soviet space efforts. Slayton's perseverance and eventual return to flight exemplified the resilience and determination of the Mercury Seven. His role in the Apollo-Soyuz mission symbolized a moment of détente during the Cold War and highlighted the potential for international cooperation in space.

Donald K. "Deke" Slayton, one of the original Mercury Seven astronauts, faced an unexpected setback when a heart condition grounded him from active spaceflight. Despite this, Slayton's influence within NASA remained significant. As NASA's first Chief of the Astronaut Office, he was crucial in selecting and managing astronaut crews for subsequent missions, shaping the course of American space exploration.

Despite undergoing extensive physical examinations, Deke Slayton had an undiagnosed condition of atrial fibrillation, which ultimately led to his grounding before his first space flight. This unexpected medical issue was a significant setback, but Slayton played a crucial role within NASA, contributing to the space program's success from a leadership position.

Slayton's responsibilities included overseeing the training, performance, and assignments of astronauts for the Gemini, Apollo, and early Space Shuttle programs. His deep understanding of spaceflight's technical and human aspects ensured that the best candidates were chosen for each mission, contributing to the success and safety of NASA's endeavors.

Although grounded, Slayton never relinquished his dream of spaceflight. His perseverance paid off in 1975 when medical advancements allowed him to return to flight status. He flew on the Apollo-Soyuz Test Project, the first international space mission, marking a historic moment in space exploration. This mission, launched on July 15, 1975, symbolized a thawing of Cold War tensions as American and Soviet spacecraft docked in orbit, and crews conducted joint scientific experiments and shared meals. Slayton's participation was a testament to his resilience and dedication.

Donald Kent (Deke) Slayton

In the years following his spaceflight, Slayton continued contributing to NASA's efforts, particularly the Space Shuttle program. His expertise and leadership helped guide the development and implementation of this new era of reusable spacecraft. Slayton retired from NASA in 1982, capping a distinguished career spanned the most formative years of human spaceflight.

Born in 1924, Major Donald K. Slayton of the US Air Force (USAF) was a pioneer whose impact on space exploration extended far beyond his time in space. His legacy is one of determination, leadership, and an unwavering commitment to advancing human space exploration. Slayton passed away in 1993, but his contributions continue to be felt in the ongoing journey to explore the cosmos.

Slayton's career highlights the importance of adaptability and perseverance in facing challenges. His work as Chief of the Astronaut Office ensured the success of numerous missions, and his eventual flight on the Apollo-Soyuz Test Project demonstrated that dreams deferred can still be realized. Astronaut Training Regimen

Personal Lives and Commitment

The Mercury Seven were more than just astronauts; they were symbols of American ingenuity and determination. They gave interviews, made public appearances, and became symbols of hope and progress during intense competition with the Soviet Union. Their commitment to the mission was unwavering, knowing the risks involved in pioneering human spaceflight but driven by a profound sense of duty and the desire to push the boundaries of human knowledge and capability.

Scott Carpenter's selection for Project Mercury tested the Navy's commitment to the program when his ship's skipper initially refused to release him for the mission, requiring intervention from Admiral Arleigh Burke. Gordon Cooper faced personal challenges when his wife, Trudy, left him after an affair, only to return to uphold the image of stability NASA desired for its astronauts. These personal sacrifices underscored the dedication and resilience of the Mercury Seven and their families.

Critical tests were conducted on two escape configurations for the Mercury spacecraft on April 9-10, 1959, at the Arnold Engineering Development Center in Tullahoma, Tennessee. These tests evaluated the escape systems' static stability and drag characteristics, essential for ensuring astronaut safety during launch emergencies.

The two-day testing regimen was crucial in refining the Mercury spacecraft's design. Engineers meticulously assessed how each escape configuration performed under simulated flight conditions. Static stability tests aimed to determine how the spacecraft would behave aerodynamically when subjected to different forces, ensuring that it remained controllable during an abort scenario. Drag characteristics were also scrutinized, as minimizing aerodynamic drag was vital for achieving the necessary speed and trajectory for a safe escape.

These tests were integral to the Mercury program's iterative design process. By rigorously evaluating and comparing the two configurations, engineers could identify the most reliable and efficient design. The data gathered from these tests would inform further refinements, enhancing the escape system's overall safety and effectiveness.

The dedication to exhaustive testing and validation underscored NASA's commitment to astronaut safety. Ensuring that the escape system could function flawlessly under extreme conditions was paramount. The lessons learned from these tests would benefit the Mercury missions and contribute to the broader body of knowledge in spacecraft design and safety protocols.

The work done at the Arnold Engineering Development Center during these two days was a testament to the meticulous planning and engineering excellence that characterized Project Mercury. Each successful test brought NASA closer to its goal of manned spaceflight, with the assurance that every possible measure had been taken to protect the lives of the astronauts.

As the program advanced, the data and insights gained from these early tests would continue to play a crucial role in developing and refining the Mercury spacecraft. The unwavering focus on safety and reliability laid a solid foundation for the following historic missions, ultimately contributing to the United States' achievements in human space exploration.

On April 10, 1959, engineers at Wallops Island completed critical tests on the escape motor canting angles for the Mercury spacecraft. These tests were designed to evaluate how different canting angles—angles at which the escape motors were tilted—affected the stability of the spacecraft during an emergency escape scenario.

The results of these tests indicated that stability improved with larger canting angles. The engineers meticulously documented this finding, understanding the paramount importance of these tests for the safety of the astronauts. Ensuring the spacecraft could maintain stability during a high-stress abort was crucial for protecting the crew's lives.

The tests involved a series of controlled simulations in which the escape motors were fired at various angles. By carefully observing the behavior of the spacecraft under these conditions, the engineers determined the optimal canting angle that would provide the best stability and control during an emergency escape. This information was essential for refining the design of the escape system and ensuring its reliability.

The work done on April 10th was part of the broader effort to develop a robust and effective escape mechanism for the Mercury spacecraft. Each test and subsequent analysis contributed to a deeper understanding of the spacecraft's dynamics and helped identify the configurations that would maximize safety.

The dedication and precision of the engineering team at Wallops Island reflected the meticulous approach that defined Project Mercury. Every detail was carefully considered, and every test was a step closer to ensuring the success and safety of America's first manned space missions.

As NASA progressed toward launching astronauts into space, the insights gained from these tests would play a crucial role in finalizing the design of the escape system. The commitment to thorough testing and validation was instrumental in building a spacecraft capable of safely carrying humans into space and returning them home.

The findings from Wallops Island on escape motor canting angles were another piece in the complex puzzle of human spaceflight. They demonstrated the relentless pursuit of excellence that would eventually lead to the historic successes of Project Mercury and beyond.

On April 12, 1959, significant advancements were made in the safety and durability of the Mercury spacecraft through innovative testing and successful demonstrations. At Langley Research Center, tests were underway on an aluminum honeycomb structure designed to absorb the impact loads experienced by the spacecraft during reentry and landing. This cutting-edge approach aimed to enhance the structural integrity of the spacecraft, ensuring it could withstand the harsh conditions of spaceflight and protect the astronaut inside.

The aluminum honeycomb structure represented a pioneering effort in spacecraft design. By distributing impact forces across a larger area, the honeycomb material could reduce the stresses on the spacecraft's body, minimizing the risk of damage upon landing. The tests at Langley were crucial for validating this concept and refining the design to maximize its effectiveness.

The second full-scale beach abort test was conducted on the same day at Wallops Island. This test was a critical evaluation of the Mercury spacecraft's escape system, designed to ensure that astronauts could be safely extracted during an emergency during launch. Despite a deliberate thrust misalignment introduced to test the system's resilience, the escape mechanism performed successfully.

The beach abort test simulated an emergency scenario where the spacecraft must be rapidly propelled away from the launch vehicle. Engineers meticulously monitored the test, noting the escape motors' performance and the spacecraft's stability. The success of this test, even under the challenging condition of a thrust misalignment, demonstrated the robustness and reliability of the escape system.

The combination of these two testing efforts on April 12th underscored the comprehensive and methodical approach that defined Project Mercury. By simultaneously advancing the spacecraft's structural safety and validating its emergency escape capabilities, NASA was making significant strides toward ensuring the overall mission success and astronaut safety.

These efforts were part of the broader framework of Project Mercury's development, where each test and experiment contributed to the refinement and optimization of the spacecraft. The dedication to rigorous testing and innovation at Langley and Wallops Island exemplified the relentless pursuit of excellence essential for the United States to achieve its goals in human space exploration.

The progress made on this day reflected the unwavering commitment of NASA's engineers and scientists to overcome the challenges of spaceflight. Their work not only paved the way for Project Mercury's successful missions but also laid the groundwork for future achievements in space exploration.

As the spring of 1959 unfolded, Project Mercury's development phase was characterized by rigorous testing, strategic planning, and the foundational steps for the astronauts who would soon venture into the unknown. Each decision and test brought NASA closer to the monumental goal of manned spaceflight, propelled by the collaborative efforts of engineers, scientists, and military personnel.

On April 13, 1959, two small-scale spacecraft escape-tower combinations were successfully launched at Wallops Island. These tests were essential for validating the design and functionality of the escape system, ensuring it could perform reliably under various conditions. The success of these small-scale tests set the stage for a full-scale launch conducted the following day, which proceeded flawlessly from the firing of the escape system to the recovery by helicopter. This sequence of successful tests demonstrated the robustness of the escape mechanism, providing confidence in its ability to safeguard astronauts during an emergency.

In addition to these technical advancements, NASA took significant steps to support the training and preparation of the Mercury astronauts. NASA requested the Navy's Aviation Medical Acceleration Laboratory in Johnsville, Pennsylvania, to be used for astronaut training. This facility was equipped to simulate the high-acceleration forces astronauts would experience during launch and reentry, making it an ideal location for preparing them for the physical demands of spaceflight.

Furthermore, Rear Admiral J.W. Gannon was appointed by Deputy Secretary of Defense Donald A. Quarles to lead a Department of Defense group focused on studying the recovery aspects of Project Mercury. This appointment underscored the importance of a well-coordinated recovery operation, ensuring that astronauts and spacecraft could be safely retrieved after their missions. The involvement of the Department of Defense highlighted the collaborative nature of Project Mercury, with multiple agencies working together to address the myriad challenges of human spaceflight.

This period's meticulous planning and rigorous testing were crucial for advancing Project Mercury. The successful launches at Wallops Island, the establishment of training protocols, and the strategic focus on recovery operations all contributed to the project's momentum.

As NASA and its partners continued to refine and validate the various components of the Mercury program, each step brought the United States closer to achieving its goal of sending humans into space. The dedication and expertise of all involved laid a strong foundation for the following historic missions, marking the beginning of a new era in space exploration.

On April 15, 1959, the offices of the Space Task Group were abuzz with activity as engineers and scientists meticulously drafted the ground-instrumentation requirements for firing the Little Joe test vehicles at Wallops Island. These detailed specifications were essential for ensuring the success and accuracy of the upcoming tests, designed to validate various aspects of the Mercury spacecraft and its systems.

The instrumentation requirements included a range of sophisticated equipment:

Pulse Radars: To track the precise position and trajectory of the test vehicles during flight.

Cameras: To capture high-resolution images and video of the launches, providing visual data for analysis.

Doppler Radar: To measure the velocity and acceleration of the vehicles, offering crucial information about their performance.

Wind-Monitoring Instruments: To assess atmospheric conditions and their impact on the test flights.

Telemetry Equipment: To transmit real-time data from the spacecraft to ground control, allowing for immediate monitoring of the vehicle's systems and performance.

Ground Destruct System: To safely terminate the flight if necessary, ensuring the safety of the test range and personnel.

The precise coordination and integration of these instruments were critical. Each piece of equipment was vital in collecting data, monitoring the vehicle's performance, and ensuring the overall success of the tests. The engineers worked tirelessly to ensure that all systems were correctly configured and that the data collected would be comprehensive and reliable.

The Little Joe test vehicles were integral to the development of the Mercury spacecraft. These suborbital rockets simulated various mission phases, including launch, ascent, and abort scenarios. By meticulously planning and implementing the ground-instrumentation requirements, the Space Task Group aimed to gather invaluable data that would inform the design and refinement of the Mercury spacecraft.

This meticulous attention to detail and rigorous planning were emblematic of Project Mercury's approach to problem-solving and innovation. Every test, data, and every analysis contributed to building a safer, more reliable spacecraft capable of carrying American astronauts into space and back.

As the Space Task Group and its partners prepared for the Little Joe tests, their efforts exemplified the dedication and expertise that would ultimately lead to Project Mercury's success. The groundwork laid on days like April 15, 1959, was crucial in paving the way for the United States' historic achievements in human space exploration.

On April 16, 1959, key meetings and discussions occurred that would shape the future of Project Mercury and its associated missions. These discussions underscored the collaborative efforts between NASA and the military, ensuring that every aspect of the project was meticulously planned and executed.

In a significant meeting led by Admiral J.W. Gannon, NASA, and military officials convened to discuss the search and recovery aspects of Project Mercury. This exploratory meeting highlighted the Navy's and other military services' commitment to supporting the project. The discussions focused on strategies and logistics for safely recovering the Mercury spacecraft and its crew after the splashdown. Ensuring a reliable and efficient recovery process was vital for the success of manned missions, as it guaranteed the astronauts' safe return after their journey into space.

Simultaneously, personnel from the Space Task Group, Langley Research Center, and Lewis Research Center met to discuss the development plans for constructing and instrumenting the Big Joe Number I reentry spacecraft test vehicle. The Big Joe test vehicle was designed to validate the Mercury spacecraft's heat shield and reentry procedures, a critical component for ensuring the spacecraft could withstand the intense heat of reentry and protect its occupants. Milestone objectives were meticulously drafted during the meeting, and clear goals and timelines for the development and testing phases were set. This planning was crucial for maintaining the project's momentum and ensuring that each milestone was achieved on schedule.

In addition to these discussions, NASA requested the Air Force to provide two TF-102B and two T-33 aircraft for the Mercury astronauts to maintain their proficiency in high-performance aircraft. These aircraft were essential for the astronauts to practice and hone their piloting skills, ensuring they could handle high-speed, high-altitude flights. Proficiency in these aircraft was crucial for the astronauts' overall training and preparedness for their space missions.

The April 16, 1959, events demonstrated the comprehensive and collaborative efforts required to advance Project Mercury. The detailed planning and coordination evidently demonstrated NASA's and the military's commitment to supporting and enhancing the project's various aspects. These efforts ensured that every component, from recovery operations to reentry tests and astronaut training, was thoroughly addressed and optimized.

As Project Mercury progressed, the dedication and expertise of all involved continued to drive the project forward. The milestones and objectives set during these meetings would pave the way for the successful execution of the program, ultimately leading to the United States' first manned spaceflights and solidifying its position as a leader in space exploration.

On April 22, 1959, NASA officials made a critical decision in developing the Mercury spacecraft: the tower configuration was the optimal escape system. This conclusion resulted from thorough discussions and rigorous testing, reflecting NASA's unwavering commitment to astronaut safety.

The tower escape system, consisting of a solid-fuel rocket mounted above the Mercury capsule, was designed to rapidly propel the spacecraft away from its launch vehicle in an emergency. This configuration was chosen for its effectiveness in quickly distancing the spacecraft from any potential explosion or malfunction of the launch vehicle, safeguarding the astronaut's life.

Adopting the tower configuration was a significant milestone in the Mercury program. It provided a clear path forward for the development and integration of the escape system, ensuring that one of the spacecraft's most crucial safety components was reliably designed and tested.

Despite this definitive choice, NASA officials acknowledged the importance of continuous innovation and improvement. As such, tentative studies of alternate escape configurations would continue. This approach ensured that any potential advancements or improvements in escape technology could be considered and integrated if proven superior.

Adopting the tower configuration marked a key advancement in Project Mercury's goal of achieving safe manned spaceflight. It underscored NASA's methodical approach to problem-solving and its dedication to the safety of the astronauts. Like this one, each decision brought the United States closer to realizing its ambition of sending humans into space and ensuring their safe return.

On April 23-24, 1959, a pivotal coordination meeting occurred at NASA's Langley Research Center, a cornerstone of the United States' burgeoning space program. The meeting's focus was to strategize on the handling, reduction, and analysis of data collected from the upcoming Big Joe spacecraft mission. This mission was integral to the Mercury Project, aimed at putting an American astronaut into orbit and ensuring their safe return.

During the two-day session, detailed data pickup and information dissemination procedures were established. These procedures were crucial for ensuring that the data collected from Big Joe would be efficiently processed and analyzed. The bulk of the data reduction workload was assigned to the Lewis Research Center (now known as NASA Glenn Research Center) and the Space Task Group. These entities were at the forefront of America's space efforts, tasked with turning raw data into actionable insights that could advance the nation's knowledge and capabilities in space exploration.

The Big Joe mission was designed to test the heat shield of the Mercury capsule, a critical component for the safe re-entry of astronauts from space. The success of this mission would pave the way for future manned flights, contributing significantly to the United States' progress in the Space Race. The coordination meeting at Langley exemplified the collaborative effort and meticulous planning underpinning NASA's early achievements, laying the groundwork for the eventual success of the Mercury missions and the broader Apollo program. This era of innovation and discovery propelled America to the forefront of space exploration and inspired generations of scientists, engineers, and astronauts who would continue to push the boundaries of human knowledge and capability.

On April 27, 1959, Project Mercury achieved a significant milestone by receiving the DX priority procurement rating. This designation ensured that the project would receive the highest national priority for all necessary materials and components, reflecting its critical importance to the United States' efforts in the Space Race. This priority status expedited the acquisition of essential resources, underscoring the nation's commitment to advancing its space exploration capabilities.

The same day also marked a momentous occasion as the seven Mercury astronauts reported for duty. These pioneering men, selected from a pool of military test pilots, were about to embark on a journey to test human endurance and ingenuity limits. Their rigorous training program began immediately, designed to prepare them for the challenges of spaceflight.

During their first week, the astronauts' schedule was packed with general briefings that provided an overview of the mission objectives and expectations. They delved into the specifics of spacecraft configuration, learning about the intricate systems that would keep them safe and operational in the harsh space environment. Training on escape methods and using support and restraint systems was critical, as these skills would be vital for their survival in an emergency.

Operational concepts were also a key focus, ensuring that the astronauts understood every aspect of the mission and their roles within it. This comprehensive approach was essential for building the confidence and competence required for the groundbreaking flights ahead.

In addition to technical training, the astronauts engaged in flying time and athletic activities. Flying time helped them maintain and enhance their piloting skills, while athletic activities ensured they remained in peak physical condition. This blend of intellectual and physical preparation was crucial for the demanding nature of space missions.

The arrival of the Mercury astronauts and the initiation of their training program marked the beginning of an era of intense preparation and dedication. Their efforts would lay the foundation for America's first manned spaceflights, ultimately leading to the historic achievements of Project Mercury. These early days were characterized by a spirit of innovation and determination that would drive the United States to new heights in space exploration, inspiring future generations to reach for the stars.

On April 27-28, 1959, the Department of Defense working group on Mercury search and recovery operations convened at Patrick Air Force Base. The primary goal of this meeting was to delineate service responsibilities for the first two Mercury-Atlas ballistic flights. This coordination was essential to ensure that all aspects of search and recovery were meticulously planned and executed. By defining these roles, the meeting solidified the military's integral support for Project Mercury, highlighting the collaboration between NASA and the Department of Defense. This partnership was critical to the success of the missions, as it provided the necessary logistical and operational support to recover astronauts and spacecraft safely after their missions, marking a significant step forward in the United States' space exploration efforts.

Significant advancements were made in the spacecraft's recovery landing system throughout April. The original design, which featured an extended-skirt main parachute, was tested and found wanting during a critical drop test. This failure highlighted the potential dangers and unreliability of the extended-skirt parachute in ensuring a safe descent and landing for the spacecraft.

In response to these findings, engineers swiftly replaced the problematic parachute with a more reliable ring-sail parachute. The ring-sail design, known for its stability and effectiveness in managing descent velocities, was rigorously tested and proved a superior alternative. This change was not merely a technical adjustment but a crucial enhancement that significantly improved the safety and reliability of the spacecraft's landing system.

On May 1, 1959, a pivotal coordination meeting for the Little Joe Project occurred, gathering key personnel from the Space Task Group, McDonnell Aircraft Corporation, and Wallops Island. This crucial assembly evaluated the project's various developmental phases and aimed to synchronize efforts across multiple related initiatives.

The participants meticulously reviewed progress, addressing technical challenges and assessing readiness levels. The goal was to establish a realistic and efficient launch schedule that would align seamlessly with other significant projects in the burgeoning space program, such as Big Joe, Mercury-Atlas, Mercury-Redstone, and Mercury-Jupiter. These projects represented a vital component of the nation's efforts to achieve manned spaceflight, with Little Joe playing a critical role in testing and validating launch and abort systems.

On May 5, 1959, a comprehensive meeting was convened to plan the spacecraft's complete recovery test program meticulously. This gathering brought together experts and key stakeholders to discuss crucial elements necessary for the recovery process's success.

The agenda covered a wide range of topics, including the availability of model spacecraft needed for testing, identifying optimal test locations, and establishing a detailed time schedule for the entire program. Each aspect of the spacecraft's recovery process, from the initial descent to the final retrieval after landing, was scrutinized to ensure nothing was overlooked.

On May 6, 1959, a significant decision was made regarding using test subjects for the Little Joe flights. Initially considered for these tests, pigs were eliminated due to their inability to survive extended periods on their backs, which is a necessary condition for flight tests.

Despite this setback, McDonnell's innovative approach using a pig named "Gentle Bess" yielded promising results. "Gentle Bess" was employed to test the impact crushable support system, a crucial component designed to absorb shock and protect the spacecraft's occupants during landing. The test was a success, demonstrating the effectiveness and viability of the crushable support system.

On May 11, 1959, two significant developments marked the progress of the space program. First, a spacecraft recovery study contract was awarded to Grumman Aircraft Corporation. This contract entrusted Grumman with the critical task of developing and refining the systems necessary to ensure the spacecraft's and its occupants' safe recovery after missions. Grumman's expertise in aerospace engineering was expected to provide innovative solutions and robust designs, further enhancing the reliability of the recovery process.

In parallel, NASA issued an essential policy concerning the Mercury astronauts. According to this new directive, the astronauts were officially subject to NASA regulations, ensuring their activities and conduct aligned with the agency's standards and protocols. Moreover, the policy stipulated that any unclassified information reported by the astronauts would be made publicly available. This commitment to transparency was intended to foster public trust and support for the space program, allowing the general public to stay informed about the progress and achievements of the Mercury missions.

These developments underscored NASA's dedication to space exploration's technical and administrative aspects. By securing Grumman's expertise for spacecraft recovery and establishing clear policies for astronaut conduct and information dissemination, NASA laid a strong foundation for the success of the Mercury program and future space endeavors.

On May 17, 1959, the Langley Research Center prepared a one-fourteenth-scale model of the Mercury spacecraft for an ambitious test launch from Wallops Island. This launch was designed to achieve a speed of Mach 18 using a five-stage rocket. The creation and testing of this scale model were integral to the ongoing efforts to test and refine the spacecraft's capabilities, ensuring it could withstand the extreme conditions of high-speed flight.

The meticulous preparation in this project underscored the rigorous approach taken by engineers and scientists at Langley. The scale model allowed the simulation of various flight dynamics and aerodynamic stresses that the full-sized Mercury spacecraft would encounter. By achieving Mach 18 speeds, the team aimed to gather critical data on the spacecraft's performance at hypersonic velocities, including its structural integrity, thermal protection systems, and aerodynamic stability.

This test was a crucial step in the iterative process of spacecraft development, providing valuable insights that would inform the design and construction of the actual Mercury spacecraft. The efforts at Langley were part of a broader strategy to ensure that when the time came for manned missions, the Mercury spacecraft would be fully equipped to meet the challenges of spaceflight, paving the way for the United States' early ventures into human space exploration.

On May 21, 1959, Langley Research Center issued Specification Number S-45, a comprehensive document detailing the tracking and ground instrumentation system required for Project Mercury. This specification was critical to developing a reliable infrastructure to monitor and support the spacecraft throughout its missions.

In response to this detailed specification, proposals from seven contractor teams were submitted by June 22, 1959. Each proposal was carefully crafted to meet Langley's stringent requirements, encompassing various aspects of tracking, communication, and data acquisition necessary for the successful execution of Project Mercury missions.

The subsequent technical evaluations of these proposals marked the beginning of a meticulous selection process. Experts at Langley assessed each proposal for its technical merit, feasibility, and alignment with Project Mercury's objectives. This evaluation phase was crucial, as the chosen tracking and ground instrumentation system would play a vital role in ensuring the astronauts' and their spacecraft's safety and success.

These efforts underscored the complexity and precision required in orchestrating manned space missions. By issuing Specification Number S-45 and initiating the proposal review process, Langley Research Center laid the groundwork for a robust support system essential for Project Mercury's groundbreaking achievements and the future of American space exploration.

On May 22, 1959, negotiations with the Army Ordnance Missile Command regarding the costs of Redstone and Jupiter boosters continued. These discussions focused on refining funding estimates to ensure that the necessary financial resources were allocated efficiently for the boosters, which were crucial for Project Mercury's launch operations. The Redstone and Jupiter boosters were integral to the program, providing the necessary thrust for early manned missions and suborbital tests.

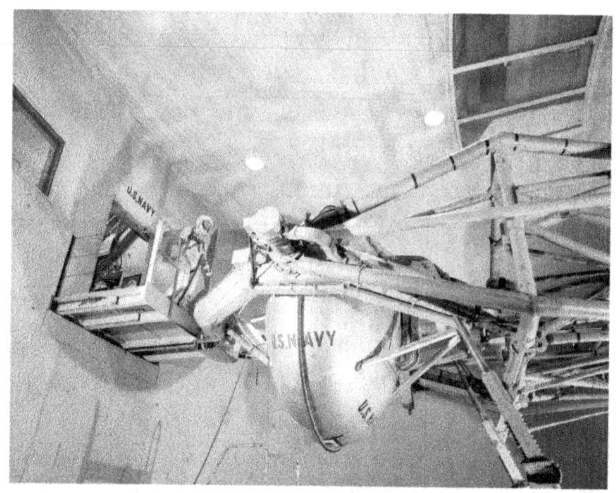

G-force training, Johnsville, 1960

In a parallel development, the decision was made to cancel the balloon flight test program for Project Mercury. Initially considered for high-altitude testing, the balloon flights were deemed unnecessary. Instead, the spacecraft could undergo comprehensive environmental testing at the Lewis Research Center's altitude wind tunnel. This advanced facility could simulate the exact temperature and altitude conditions up to 80,000 feet, providing a controlled environment to test the spacecraft's performance rigorously.

The altitude wind tunnel allowed engineers to conduct detailed assessments of the spacecraft's systems under conditions that closely mimicked those of the upper atmosphere. This decision not only streamlined the testing process but also ensured more precise and reliable data, contributing to the overall safety and readiness of the spacecraft for upcoming missions.

On May 25, 1959, a crucial meeting was held at Johnsville, Pennsylvania, focusing on developing astronaut training programs using the centrifuge. This meeting brought together experts to discuss various training methods and the potential duration of these training sessions. The goal was to ensure that astronauts would be adequately prepared for the intense conditions they would face during space travel.

During the discussions, tentative training periods were established for August 1959 and January 1960. These sessions would subject astronauts to the high G-forces they would experience during launch, re-entry, and other critical phases of their missions. Centrifuge training was a vital component of the overall astronaut preparation program. It was designed to condition astronauts' bodies to withstand spaceflight's physical stresses and train them to handle various emergencies.

This meeting was part of a broader, meticulously planned effort to ensure the success of Project Mercury. Each step in the project's development, from technical specifications and booster negotiations to training and environmental testing, brought NASA closer to achieving its goal of manned spaceflight. The collaborative efforts, rigorous testing protocols, and strategic planning underscored the nation's unwavering commitment to exploring the final frontier

On May 28, 1959, North American Aviation marked a significant milestone in the Little Joe project by delivering the first two booster airframes. Fabrication of the remaining four airframes was progressing on schedule, indicating a steady advancement towards upcoming test flights. At the designated test flight site at Wallops Station, Virginia, rocket motors for the initial flight were already in place. Concurrently, the procurement of test spacecraft, which incorporated critical Mercury flight components, was moving smoothly. Personnel from the Space Task Group were busy instrumenting the first spacecraft. At the same time, engineers meticulously worked on other test units, ensuring that every component met the stringent standards necessary for mission success.

On the same day, another momentous event occurred with the launch of Primates Able and Baker aboard an Army Jupiter missile nose cone. The mission saw the primates travel 300 miles into space, landing 1,700 miles downrange from Cape Canaveral. Telemetry data revealed that the animals' physiological responses remained within normal limits despite the extreme conditions. During the boost phase, there were increases in their body temperature, respiration, pulse rate, and heartbeat, all of which were tolerable. During the weightless period, their physiological responses normalized, with Baker appearing to doze. Upon reentry, their physiological indicators rose again but stabilized upon landing. This successful mission demonstrated that life could be sustained in a space environment, marking a crucial step forward in human spaceflight.

In parallel, a new quick-release side exit hatch was designed for the Mercury spacecraft. This hatch featured a continuous double explosive train, ensuring all bolts would break upon activation, providing a reliable and swift exit mechanism.

Moreover, astronauts and NASA personnel dedicated significant effort to studying the Mercury spacecraft cockpit. Their focus was on mastering both routine and emergency flight procedures, refining the cockpit layout, and ensuring that astronauts could reach and operate any control under all conditions. This thorough preparation was essential to guarantee the safety and effectiveness of future manned missions.

These developments collectively underscored Project Mercury's relentless progress. The delivery of the booster airframes, successful biological experiments in space, and meticulous attention to spacecraft design and astronaut training highlighted NASA's comprehensive approach to overcoming the challenges of space exploration. Each achievement brought the United States closer to realizing its goal of manned spaceflight, laying a robust foundation for the following historic missions.

On June 1, 1959, the personnel strength supporting Project Mercury reached 363 individuals. This team was strategically distributed across several key locations, each playing a crucial role in the mission's progress. The Space Task Group, the project's core team, comprised 204 personnel. These individuals were responsible for overseeing and coordinating the multifaceted aspects of Project Mercury, ensuring that every detail aligned with the mission's stringent requirements.

At the Langley Research Center, 98 experts contributed their specialized knowledge and skills to advance the project. Langley's role was pivotal, providing essential research and development support that underpinned the technical and engineering breakthroughs necessary for manned spaceflight.

Meanwhile, 44 personnel at the Lewis Research Center focused on propulsion and power systems, critical components that would ensure the spacecraft could perform reliably during its missions. Their work was instrumental in addressing the challenges of space travel, from launch to reentry.

Lastly, the Mercury tracking network, vital for maintaining communication and monitoring the spacecraft's trajectory, was supported by 21 dedicated individuals. This network ensured data was accurately relayed to mission control, allowing real-time tracking and adjustments.

The collective efforts of these 363 personnel exemplified the collaborative spirit and technical expertise driving Project Mercury forward. Their dedication and hard work were essential in overcoming the myriad challenges of early space exploration, paving the way for the United States to achieve its goal of manned spaceflight.

On June 5, 1959, several key modifications and proposals were made to enhance Project Mercury's safety and performance. The drogue parachute configuration was altered to a 28 percent porosity, 30-degree conical canopy. This adjustment improved the parachute's stability and effectiveness during the spacecraft's descent, ensuring a smoother and safer landing.

The Army Ballistic Missile Agency proposed a Mercury-Redstone inflight abort sensing system in a significant development for inflight safety. This innovative system was designed to monitor critical performance parameters during the flight. If the system detected that operational limits were exceeded, it would automatically initiate spacecraft ejection, thereby protecting the astronaut in an emergency.

Space Technology Laboratories and Convair also recommended a lightweight telemetry system tailored for the Mercury-Atlas program. Weighing in at 270 pounds, this telemetry system was considerably lighter than previous versions, offering a more efficient means of transmitting vital data back to mission control. NASA promptly approved this recommendation, recognizing the benefits of reducing the overall weight of the spacecraft without compromising the quality and reliability of data transmission.

On June 8, 1959, the Big Joe spacecraft, intended for a crucial reentry test, was successfully delivered to Cape Canaveral. This delivery marked a significant milestone in Project Mercury's development, as the Big Joe mission aimed to validate the spacecraft's heat shield and reentry capabilities, which are critical components for the safety of future manned missions.

In preparation for these missions, the Space Task Group reached out to the Navy's Bureau of Aeronautics regarding the procurement of essential government-furnished survival items. These items included desalter kits, distress signals, and survival rations, all vital for ensuring the safety and well-being of astronauts in the event of an emergency landing. The Space Task Group requested the Navy's assistance in obtaining these crucial survival tools, highlighting the collaborative effort between different branches of the military and NASA to support the Mercury program.

Additionally, discussions with the School of Aviation Medicine led to the finalization of bio-pack details for the NASA Little Joe Flight program. The bio-pack was designed to gather comprehensive life support data, essential for understanding space travel's physiological impacts on humans. By simulating the conditions astronauts would experience, these tests aimed to ensure that the life support systems were reliable and effective for manned flights.

On June 12, 1959, Langley Research Center took a critical step in advancing Project Mercury by organizing a Source Selection Panel and a Technical Evaluation Board. These bodies were established to meticulously evaluate the Mercury tracking and ground instrumentation systems proposals, essential for ensuring reliable communication and data collection during space missions.

The Technical Evaluation Board began its assessments on June 23, considering proposals from seven prominent companies. Among these were Western Electric, Aeronutronics, and RCA, each presenting their designs and solutions to meet NASA's rigorous requirements. The evaluations focused on the technical merits, feasibility, and potential for integrating each company's proposed systems into the broader Mercury program.

This selection process was crucial for identifying the best possible tracking and ground instrumentation solutions, which would play a vital role in monitoring spacecraft trajectories, ensuring accurate data transmission, and supporting mission control operations. Langley Research Center's meticulous approach to organizing these panels and boards underscored the importance of thorough and objective evaluations to advance the nation's space exploration efforts.

The formation of these evaluation bodies and the commencement of technical reviews marked another significant milestone in Project Mercury. By carefully scrutinizing and selecting the most promising proposals, NASA ensured that the tracking and ground instrumentation systems would be robust, reliable, and capable of supporting the ambitious goals of human spaceflight. This process highlighted the collaborative and innovative spirit driving the program, bringing the United States closer to achieving its historic achievements in space exploration.

Between June 14 and 27, 1959, a visit to McDonnell Aircraft Corporation provided valuable insights into the ongoing development of the Mercury spacecraft. It was revealed that the spacecraft was designed to withstand noise levels up to 149 decibels, significantly higher than the anticipated levels of 128 decibels. This conservative design approach ensured the spacecraft could handle the extreme acoustic environments experienced during launch and flight.

Recognizing the potential impact of such high noise levels on astronauts, Space Task Group personnel initiated extensive research on pilot comfort. This research aimed to develop effective strategies and technologies to mitigate the adverse effects of high decibel levels on astronauts. This included studying materials and engineering solutions for noise insulation and assessing the potential psychological and physiological impacts of prolonged exposure to high noise levels.

This period of intensive research and development underscored the commitment to ensuring not only the Mercury spacecraft's structural integrity but also its occupants' well-being and comfort. By addressing these challenges, NASA and McDonnell were taking proactive steps to create a safer and more conducive environment for astronauts, enhancing the overall safety and success of manned space missions.

These efforts reflected the broader goals of Project Mercury, which aimed to push the boundaries of human spaceflight while prioritizing astronaut safety and mission success. Each aspect of the spacecraft's design, from noise insulation to life support systems, was meticulously crafted to meet the rigorous demands of space exploration, paving the way for the United States' pioneering achievements in sending humans into space.

On June 18, 1959, the centrifuge program at Johnsville, Pennsylvania, critically investigated the pilot's role during a multi-stage vehicle launch. This program aimed to understand how astronauts could perform under the extreme conditions of space travel, particularly focusing on the boost-control tasks essential during launch. Test subjects, including future astronaut Neil Armstrong, were subjected to intense accelerations of up to 15 g-forces. Remarkably, none of the participants reached their control capability limits, showcasing the potential of human endurance and adaptability.

Neil Armstrong's participation in this program was a significant milestone in his illustrious career. This early experience under extreme conditions likely contributed to his selection for the Gemini program, where he would further hone his skills and demonstrate exceptional piloting abilities. The insights from the Johnsville centrifuge tests were invaluable, shaping the training protocols and safety measures for future space missions.

This centrifuge program not only highlighted the physical and mental resilience of astronauts but also played a crucial role in the United States' progress in space exploration. By understanding the limits of human performance, NASA could design more effective training regimens and spacecraft controls, ultimately leading to the successful manned missions of the Gemini and Apollo programs. These advancements paved the way for the United States to achieve its ambitious goal of landing a man on the Moon, solidifying its leadership in space exploration.

On June 19, 1959, the Mercury Capsule Coordination Office was established within NASA's Space Task Group, marking a significant step in organizing the burgeoning U.S. space program. J.A. Chamberlin was appointed as the head of this office, overseeing the comprehensive development and management of the Mercury capsule.

The Mercury Capsule Coordination Office's responsibilities were meticulously categorized into several critical areas: loads, thermodynamics, structures, aerodynamics, cabin life support, controls, electronics, recovery, sequencing, transportation, handling, schedules, and testing. This structured approach ensured that each aspect of the capsule's design and function was focused on creating a safe and reliable spacecraft.

In conjunction with establishing the Coordination Office, a Capsule Review Board, chaired by Paul E. Purser, was formed. This board was responsible for reviewing the actions and decisions made by the Capsule Coordination Office, providing an additional layer of oversight and ensuring that all components of the Mercury capsule met the stringent standards required for manned spaceflight.

The formation of the Mercury Capsule Coordination Office and the Capsule Review Board was pivotal in the success of the Mercury program. This structured and systematic approach to spacecraft development set the foundation for the United States' achievements in human space exploration, ultimately leading to the historic manned missions of the 1960s and the realization of America's space aspirations.

On June 24, 1959, the final cost for eight Redstone launch vehicles was negotiated and settled at $20.1 million, significantly higher than the original estimate of $15.5 million. This increase reflected the complexities and unforeseen challenges of developing and producing these pioneering rockets.

Name	Launch	Rank	Unit	Born	Died
M. Scott Carpenter	1962/5/24	Lieutenant	USN	1925	2013
L. Gordon Cooper	1963/5/15	Captain	USAF	1927	2004
John H. Glenn, Jr.	1962/2/20	Major	USMC	1921	2016
Virgil I. Grissom	1961/7/21	Captain	USAF	1926	1967
Walter M. Schirra, Jr.	1962/10/3	Lt Commander	USN	1923	2007
Alan B. Shepard, Jr.	1961/5/5	Lt Commander	USN	1923	1998
Donald K. Slayton		Major	USAF	1924	1993

Chapter 7 - The Rockets

The Redstone launch vehicles played a crucial role in the early stages of the United States' space program. Initially designed as ballistic missiles, they were adapted for space exploration and used in several key missions, including launching America's first astronaut, Alan Shepard, on a suborbital flight in 1961.

On June 25, 1959, a coordinated recovery operation by Navy surface vessels and aircraft successfully retrieved a spacecraft off the coast of Jacksonville, Florida. The recovery occurred 2.5 hours after the spacecraft was airdropped, demonstrating the effectiveness and precision of the recovery procedures essential for future manned space missions.

This successful recovery operation was a critical milestone in the development of the United States' space program. It validated the procedures and technologies necessary to retrieve spacecraft and astronauts after reentry safely. The exercise also provided valuable experience and confidence for the recovery teams, who would later play pivotal roles in the Mercury, Gemini, and Apollo missions.

Between June 28 and July 11, 1959, a series of twelve heat-transfer tests were conducted at Wallops Island. These tests focused on various ablation materials intended for use in the Little Joe test flights. These tests were crucial for evaluating the effectiveness of materials that protect spacecraft from the intense heat generated during reentry into the Earth's atmosphere.

Among the materials tested, triester polymer and thermolag emerged as particularly effective in providing heat protection. These materials demonstrated their ability to withstand extreme temperatures, making them suitable candidates for the thermal protection systems of the spacecraft.

On June 29, 1959, a longitudinal static stability investigation was conducted at the Arnold Engineering Development Center for the Mercury manned orbital spacecraft model. This critical assessment focused on evaluating the spacecraft's aerodynamic stability, an essential factor in ensuring the safety and control of the vehicle during its flight through the Earth's atmosphere and into space.

The investigation involved testing the spacecraft model in various simulated flight conditions to determine its stability characteristics. Longitudinal static stability is crucial for maintaining the correct pitch attitude of the spacecraft, preventing uncontrolled rotations, and ensuring a smooth and predictable trajectory.

Throughout June 1959, significant progress was made in refining the recovery techniques for Project Mercury. Several boilerplate spacecraft were provided to Destroyer Flotilla Four (DesFlotFour), where detailed recovery procedures were developed and practiced. This collaboration ensured recovery operations could be conducted efficiently and safely, an essential component of manned space missions.

In parallel, McDonnell Aircraft Corporation selected Northrop as the subcontractor to design and fabricate the landing system for Project Mercury. Northrop, leveraging its extensive experience in parachute recovery systems dating back to 1943, developed the 63-foot ring-sail main parachute. This advanced parachute system was crucial for the safe descent and landing of the Mercury spacecraft, ensuring the astronauts' return to Earth.

As spring transitioned into summer, Project Mercury continued to advance through rigorous testing, strategic planning, and coordination among NASA, the military, and various contractors. Each step, from developing recovery techniques to fabricating critical components, brought NASA closer to realizing the goal of manned spaceflight. This period highlighted the collaborative effort and meticulous attention to detail driving the mission forward, showcasing the united dedication to pushing the boundaries of human exploration and achieving unprecedented milestones in space.

On July 1, 1959, NASA strategically canceled the order for Jupiter launch vehicles intended to support Project Mercury. The assessment concluded that the necessary data for the program could be obtained more effectively from Atlas flights. This decision streamlined Project Mercury, allowing NASA to concentrate its resources and efforts on the Atlas launch vehicle as the primary platform for the upcoming manned missions.

With its greater payload capacity and advanced technology, the Atlas rocket offered more reliable and comprehensive data than the Jupiter launch vehicles. By focusing on the Atlas, NASA could ensure a more efficient and cohesive approach to developing the Mercury spacecraft and its associated systems.

Between July 1 and 2, 1959, a crucial pressure suit compatibility evaluation occurred using the Mercury spacecraft mock-up. This evaluation aimed to test the fit and functionality of pressure suits from three different manufacturers: the David Clark Company, B.F. Goodrich Company, and the International Latex Company.

During the evaluation, four test subjects donned various pressure suits and operated within the tight confines of the spacecraft mock-up. This rigorous testing was essential to ensure that the suits provided mobility, comfort, and protection for astronauts during their missions. The suits needed to accommodate the astronauts' movements and tasks without compromising safety or performance.

The results of these evaluations informed NASA's decision on which pressure suit design would be most suitable for the Mercury mission. This process was critical in selecting equipment supporting astronauts in extreme space conditions. The thorough testing and attention to detail in selecting the right

pressure suit underscored NASA's commitment to astronaut safety and mission success, contributing to the overall progress and readiness of Project Mercury.

On July 6, 1959, Maxime A. Faget of the Space Task Group and John E. Naugle of NASA Headquarters engaged in discussions that yielded a significant scientific opportunity for Project Mercury. They concluded that valuable experiments on energetic particles could be carried out using packets of emulsion within the Mercury spacecraft. This innovative approach promised to enhance the scientific return of the missions by enabling the study of high-energy particles in space.

Following their discussions, work began on determining the most suitable location within the spacecraft for these emulsion packets. Considerations included ensuring minimal interference with other spacecraft systems and maintaining the integrity of the scientific data. Additional experimental details, such as the precise configuration and handling of the emulsion packets, were also meticulously planned.

On July 12, 1959, an agreement was reached with the Air Force that permitted the Space Task Group to install microphone pickups on the skin of the Atlas launch vehicle. This setup was designed to measure noise levels during the Big Joe-Atlas launch, providing critical data on the acoustic environment experienced by the spacecraft and its components.

The configuration included five microphones strategically placed to capture comprehensive noise data. One microphone was installed inside the Mercury spacecraft to measure the internal noise levels that astronauts would experience. Three microphones were mounted externally about midway up the launch vehicle to gauge the noise generated along its body during ascent. The final microphone was positioned on the Atlas skirt to record the noise levels at the rocket's base.

This arrangement allowed for a detailed analysis of the acoustic conditions during the launch, contributing to the understanding of how noise could affect the spacecraft's structural integrity and the astronauts' comfort and safety. The data collected from these measurements would inform design improvements and operational procedures, enhancing future Mercury missions' overall reliability and success. This collaborative effort between NASA and the Air Force underscored the importance of interagency cooperation in advancing space exploration objectives.

On July 13, 1959, two significant developments marked the progress of Project Mercury. First, qualification tests for the spacecraft horizon scanner commenced. The horizon scanner was a crucial navigational instrument designed to help the spacecraft maintain its proper orientation relative to the Earth's horizon. These tests aimed to validate the scanner's performance and reliability under simulated space conditions, ensuring it could accurately guide the spacecraft during its missions.

In parallel, it was announced that the Western Electric Company and its associates had won the competition to construct the Mercury tracking network. This network was essential for monitoring and communicating with the Mercury spacecraft throughout its flight. It comprised a series of ground stations strategically positioned around the globe to maintain continuous contact with the spacecraft. This capability was vital for tracking the spacecraft's trajectory, relaying telemetry data, and ensuring the safety and success of manned missions.

On July 20, 1959, NASA negotiated with Western Electric and its subcontractors to construct the Mercury tracking network. This network was essential for ensuring the safety and success of the Mercury missions by providing continuous monitoring and communication with the spacecraft.

The negotiations culminated in a letter contract signed on July 30, 1959, covering the comprehensive scope of the tracking network. This contract included the development and implementation of radar tracking systems, telemetry receiving and recording equipment, and display capabilities. It also encompassed spacecraft and surface station communications, as well as the necessary computing and control facilities.

Establishing this extensive tracking network was a critical step in supporting the Mercury program. It enabled precise tracking of the spacecraft's trajectory, real-time telemetry data analysis, and effective communication between the spacecraft and mission control. These capabilities were vital for monitoring the spacecraft's status, ensuring the astronauts' safety, and responding promptly to potential issues during the missions.

On July 21, 1959, discussions occurred regarding the necessary alterations to Building "S" at Cape Canaveral to support Project Mercury. These modifications were essential to accommodate the unique requirements of the Mercury program, ensuring that the facility could effectively support the preparation and launch of the Mercury spacecraft.

The target completion date for these alterations was December 1, 1959. To meet this deadline, the existing Vanguard activities from Building "S" had to be phased out. The Vanguard program, which had previously utilized the facility, needed to relocate its operations to make way for the incoming Mercury project.

These planned changes to Building "S" reflected Cape Canaveral's shifting focus towards the burgeoning manned spaceflight program. The adaptations were a critical component of the infrastructure development required to support NASA's ambitious goals for Project Mercury. By ensuring that the facilities were properly equipped and tailored to the needs of the Mercury missions, NASA could better manage the preparation, testing, and launch processes, ultimately contributing to the success of America's first manned spaceflights.

On July 22, 1959, NASA selected the B.F. Goodrich Company to design and develop the Mercury astronaut pressure suit. B.F. Goodrich was a natural choice, given its extensive history of developing high-altitude suits, a specialization that dated back to 1934. Their expertise was crucial in ensuring that the pressure suits would provide the necessary protection and mobility for astronauts operating in harsh space conditions.

On the same day, a significant milestone was achieved with successfully executing a pad abort flight. This test involved a boilerplate spacecraft outfitted with a production version of the escape tower and rocket. The pad abort flight marked the first operational test of this critical safety component, designed to quickly propel the spacecraft away from the launch pad in an emergency. This test's success demonstrated the escape system's effectiveness and provided confidence in the safety measures in place for future manned missions.

These developments underscored the meticulous planning and rigorous testing that characterized Project Mercury. The selection of B.F. Goodrich for the pressure suits and the successful pad abort test were pivotal steps in advancing the program. Together, they contributed to the robust safety framework and technological readiness required to achieve the historic goal of sending humans into space and safely bringing them back.

On July 28, 1959, a crucial test was conducted involving a boilerplate spacecraft equipped with instruments to measure sound pressure levels and vibration. This second beach abort test aimed to collect detailed data on the vibration and acoustic environment experienced during the firing of the Grand Central abort rocket.

The primary objective of this test was to understand the conditions the spacecraft and its systems would endure in an actual abort scenario. By accurately measuring sound pressure levels and vibrations, engineers could assess these forces' impact on the spacecraft's structural integrity and the astronauts' safety and comfort.

The data gathered from this test was invaluable in refining the design and operational procedures for the Mercury spacecraft. Ensuring that the spacecraft could withstand the intense forces generated during an abort was critical for the success of manned missions. This rigorous testing and analysis underscored NASA's commitment to astronaut safety and mission reliability, key elements in the broader effort to achieve human spaceflight.

On July 30, 1959, NASA awarded a letter contract to the Western Electric Company to construct the Mercury tracking and ground instrumentation system. This system was integral to the Mercury program, ensuring that NASA could maintain continuous communication with and precise tracking of the Mercury spacecraft throughout its missions.

The contract covered developing and implementing a comprehensive network of ground stations equipped with advanced radar tracking, telemetry receiving, and data recording capabilities. These ground stations would relay vital information about the spacecraft's status, trajectory, and telemetry data to mission control, facilitating real-time monitoring and decision-making.

Establishing this tracking and ground instrumentation system was a critical step in ensuring the success of the Mercury missions. It provided the necessary infrastructure to support the complex operations of human spaceflight, marking a significant advancement in NASA's ability to manage and execute space missions. The collaboration with Western Electric underscored the importance of leveraging industrial expertise to achieve the ambitious goals of Project Mercury and advance the United States' position in space exploration.

On July 31, 1959, personnel from the Aeromedical Field Laboratory inspected the first animal couch fabricated by McDonnell for the Mercury animal flight program. This inspection was a crucial step in preparing for the upcoming test flights involving animal subjects, which aimed to ensure the spacecraft's conditions were suitable for human astronauts.

The Mercury animal flight program was designed to verify that the environment within the spacecraft was safe and that life support systems were functioning correctly. By sending animals into space, NASA could gather essential data on the physical and mental stresses associated with space travel. This information would help scientists and engineers understand how spaceflight conditions—such as microgravity, acceleration forces, and confined spaces—would impact living organisms.

The animal couch was a key component in these experiments, specifically designed to accommodate and protect the test animals. Aeromedical Field Laboratory personnel inspected the couch to ensure that it met the necessary standards for safety and functionality, providing a secure environment for the animals during the flight.

These tests were a critical precursor to manned missions, allowing NASA to refine spacecraft systems and procedures based on empirical data. The successful execution of the animal flight program would build confidence in the spacecraft's ability to support human life, paving the way for the historic manned missions of Project Mercury.

Throughout July 1959, several key activities and milestones were achieved in the Mercury program, marking significant progress towards manned spaceflight.

The Mercury astronauts completed a series of disorientation flights using the three-axis space simulator at the Lewis Research Center. These flights were designed to prepare the astronauts for the disorienting conditions of space travel and help them develop the skills needed to maintain control and orientation in a weightless environment.

During this same period, Minneapolis-Honeywell delivered the Mercury spacecraft's first automatic stabilization and control system. This system was crucial for maintaining the spacecraft's orientation and stability during flight, ensuring that the astronauts could focus on their mission tasks without needing constant manual adjustments.

At Wallops Island, the Pilotless Aircraft Research Division launched a 1/14th-scale model of the Mercury spacecraft. Although this test flight resulted in the model entering a continuous tumble from separation to landing, the data collected provided valuable insights into the spacecraft's aerodynamic behavior. This information was essential for refining the design and improving the stability of future models.

Additionally, the Mercury astronauts were assigned specific specialty areas within the program. This specialization ensured that each astronaut could contribute expertise to different aspects of the mission, such as navigation, communications, or life support systems. Their active participation in coordination meetings and their respective areas of expertise helped streamline the development process and enhance the program's overall effectiveness.

These efforts throughout July 1959 underscored the comprehensive and collaborative nature of the Mercury program, highlighting the dedication and meticulous planning required to achieve the goal of sending humans into space.

The vision for Project Mercury initially encompassed a series of missions of increasing complexity, aiming for both suborbital and extended orbital flights, including missions lasting one and three days. However, the project's trajectory shifted with the success of early Mercury missions and Project Gemini's emergence, which aligned more closely with NASA's ambitious goals for lunar exploration. This strategic realignment led to the cancellation of several planned Mercury missions.

Among those canceled missions were Mercury-Jupiter 1 and Mercury-Jupiter 2.

Mercury-Jupiter 1:

Pilot: N/A

Planned Launch: July 1, 1959

Cancellation Date: July 1, 1959

Details: This mission was intended as an early test to evaluate the spacecraft's capabilities. However, it was canceled on the very day it was scheduled to launch due to shifting program priorities and concerns that the Jupiter rocket might not adequately meet the evolving needs of the Mercury program.

Mercury-Jupiter 2:

Pilot: Chimpanzee

Planned Launch: First quarter, 1960

Cancellation Date: July 1, 1959

Details: Similar to its predecessor, Mercury-Jupiter 2 was canceled shortly after NASA decided to stop using the Jupiter rocket for Mercury missions. This decision made early in the program, reflected NASA's strategic pivot towards focusing resources on more suitable launch vehicles that could better support the goals of manned space exploration.

On August 3, 1959, Major General Donald N. Yates was appointed Department of Defense representative for Project Mercury support operations. This appointment underscored the critical role of military collaboration in the success of NASA's ambitious space program.

Major General Yates' involvement brought significant expertise and resources from the Department of Defense to Project Mercury. His leadership ensured that military assets, such as tracking stations, recovery operations, and logistical support, were effectively integrated into the overall mission framework. This collaboration was essential for coordinating the various elements required for the program's success, including launch operations, in-flight tracking, and post-mission recovery of the spacecraft and astronauts.

The appointment of Major General Yates highlighted the strategic importance of Project Mercury and the recognition that the pioneering efforts in space exploration required a concerted effort across both civilian and military domains. His role facilitated smoother coordination between NASA and the Department of Defense, contributing to the robustness and efficiency of the support operations necessary for America's first manned spaceflights.

On August 4, 1959, tests commenced to evaluate the operation of the redesigned Mercury drogue parachute. These tests were a crucial step in ensuring the safe descent and landing of the Mercury spacecraft, particularly during the final stages of reentry.

The drogue parachute was a vital component of the spacecraft's landing system, designed to stabilize and decelerate the capsule before the main parachute deployed. The redesign aimed to improve its reliability and performance under harsh reentry conditions.

During the tests, the drogue parachute's deployment, stability, and deceleration capabilities were rigorously assessed. Engineers analyzed its behavior in various simulated conditions to confirm that it met safety and performance standards. These evaluations were critical in validating the changes made to the parachute design and ensuring that it would function correctly during an actual mission.

On August 6, 1959, four F-102 aircraft were made available to the Mercury astronauts to help them maintain their proficiency in piloting high-performance vehicles. The F-102 Delta Dagger, a supersonic interceptor aircraft, provided an excellent platform for the astronauts to hone their flying skills and stay sharp for their upcoming space missions.

Flying the F-102s allowed the astronauts to experience the high speeds, quick reactions, and intense focus required for piloting advanced aircraft, similar to the demands they would face during spaceflight. Regular flights in these aircraft ensured that the astronauts' reflexes, situational awareness, and handling of dynamic conditions remained at peak levels.

On August 14, 1959, NASA Headquarters approved a proposal from the Space Task Group to enter negotiations with McDonnell to fabricate six additional Mercury spacecraft. This decision was driven by the need to ensure an adequate supply of spacecraft for the expanding scope of the Mercury program, which aimed to achieve a series of increasingly complex manned space missions.

The approval to negotiate for additional spacecraft indicated NASA's commitment to thoroughly testing and perfecting the Mercury capsule through multiple missions. Each spacecraft would undergo rigorous testing and evaluation, providing valuable data to enhance the capsules' design and functionality.

By expanding the fleet of Mercury spacecraft, NASA aimed to address potential challenges, mitigate risks, and ensure the readiness of backup vehicles in case of any unforeseen issues. This proactive approach was essential for maintaining Project Mercury's momentum and achieving its goal of sending humans into space and returning them safely to Earth.

The collaboration with McDonnell, a key contractor in the Mercury program, reinforced the importance of industrial partnerships in advancing space exploration. McDonnell's expertise in spacecraft fabrication played a critical role in successfully developing and deploying the Mercury capsules, contributing to the United States' early achievements in human spaceflight.

On August 15, 1959, the Mercury astronauts began their initial centrifuge training at the Aviation Medical Acceleration Laboratory. This training was critical to their preparation for the intense accelerations they would experience during launch, orbit, and reentry.

In preparation for this training, personnel from the Space Task Group had installed and thoroughly checked out Mercury spacecraft simulation equipment at the laboratory. This equipment was designed to replicate the conditions and forces the astronauts would face, providing a realistic and controlled environment for their training.

The centrifuge training allowed astronauts to experience high g-forces and learn how to manage the physiological stresses of rapid acceleration and deceleration. This training was essential for developing the skills and endurance needed to maintain control of the spacecraft and perform mission-critical tasks under extreme conditions.

Commencing centrifuge training marked a significant step in the comprehensive training regimen to prepare the Mercury astronauts for their pioneering journeys into space. It demonstrated NASA's commitment to ensuring the astronauts were fully equipped to handle the demands of spaceflight, thereby enhancing the overall safety and success of the Mercury program.

On August 21, 1959, during the first programmed Little Joe launch (LJ-1 beach abort test) at Wallops Island, an unexpected event occurred when the escape rocket fired prematurely 31 minutes before the scheduled launch. The unexpected activation caused the spacecraft to ascend to 2,000 feet before descending and landing approximately 2,000 feet from the launch site.

The premature firing of the escape rocket was a significant anomaly, prompting an immediate investigation. The cause was identified as a faulty escape circuit, which had triggered the escape sequence ahead of time. Although the incident resulted in an unplanned launch, it provided valuable data on the performance of the escape system and highlighted the importance of rigorous pre-flight testing and troubleshooting.

This early setback underscored the challenges inherent in developing reliable spacecraft systems. However, it also demonstrated NASA's ability to respond to and learn from such incidents, improving the design and functionality of the escape system to ensure the safety of future manned missions. The insights from the LJ-1 test contributed to the refinement of Project Mercury's safety protocols, ultimately enhancing the program's overall robustness and reliability.

On August 25, 1959, testing concluded on the effectiveness of the drogue parachute as a stabilizing device for the Mercury spacecraft. These tests were essential to ensure that the drogue parachute could reliably stabilize the spacecraft during reentry, thereby preparing it for a safe landing.

The tests demonstrated that the drogue parachute was fully qualified for deployment at speeds up to Mach 1.5 and altitudes up to 70,000 feet. This qualification confirmed that the parachute met all operational requirements, effectively stabilizing the spacecraft under high-speed and high-altitude conditions.

The successful qualification of the drogue parachute marked a significant milestone in the Mercury program. It provided confidence that the parachute system would function as intended during actual missions, enhancing the overall safety and reliability of the reentry and landing phases. This achievement was a crucial step forward, contributing to Project Mercury's readiness for its forthcoming manned spaceflights and ensuring that astronauts could return safely to Earth.

On August 28, 1959, NASA Headquarters authorized the Space Task Group to negotiate with the Air Force Ballistic Missile Division to procure additional Atlas launch vehicles to support Project Mercury. This authorization was to be formalized within Contract No. HS-36, ensuring the necessary resources were secured for the upcoming manned missions.

Throughout the summer of 1959, each milestone, discussion, and test brought Project Mercury closer to achieving its goal of manned spaceflight. The collaborative efforts of NASA personnel, military support, and various contractors were pivotal in meticulously planning and executing every aspect of the mission. This period was marked by significant progress, including the qualification of the drogue parachute, advancements in astronaut training, and the integration of critical systems and technologies.

The dedication and determination of everyone involved in Project Mercury reflected America's commitment to advancing space exploration. Procuring additional Atlas launch vehicles was a key step in ensuring the program's success, providing the required lift capacity for the Mercury spacecraft. These concerted efforts underscored the nation's ambition to lead in space exploration and set the stage for the historic manned missions that would soon follow.

Throughout August 1959, significant strides were made in the qualification tests for the 63-foot ringsail main parachute, a critical component for the safe landing of the Mercury spacecraft. These tests, which had commenced in May, reached a successful conclusion, marking a major milestone in developing the United States' first manned space program.

After completing the parachute qualification tests, comprehensive parachute landing tests began. These tests involved dropping spacecraft models from a C-130 aircraft over the Salton Sea in California. The objective was to simulate actual landing conditions and ensure that the parachutes would perform reliably during re-entry and descent, providing a safe landing for the astronauts.

In conjunction with these technical advancements, McDonnell Aircraft

Corporation, the prime contractor for the Mercury spacecraft, submitted its first monthly reliability report. This report was a comprehensive summary of the efforts undertaken by McDonnell and its subcontractors to enhance the reliability of the Mercury spacecraft's design and development. The reliability report highlighted the meticulous attention to detail and rigorous testing protocols implemented to ensure the spacecraft's systems would function flawlessly during the mission.

These developments in parachute testing and reliability reporting were pivotal in the broader context of the United States' space exploration efforts. The successful completion of the parachute tests validated the design and instilled confidence in manned space missions' overall safety and feasibility. The rigorous reliability measures adopted by McDonnell set a new standard for spacecraft development, emphasizing the importance of systematic testing and quality assurance.

Chimpanzee Mission Little Joe 1:

Little Joe 1
Spacecraft No.: Boilerplate
Launch Date: August 21, 1959
Duration: 20 seconds
Purpose: Test of launch escape system during flight.
Result: Failure

The first flight of Little Joe 1 marked a significant milestone in the early days of space exploration, particularly in testing critical safety systems for manned missions. Launched on August 21, 1959, this mission aimed to evaluate the effectiveness of the launch escape system designed to protect astronauts in the event of a launch vehicle failure.

Regrettably, the test did not achieve its intended objectives, failing the launch escape system during the 20-second flight. Despite the setback, the mission provided invaluable insights and lessons that would inform subsequent tests and improvements to ensure the safety and reliability of future manned missions.

This early setback underscored the complexities and challenges inherent in space exploration, reinforcing NASA's commitment to rigorous testing and continuous improvement as it pursued ambitious goals in human spaceflight.

Strategic Relocation and Mission Planning at Cape Canaveral

Between September 1-7, 1959, McDonnell Aircraft Corporation took a significant step forward in the Mercury program by relocating a segment of its efforts to Cape Canaveral. This strategic move aimed to gear up for the program's operational phase, bringing the company closer to the heart of America's space launch activities.

Upon arrival, McDonnell personnel were promptly integrated into various committees to develop detailed plans for the upcoming Mercury-Redstone and Mercury-Atlas missions. These missions represented crucial stages in the Mercury program, with Mercury-Redstone focusing on suborbital flights and Mercury-Atlas targeting orbital missions. The collaborative planning efforts at Cape Canaveral were essential for addressing the technical and logistical challenges associated with launching and safely returning a manned spacecraft.

McDonnell established its office in Hangar S, a historic site that would become synonymous with the early days of American space exploration. Hangar S served as the central hub for McDonnell's activities, facilitating close coordination with NASA and other key stakeholders involved in the Mercury program.

The relocation to Cape Canaveral and the establishment of mission planning committees underscored the accelerating momentum of the Mercury program. This period transitioned from preliminary testing and development to focused preparation for actual manned spaceflights. The groundwork laid during this time would be instrumental in achieving the program's objectives, ultimately leading to America's first human spaceflights and advancing the nation's capabilities in space exploration.

These efforts at Cape Canaveral highlighted the collaborative spirit between McDonnell and NASA and demonstrated the meticulous planning and dedication required to make human spaceflight a reality. The work done during this early September period contributed significantly to the success of the Mercury-Redstone and Mercury-Atlas missions, paving the way for future achievements in the United States' space endeavors.

Establishing Ground Rules for Prelaunch Preparations

On September 3, 1959, the Space Task Group took a pivotal step in standardizing the Mercury program's prelaunch preparations by forwarding a comprehensive set of ground rules to McDonnell Aircraft Corporation. These guidelines were crucial for designing and developing Mercury checkout equipment, ensuring that all systems and processes operated seamlessly during the critical prelaunch phase.

The ground rules provided detailed instructions for various equipment essential for the Mercury missions. This included blockhouse equipment, encompassing the control panels and monitoring systems used by mission control to oversee the launch process. The guidelines ensured that these systems would be capable of handling the complex tasks required for a successful launch and spacecraft operation.

Additionally, the ground rules addressed the design of checkout trailers. These mobile units had the tools and systems to conduct thorough preflight inspections and tests on the Mercury spacecraft. The guidelines aimed to standardize the procedures for these inspections, enhancing the reliability and safety of the spacecraft before launch.

Telemetry trailers, another critical component outlined in the ground rules, were designed to track and communicate with the spacecraft during prelaunch preparations and the actual mission. These trailers housed advanced telemetry equipment that would capture and transmit data from the spacecraft, providing real-time insights into its performance and status.

The Space Task Group and McDonnell established these ground rules to ensure that all prelaunch preparations were meticulously planned and executed. This level of detailed planning was vital for the success of the Mercury program, as it minimized the risks associated with human spaceflight and maximized the chances of mission success.

The issuance of these guidelines on September 3, 1959, marked a significant advancement in the Mercury program's operational readiness. It demonstrated the program's commitment to rigorous standards and thorough preparation, laying a solid foundation for the upcoming Mercury-Redstone and Mercury-Atlas missions. The careful planning and adherence to these ground rules played a crucial role in the eventual success of America's first manned space missions, contributing to the nation's progress in space exploration and its emerging leadership in the space race.

The Big Joe Atlas Test Flight and Its Achievements

On September 9, 1959, the Mercury program achieved a significant milestone with the successful launch of a Big Joe Atlas boilerplate Mercury spacecraft model from Cape Canaveral. Despite the booster-engine separation not occurring as planned, this test flight accomplished most of its primary objectives. It marked a pivotal step forward in the United States' efforts to master manned spaceflight.

The Big Joe Atlas test flight was designed to evaluate several critical aspects of the Mercury spacecraft's performance. One of the key objectives was to test the ablation shield, a crucial component that protects the spacecraft from the intense heat generated during reentry into Earth's atmosphere. During the flight, the heat shield reached a peak temperature of 3,500 degrees Fahrenheit, successfully demonstrating its capability to withstand extreme thermal conditions.

Another important aspect of the test was measuring afterbody heating, which refers to the thermal stress experienced by the spacecraft's structure behind the heat shield. The data collected from this test provided valuable insights into the spacecraft's thermal dynamics, informing future design improvements and ensuring the safety of manned missions.

The spacecraft's flight dynamics during reentry were also assessed, providing essential information about its behavior and stability as it descended back to Earth. This evaluation helped engineers refine the spacecraft's design to ensure it could safely and predictably return from orbit.

Additionally, the test flight aimed to test the recovery system and procedures. The spacecraft was successfully recovered about eight hours after liftoff, validating the effectiveness of the recovery protocols and equipment. This successful recovery was crucial for the future retrieval of manned spacecraft, ensuring the astronauts' safe return.

The success of the Big Joe Atlas test flight rendered a similar subsequent launch unnecessary, leading to its cancellation. This decision highlighted the test's comprehensive success in achieving its objectives, allowing the Mercury program to advance to the next stages of development with increased confidence.

Overall, the achievements of the Big Joe Atlas test flight on September 9, 1959, represented a major leap forward in the Mercury program. The insights gained from this test validated critical components and procedures and propelled the United States closer to its goal of manned spaceflight. This milestone was a testament to the rigorous testing and innovation that defined the early days of America's space exploration efforts, laying the groundwork for future successes in the space race.

September 10-11, 1959: Astronauts' Recommendations During the Spacecraft Mock-Up Review

A critical spacecraft mock-up review occurred on September 10-11, 1959. During this review, the astronauts provided invaluable feedback and recommended several key changes to the Mercury spacecraft's design. Their insights were instrumental in refining the spacecraft to meet the mission's needs better and ensure the astronauts' safety and effectiveness during spaceflight.

One of the primary recommendations was the redesign of the instrument panel. The astronauts advocated for a new layout that would improve accessibility and readability of the controls and displays. This change was aimed at enhancing the ergonomics and functionality of the cockpit, allowing the astronauts to monitor and manage the spacecraft's systems during flight more efficiently.

Another significant recommendation was the inclusion of a forward centerline window. This addition would provide the astronauts with a direct line of sight forward, improving their ability to observe and navigate during critical phases of the mission. The forward centerline window would enhance situational awareness, making it easier for the astronauts to perform tasks that required visual confirmation.

The astronauts also suggested implementing an explosive side egress hatch. This feature was designed to allow for rapid and reliable emergency exit from the spacecraft. In an emergency, the explosive hatch would enable the astronauts to evacuate quickly, increasing their chances of survival. This recommendation underscored the importance of safety and quick response capabilities in the design of the spacecraft.

These recommendations from the astronauts during the mock-up review reflected their practical insights and firsthand understanding of the challenges they would face in space. Their input was crucial in making the Mercury spacecraft more user-friendly, safer, and better suited to the demands of manned space missions.

Incorporating these changes improved the overall design of the Mercury spacecraft and highlighted the collaborative effort between engineers and astronauts. This collaboration was essential for addressing the practical realities of space travel and ensuring that the spacecraft was fully equipped to handle the rigors of spaceflight.

The spacecraft mock-up review on September 10-11, 1959, was pivotal in the Mercury program. It brought together the astronauts' expertise and the engineers' technical knowledge. The resulting design improvements significantly impacted the program's success, contributing to the safety and effectiveness of America's first manned space missions.

September 11, 1959: Estimating Astronaut Needs in the Mercury Environment

On September 11, 1959, Dr. Douglas H.K. Lee conducted a preliminary study of the environmental conditions the astronauts would face during the Mercury missions. His research focused on the physiological needs of astronauts, providing crucial data to support their health and well-being in space.

Dr. Lee estimated that each astronaut would require approximately 500 cubic centimeters (cc) of water per hour. This calculation considered the various ways in which the astronauts' bodies would use water, including hydration, metabolic processes, and temperature regulation in the spacecraft's closed environment. Ensuring an adequate water supply was vital to maintaining the astronauts' physical health and performance during the mission.

In addition to water requirements, Dr. Lee estimated the astronauts' daily caloric intake needs at about 3,200 calories. This high caloric intake was necessary to sustain the astronauts' energy levels and support their bodily functions in the unique conditions of space. The caloric estimate included considerations for the increased metabolic demands posed by the microgravity environment and the physical and mental exertion required for mission tasks.

These estimates were critical for designing the Mercury spacecraft's life support systems. They informed the development of food and water storage solutions, ensuring that the spacecraft could adequately supply the astronauts with their essential needs. Dr. Lee's research provided a scientific basis for planning the nutritional and hydration strategies necessary for the success of the Mercury missions.

By establishing these fundamental requirements, Dr. Lee's study played a pivotal role in the overall mission planning. His work ensured that the Mercury program could provide a sustainable and safe environment for the astronauts, addressing their basic physiological needs and enabling them to focus on the demanding tasks of space exploration.

Dr. Lee's September 11, 1959, contributions underscored the importance of thorough scientific research and planning in developing manned space missions. His findings helped to lay the foundation for the life support systems that would become a critical component of the Mercury spacecraft, ensuring the health and safety of America's first astronauts as they ventured into space.

September 15, 1959: Appointment of Walter C. Williams as Associate Director for Project Mercury Operations

On September 15, 1959, Walter C. Williams was appointed as the Associate Director for Project Mercury Operations, a key leadership position that underscored the significance of his role in the mission's success. Williams was also designated as the primary contact between NASA and the Department of Defense (DoD) for Mercury flight operations, highlighting the collaborative nature of the project.

Williams brought extensive experience and expertise to this critical role. His appointment was a strategic move to ensure the smooth coordination and execution of the Mercury missions. As the Associate Director for Operations, Williams was responsible for overseeing the day-to-day activities related to preparing, launching, and recovering the Mercury spacecraft. His leadership was essential in managing manned spaceflight's complex logistics and technical challenges.

Williams facilitated effective communication and cooperation between the two agencies as the primary NASA-DoD contact. The Department of Defense provided crucial support for the Mercury program, including using military facilities, aircraft, and personnel. Williams' ability to navigate the interagency dynamics was vital for ensuring that all flight operations were well-coordinated and aligned with the mission objectives.

Walter C. Williams' appointment on September 15, 1959, marked a significant step in strengthening the operational framework of Project Mercury. His leadership and the close collaboration between NASA and the Department of Defense were instrumental in advancing the United States' efforts to achieve manned spaceflight. Williams' contributions played a pivotal role in the successful execution of the Mercury missions, paving the way for future advancements in space exploration and solidifying America's position in the space race.

September 16, 1959: Continued Ablation Heat-Shield Testing at Langley Research Center

On September 16, 1959, Langley Research Center continued its rigorous testing of ablation heat shields, examining nine model shields to validate their effectiveness further. These tests continued even though the recent Big Joe test flight had already successfully demonstrated the feasibility of the ablation heat shield concept and confirmed the suitability of the selected materials.

The ongoing tests at Langley were part of a comprehensive effort to ensure the highest safety and reliability for the Mercury spacecraft. Ablation heat shields protect the spacecraft from the extreme temperatures encountered during reentry into Earth's atmosphere. As the shield heats up, its material gradually erodes, carrying the heat and preventing the spacecraft's structure from overheating.

The Big Joe test, conducted in September, provided significant proof of concept by validating the heat shield's performance under real flight conditions. The shield reached temperatures of up to 3,500 degrees Fahrenheit and successfully protected the spacecraft, confirming that the chosen materials and design could withstand the intense heat of reentry.

Despite this success, Langley's continued testing aimed to gather more data and refine the heat shield's design. Testing multiple models allowed engineers to assess various aspects of the heat shield's performance, including its durability, the efficiency of heat dissipation, and the behavior of different materials under extreme conditions. This thorough approach was crucial for identifying potential weaknesses and ensuring the heat shields would perform reliably during actual manned missions.

The efforts at Langley Research Center reflected the meticulous attention to detail and commitment to safety that characterized the

Mercury program. By continuing to test and improve the heat shield, NASA ensured that the Mercury spacecraft would provide maximum protection for the astronauts, increasing the likelihood of mission success.

The work done on September 16, 1959, was a testament to the rigorous scientific and engineering standards upheld by the Mercury program. These ongoing tests contributed to the overall robustness of the spacecraft, enhancing its capability to return astronauts from space safely and reinforcing the United States' position as a leader in space exploration.

September 19, 1959: Conclusion of Project Mercury Drogue Parachute Testing

On September 19, 1959, the NASA Flight Research Center successfully conducted the final development and qualification test for the Project Mercury drogue parachute, marking the 15th and concluding test in the series. This milestone represented the culmination of an extensive testing program designed to ensure the reliability and effectiveness of the drogue parachute system, a critical component for the safe descent and landing of the Mercury spacecraft.

The drogue parachute was vital in the reentry and landing sequence of the Mercury missions. It was responsible for stabilizing and decelerating the spacecraft after reentry into Earth's atmosphere, preparing it to deploy the main parachutes. Ensuring the drogue parachute's performance was essential to safeguard the astronauts and the spacecraft during the final stages of their journey.

Throughout the 15-test series, various aspects of the drogue parachute's design and functionality were evaluated under different conditions. These tests included assessing the parachute's deployment mechanisms, its behavior in various atmospheric conditions, and its ability to stabilize the spacecraft during descent reliably. The rigorous testing regimen aimed to identify and address potential issues, ensuring that the parachute would perform flawlessly in actual mission scenarios.

The September 19, 1959, final test validated that the drogue parachute system met all design specifications and performance requirements. The success of this test series gave NASA and its contractors the confidence that the drogue parachute would function as intended, contributing to the overall safety and success of the Mercury missions.

This achievement marked a significant step forward in the preparation of manned spaceflights. The successful qualification of the drogue parachute system was a testament to the meticulous planning, rigorous testing, and collaborative efforts of the engineers and scientists involved in Project Mercury. Their work ensured the spacecraft would have a reliable and effective descent system, crucial for the astronauts' safe return to Earth.

Testing Pilot Control Under Sustained Acceleration

Between September 21 and October 10, 1959, the Aviation Medical Acceleration Laboratory conducted a crucial research program to measure the effects of sustained acceleration on a pilot's ability to control a vehicle. This study was part of a broader effort to ensure that astronauts could effectively manage the Mercury spacecraft under demanding spaceflight conditions.

The research involved testing various side-arm controllers, allowing pilots to manage the spacecraft's movements. The focus was on determining which configuration would enable the most precise and reliable control under sustained acceleration forces, similar to those experienced during launch and reentry.

Several controller designs were evaluated, but the three-axis type, which controlled yaw, roll, and pitch, proved the most satisfactory. This configuration allowed pilots to maintain accurate control over the spacecraft's orientation and trajectory, even under the high-stress conditions simulated in the laboratory.

The three-axis side-arm controller's effectiveness was later extensively evaluated, confirming its suitability for the Mercury spacecraft's control system. This controller configuration allowed astronauts to perform precise maneuvers, maintain stable flight, and ensure a safe reentry and landing.

The findings from this research program were instrumental in refining the Mercury spacecraft's design and operational procedures. NASA increased the Mercury mission's overall safety and success potential by ensuring that astronauts could maintain control under sustained acceleration.

Adopting the three-axis side-arm controller for the Mercury spacecraft underscored the importance of rigorous testing and human factors research in space technology development. The work conducted between September 21 and October 10, 1959, demonstrated a commitment to excellence and innovation, contributing to the Mercury program's readiness for manned spaceflight and the advancement of America's capabilities in space exploration.

Study on Retrorocket Capability for Pilot Safety

On September 22, 1959, a significant paper titled "Results of Studies Made to Determine Required Retrorocket Capability" was issued, addressing a critical aspect of the Mercury program. The study focused on determining the necessary capabilities of retrorockets to ensure pilot safety during both emergency landings and the conclusion of a normal mission.

Retrorockets are small rocket engines designed to slow down the spacecraft, allowing it to reenter Earth's atmosphere and make a controlled descent. Their effectiveness is vital for astronauts' safe return, as they must reliably decrease the spacecraft's velocity to achieve a safe reentry trajectory.

The study in the paper aimed to identify the specific requirements for retrorocket performance under various conditions. This included scenarios involving potential in-flight emergencies, where a swift and reliable deceleration would be essential for an unplanned landing. Additionally, the study examined the retrorockets' role at the end of a typical mission, ensuring that the spacecraft could safely and accurately reenter the atmosphere and prepare for landing.

Key findings from the study provided crucial data on the thrust levels, burn durations, and positioning of the retrorockets needed to achieve these objectives. The study laid the groundwork for designing and testing retrorocket systems integrated into the Mercury spacecraft by establishing these requirements.

The issuance of this paper marked an important step in enhancing the safety and reliability of the Mercury missions. The detailed analysis and recommendations helped engineers to develop retrorocket systems that could be trusted to perform under the demanding conditions of spaceflight, thereby protecting the lives of the astronauts.

Enhancing Recovery Operations for the Mercury Program

Throughout September 1959, significant advancements were made in the operational analysis and planning of recovery operations for the Mercury program. Grumman Aircraft Engineering Corporation issued a comprehensive report detailing the possible recovery forces required for a three-orbit Mercury mission. This operational analysis was essential for ensuring that the recovery phase of the mission would be efficient and safe for the returning astronauts.

The Grumman report outlined the various resources and logistical considerations necessary to retrieve the Mercury spacecraft after successfully completing its mission. This included deploying naval and air assets, coordinating recovery teams, and establishing communication protocols to ensure that the spacecraft could be quickly and accurately located and retrieved.

Simultaneously, the Space Task Group continued refining the overall recovery requirements. These refinements included developing and testing satisfactory helicopter recovery techniques, crucial for swiftly retrieving the spacecraft from the ocean after splashdown. Helicopters offered the advantage of rapid deployment and flexibility, making them ideal for reaching the spacecraft quickly and ensuring the safety of the astronauts.

In addition to refining recovery techniques, the Space Task Group also focused on testing and optimizing spacecraft location aids. These aids included various beacons, flares, and tracking systems designed to make the spacecraft easily detectable by recovery forces. Effective location aids were critical for minimizing the time between splashdown and recovery, thereby reducing the risk to the astronauts and ensuring the successful completion of the mission.

The work carried out during September 1959 was pivotal in strengthening the Mercury program's recovery operations. The comprehensive analysis and rigorous testing ensured that all aspects of the recovery process were well-planned and reliable. These efforts underscored the importance of meticulous preparation and coordination in the broader context of the Mercury missions, contributing to the overall success and safety of America's first manned spaceflights.

Big Joe 1

Spacecraft No.: Big Joe Boilerplate
Launch Date: September 9, 1959
Duration: 13 minutes
Purpose: Test of heat shield and Atlas/spacecraft interface.
Result: Partial success
The Big Joe 1 mission, launched on September 9, 1959, marked a pivotal step in testing critical components for manned spaceflight. This mission focused on evaluating the performance of the heat shield and the interface between the Atlas rocket and the Mercury spacecraft, key elements crucial for the safety and success of future missions.

During its 13-minute flight, Big Joe 1 partially succeeded in its objectives. The mission provided valuable data and insights into the behavior of the heat shield under reentry conditions and the effectiveness of the Atlas rocket's integration with the Mercury spacecraft. These findings were instrumental in refining heat shield designs and optimizing the spacecraft's interface with its launch vehicle.

The collaborative efforts between NASA, Grumman Aircraft Engineering Corporation, and other stakeholders were vital in overcoming the technical challenges of early space exploration. Their combined expertise and dedication contributed significantly to the advancements made during the Mercury program, laying the foundation for subsequent achievements and the United States' ongoing progress in space exploration.

The lessons learned from Big Joe 1 set the stage for further advancements in spacecraft technology, ensuring that future missions could safely and effectively push the boundaries of human exploration.

Enhancements Approved for the Mercury Spacecraft

On October 1, 1959, NASA Headquarters approved funding for several significant upgrades to the Mercury spacecraft. These enhancements improved the spacecraft's functionality, safety, and mission success. The approved changes reflected the ongoing commitment to addressing feedback from astronauts and engineers and incorporating the latest technological advancements.

One of the key upgrades was the installation of an egress hatch. This explosive side hatch was designed to allow astronauts to quickly and safely exit the spacecraft in an emergency, both on the ground and after splashdown. The ability to rapidly evacuate the spacecraft was critical for the astronauts' safety, providing them with a reliable escape route in various scenarios.

Another important addition was the installation of an astronaut observation window. This forward centerline window would give astronauts a direct view outside the spacecraft, enhancing their situational awareness and allowing for visual observation of the Earth and space. This window was crucial for navigation and performing tasks requiring visual confirmation.

The approval also included the implementation of a rate stabilization and control system. This system was essential for maintaining the spacecraft's stability and orientation during flight, especially during the high-stress phases of reentry and splashdown. By ensuring the spacecraft remained steady and controllable, this system significantly increased the safety and success rate of the missions.

The main instrument panel was set to be redesigned to improve its layout and usability. The new design would make it easier for astronauts to access and read the instruments, enhancing their ability to monitor and control the spacecraft's systems. This improvement was based on direct feedback from astronauts who had interacted with earlier versions of the panel.

Additionally, the installation of a reefed ringsail landing parachute was approved. This type of parachute was designed to deploy in stages, initially reefed (partially deployed) to reduce the opening shock and then fully deployed to slow the spacecraft's descent for a safe landing. This upgrade aimed to enhance the landing stability and safety of the Mercury spacecraft.

Lastly, funds were allocated for various research and development flight test configurations. This included modifications to the spacecraft to support different experimental setups and testing scenarios, allowing for continuous improvement and adaptation of the Mercury program based on new findings and technological advancements.

The approval of these major changes on October 1, 1959, marked a pivotal moment in the evolution of the Mercury spacecraft. These enhancements addressed critical safety concerns and improved operational functionality. They incorporated feedback from the astronaut corps, ultimately contributing to the program's success and the broader goals of the United States' space exploration efforts.

Issuance of Specifications for the Mercury Pressure Suit

On October 2, 1959, NASA issued detailed specifications for the Mercury pressure suit, an essential component designed to protect astronauts in the harsh space environment. The procurement program for these suits was strategically divided into two distinct phases to ensure thorough testing and refinement.

The first phase focused on operational research suits intended for astronaut training and system evaluation. These suits would be used extensively in ground-based simulations and training exercises to familiarize astronauts with the suit's functionality and to test its performance in various scenarios. This phase aimed to identify any potential issues and gather feedback from the astronauts, which would be critical in refining the suit's design.

The second phase involved developing and producing the final configuration of the Mercury pressure suits. These suits would incorporate the improvements and lessons learned from the operational research suits, ensuring they met the rigorous standards for actual space missions. The final suits were needed to provide reliable life support, temperature regulation, and mobility while also being durable enough to withstand the physical demands of space travel and reentry.

The specifications detailed various features of the pressure suit, including its materials, construction, and integrated life support systems. The suit had to maintain a stable internal pressure, provide oxygen, remove carbon dioxide, and manage the astronaut's body temperature. It also needed sufficient mobility to enable the astronauts to perform their tasks both inside the spacecraft and during potential extravehicular activities.

The issuance of these specifications on October 2, 1959, marked a significant step in the preparation for manned spaceflight under the Mercury program. The careful planning and phased approach to suit development ensured that astronauts would be equipped with reliable and effective protective gear, enhancing their safety and mission success.

The Mercury pressure suit's design and development were pivotal in addressing the unique challenges of human spaceflight, setting the stage for future advancements in space suit technology. This milestone underscored NASA's commitment to rigorous testing and continuous improvement, reinforcing the agency's dedication to pioneering human space exploration.

Successful Launch of Little Joe Vehicle LJ-6

On October 4, 1959, the Little Joe launch vehicle carrying a boilerplate spacecraft, designated LJ-6, was successfully launched from Wallops Island, Virginia. This mission was an important part of the Mercury program's series of tests designed to validate various components and systems of the launch vehicle and spacecraft under realistic conditions.

The flight lasted 5 minutes and 10 seconds and had several key objectives. One of the primary goals was to check the integrity of the launch vehicle's airframe and motor system. Ensuring that the structural components and propulsion systems could withstand the stresses of launch and flight was crucial for the safety and success of future manned missions.

Another objective was to validate wind corrections. Accurate wind correction data was essential for fine-tuning the launch trajectory, ensuring that the spacecraft would follow the intended path and reach the desired altitude and range.

The mission also aimed to gather performance and drag data. This information was vital for understanding how the spacecraft and launch vehicle behaved during ascent, particularly regarding aerodynamic forces and overall flight performance. The data collected would help engineers optimize the design and improve the efficiency of the launch system.

Additionally, the test included an evaluation of the destruct system. This system was a critical safety feature designed to terminate the flight in case of a malfunction, preventing the launch vehicle from veering off course and potentially causing damage or injury.

The successful launch of LJ-6 provided valuable insights and confirmed the reliability of the Little Joe launch vehicle, contributing significantly to the overall development and readiness of the Mercury program. The data obtained from this flight helped to refine the design and operational procedures, ensuring that subsequent missions would be even more robust and reliable.

NASA's Explorer VII Satellite Achieves Orbit

On October 13, 1959, NASA's Explorer VII satellite was successfully launched into orbit, marking a significant achievement in space science and satellite technology. This mission provided invaluable geophysical data, demonstrating important advancements in satellite design and operation.

Explorer VII's primary objective was to collect and transmit data on various geophysical phenomena. The satellite was equipped with instruments designed to measure solar and Earth radiation, contributing to a better understanding of the energy balance between the Sun and our planet. This data was crucial for advancing knowledge in atmospheric science and improving climate models.

Additionally, Explorer VII monitored magnetic storms, disturbances in Earth's magnetosphere caused by solar activity. Understanding these magnetic storms was essential for predicting their impact on Earth's satellite communications, navigation systems, and power grids. The data collected by Explorer VII helped scientists study the behavior of these storms and their effects on our planet.

The satellite also provided valuable information on micrometeorite penetration. By measuring the frequency and impact of micrometeorites, Explorer VII contributed to the design of more resilient spacecraft and satellites, enhancing their ability to withstand the harsh environment of space.

One of the significant technological achievements of the Explorer VII mission was the successful demonstration of a method to control the satellite's internal temperatures. Maintaining a stable internal temperature was crucial for properly functioning the satellite's instruments and electronics. The thermal control system tested on Explorer VII ensured that the satellite's components operated within their optimal temperature range, thus improving the reliability and longevity of the mission.

Collaboration Between Space Task Group and Lewis Research Center

On October 15, 1959, key personnel from NASA's Space Task Group met with representatives from the Lewis Research Center to clarify the research support needs for Project Mercury. This meeting was crucial in defining and coordinating the various test and support activities required to ensure the success of the Mercury program.

During the meeting, several critical test and support areas were identified and agreed upon:

Separation Tests: These tests were essential for understanding and perfecting the separation mechanisms between different stages of the Mercury launch vehicle. Ensuring clean and reliable stage separation was vital for the mission's success, as any failure in this process could jeopardize the entire flight.

Pilot Techniques Development: Effective pilot techniques were crucial for astronauts operating the Mercury spacecraft. This included training astronauts in spacecraft control, navigation, and emergency procedures. The collaboration aimed to refine these techniques through simulations and practical exercises.

Retrorocket Calibration Tests: Retrorockets slow the spacecraft to enable a controlled reentry into Earth's atmosphere. Calibration tests were necessary to ensure the retrorockets would fire with the correct thrust and duration, providing the precise deceleration needed for a safe reentry and landing.

Escape Rocket Plume Studies: The escape rocket, or launch escape system (LES), was designed to quickly propel the Mercury capsule away from the launch vehicle in case of an emergency during launch. Studying the escape rocket's plume, or exhaust flow, was essential for understanding its impact on the spacecraft and ensuring that it operated effectively and safely.

The Space Task Group and the Lewis Research Center meeting highlighted the collaborative effort required to address the complex challenges of the Mercury program. By clearly defining the research support needs and test areas, both teams could focus on developing and validating the technologies and procedures necessary for manned spaceflight.

This cooperation underscored the importance of interdisciplinary collaboration in advancing space exploration. The agreed-upon test and support areas played a pivotal role in refining the Mercury spacecraft and its systems, contributing to the overall success of the missions and setting the stage for future achievements in human spaceflight.

Testing Mercury Spacecraft Afterbody Shingles

On October 20, 1959, NASA initiated requests to test the Mercury spacecraft afterbody shingles at the Navy's Dangerfield test facility. These tests were designed to evaluate the shingles' heat resistance and dynamic-pressure capabilities, crucial for protecting the spacecraft during reentry into Earth's atmosphere.

The afterbody shingles were a key component of the Mercury spacecraft's heat shield system. They withstood the extreme temperatures and aerodynamic forces encountered during reentry. Ensuring their reliability and effectiveness was essential for the astronauts' safety and the mission's overall success.

The shingles would undergo rigorous testing at the Dangerfield test facility to simulate the harsh reentry conditions. The heat resistance tests aimed to determine how well the shingles could protect the spacecraft from the intense thermal environment, preventing the underlying structure from overheating. These tests involved exposing the shingles to high temperatures to measure their thermal conductivity and durability.

In addition to heat resistance, the dynamic-pressure tests were conducted to evaluate the shingles' ability to withstand the high-pressure forces generated by the spacecraft's rapid atmospheric descent. These forces could cause structural damage or compromise the integrity of the heat shield. By simulating these conditions, engineers could assess the shingles' performance and make any necessary adjustments to enhance their strength and resilience.

The initiation of these tests on October 20, 1959, represented a critical step in validating and refining the Mercury spacecraft's thermal protection system. The data collected from these tests would provide valuable insights into the materials and design of the afterbody shingles, ensuring that they met the stringent requirements for spaceflight.

This focus on thorough testing and validation underscored NASA's commitment to the safety and success of the Mercury missions. By leveraging the specialized capabilities of the Navy's Dangerfield test facility, NASA could conduct comprehensive evaluations of the spacecraft components, contributing to the development of robust and reliable systems for human space exploration. The results from these tests would play a pivotal role in finalizing the design of the Mercury spacecraft, paving the way for America's first manned spaceflights.

Evaluating Mercury Escape-System Qualification Tests

On October 30, 1959, personnel from the Space Task Group, Wallops Station, and McDonnell Aircraft Corporation held a crucial meeting to review and evaluate the results of the Mercury escape-system qualification tests. This meeting aimed to ensure that the Mercury spacecraft's escape system, a critical safety component, met all requirements and performed reliably under test conditions.

The escape system, or launch escape system (LES), was designed to quickly propel the Mercury capsule away from the launch vehicle in an emergency during launch. Its effectiveness was paramount to the safety of the astronauts, making thorough testing and evaluation of the system essential.

During the meeting, the representatives from the three organizations carefully reviewed the test results, examining data on the system's performance, including thrust, trajectory, and reliability metrics. The discussion focused on identifying any issues or areas for improvement to ensure the escape system would function flawlessly during an emergency.

Responsibilities for achieving the test objectives and addressing any identified issues were apportioned among the Space Task Group, Wallops Station, and McDonnell. This division of duties ensured that each organization could leverage its expertise to contribute to the successful validation of the escape system.

The Space Task Group, responsible for overseeing the Mercury program, focused on ensuring the system met the overall mission requirements and safety standards. With its extensive experience in testing and launching rockets, Wallops Station provided critical support in conducting and analyzing the tests. As the prime contractor for the Mercury spacecraft, McDonnell was tasked with implementing any necessary design modifications and improvements based on the test results.

The collaborative approach taken during this meeting highlighted the importance of teamwork and clear communication in addressing the complex challenges of space exploration. By working together and sharing responsibilities, the organizations involved could efficiently and effectively resolve any issues, ensuring the escape system's readiness for future missions.

In October 1959, North American Aviation and Minneapolis-Honeywell received the go-ahead to commence the production of hardware for an air-supplied launch vehicle control system. This pivotal development marked a significant step forward in the United States' burgeoning space program.

North American Aviation, a major aerospace manufacturer, and Minneapolis-Honeywell, renowned for their expertise in control systems, collaborated on this critical project. The air-supplied launch vehicle control system was designed to enhance the stability and precision of rockets during launch, a crucial factor for the success of space missions.

The need for more reliable and accurate launch capabilities influenced the decision to develop this control system. As the space race intensified, the United States sought to improve its technological prowess to ensure the success of its space missions. The hardware produced by these two companies would become integral components in the nation's launch vehicles, providing the necessary control and stability to propel rockets into space.

This project was part of a broader effort to advance the United States' capabilities in space exploration. They were developing sophisticated control systems that allowed for more complex and ambitious missions, laying the groundwork for future achievements such as the Apollo moon landing. The collaboration between North American Aviation and Minneapolis-Honeywell exemplified the synergy between different aerospace industry sectors, driving innovation and progress.

On October 30, 1959, a crucial meeting was held to review and evaluate the qualification-test results of the Mercury escape system. Representatives from the Space Task Group, Wallops Station, and McDonnell Aircraft Corporation attended this meeting. During the session, responsibilities for achieving the test objectives were delineated among the three organizations, ensuring a coordinated approach to advancing Project Mercury.

Project Mercury made significant strides throughout the late summer and fall of 1959. Successful test flights and qualification tests underscored the project's progress, bringing NASA closer to the milestone of manned spaceflight. Each accomplishment during this period was a testament to the meticulous efforts of NASA personnel, the military, and their contractors.

The dedication and precision demonstrated in these endeavors were pivotal in the evolution of NASA's space program. Detailed planning and coordination were paramount, as the team worked tirelessly to resolve challenges and refine technologies. These efforts advanced Project Mercury and laid the groundwork for future manned space missions, ultimately contributing to the United States' leadership in space exploration.

The commitment shown by all parties involved highlighted the collaborative spirit and technical expertise driving this historic mission. As Project Mercury progressed, each successful test and meeting brought NASA closer to its goal of sending humans into space, marking the beginning of a new era in human achievement and exploration.

On October 31, 1959, McDonnell Aircraft Corporation received the first ablative heat shield, a critical component designed and manufactured by General Electric, based on the Big Joe design. This heat shield was intended for installation on Spacecraft No. 1, marking a significant milestone in the development of Project Mercury.

The ablative heat shield represented a major technological advancement. Its purpose was to protect the spacecraft and its occupants from the intense heat generated during re-entry into Earth's atmosphere. The technology worked by gradually eroding, or ablating, and carrying away heat, thereby preventing the spacecraft's structure from overheating.

The Big Joe design had previously demonstrated the effectiveness of this technology. The successful incorporation of the heat shield into Spacecraft No. 1 was a testament to the collaborative efforts between McDonnell and General Electric, and it underscored the innovative spirit driving the space program.

Receiving and installing the heat shield was a critical step forward for Project Mercury. It provided a tangible indication that the program was progressing toward its goal of manned spaceflight. The ablative heat shield not only enhanced the safety of the spacecraft but also boosted confidence in the feasibility of future missions.

Chapter 8 - Little Joe Missions

During Project Mercury, the Little Joe flights were a series of unmanned test flights conducted by NASA to evaluate the launch escape system (LES) and the performance of the Mercury spacecraft in various abort scenarios. The primary objectives of these flights were to:

Little Joe 6

Spacecraft No.: Boilerplate
Launch Date: October 4, 1959
Duration: 5 minutes 10 seconds
Purpose: Test of spacecraft aerodynamics and integrity.
Result: Partial success

The Little Joe 6 mission, launched on October 4, 1959, represented a crucial step in evaluating the aerodynamics and structural integrity of the spacecraft essential for manned spaceflight. This flight aimed to test the spacecraft's ability to withstand the stresses of launch and ascent, providing valuable data for further design enhancements.

During its 5-minute, 10-second flight, Little Joe 6 partially succeeded in its objectives. The mission contributed significant insights into the spacecraft's aerodynamic performance under real-world conditions and highlighted areas where structural adjustments were necessary.

The meticulous planning, engineering prowess, and dedication demonstrated during the Little Joe 6 mission underscored the collaborative efforts of NASA, its partners, and stakeholders. These efforts were pivotal in advancing the technological capabilities needed to safely propel humans into space, marking another milestone toward achieving NASA's ambitious goals.

The lessons learned from Little Joe 6 informed subsequent spacecraft designs and testing protocols, ensuring continuous improvement and readiness for future manned missions.

On November 1, 1959, McDonnell Aircraft Corporation published the "Handbook of Operation and Service Instructions, Horizon Scanner Test, Serial MDE 4590011." This document provided detailed guidelines for operating and maintaining the Horizon Scanner, a crucial instrument for Project Mercury. The Horizon Scanner was essential for ensuring proper orientation and stabilization of the spacecraft, contributing to the mission's overall success.

The handbook was a comprehensive resource for technicians and engineers, detailing the procedures required to test and service the Horizon Scanner. Its publication marked another step forward in the meticulous preparation for manned spaceflight. This document highlighted the importance of precise and reliable equipment in achieving mission objectives.

Recognizing the need for continuous improvement and updates, McDonnell revised and reissued the handbook on June 6, 1960. The updated version incorporated the latest findings and feedback from initial tests, ensuring that the instructions remained accurate and relevant as the project progressed.

The publication and subsequent revision of the handbook underscored the commitment to excellence and the attention to detail that characterized Project Mercury. It reflected the collaborative efforts of engineers, technicians, and scientists dedicated to overcoming the challenges of space exploration. These efforts were crucial in laying the groundwork for NASA's future successes, ultimately leading to the historic achievement of sending humans into space and safely returning them to Earth.

On November 4, 1959, the Little Joe 1-A (LJ-1A) mission was launched to test a planned abort under high aerodynamic load conditions. Although the primary mission objective was not achieved due to issues with the escape motor ignition, the flight provided valuable data by completing all other planned events.

Despite the ignition problem, the LJ-1A mission reached an altitude of 9 statute miles, a range of 11.5 statute miles, and achieved a speed of 2,021.6 miles per hour. These results demonstrated the capabilities of the Little Joe rocket, an important test vehicle designed to simulate abort scenarios and validate the Mercury spacecraft's launch escape system.

The partial success of LJ-1A offered critical insights into the performance and reliability of the escape system under real flight conditions. Each test, even those with setbacks, contributed to the iterative process of refining the spacecraft and its systems. The data collected from this flight helped engineers identify and address the issues with the escape motor, leading to improvements in subsequent tests and missions.

The LJ-1A mission exemplified the rigorous testing and troubleshooting necessary to ensure the safety and success of manned spaceflight. Regardless of its outcome, each flight test brought NASA closer to its goal of sending humans into space. The lessons learned from LJ-1A and other similar tests were instrumental in the eventual success of Project Mercury, culminating in the historic achievement of placing the first American astronaut in orbit and safely returning them to Earth.

On November 5, 1959, the astronauts of Project Mercury were fitted with pressure suits and received training on their use at the B. F. Goodrich Company in Akron, Ohio. This was a crucial step in preparing the astronauts for the challenges of spaceflight, ensuring their safety and comfort in the harsh environment of space.

The pressure suits, designed and manufactured by B. F. Goodrich, were engineered to provide life support and protection in case of cabin depressurization. These suits were essential for maintaining the astronauts' bodily functions by providing necessary pressure, oxygen supply, and thermal regulation. The meticulous fitting process ensured that each suit was tailored to the astronaut's body for maximum efficiency and comfort.

Training at B. F. Goodrich included instruction on donning and doffing the suits, operating the various life-support systems, and performing necessary tasks while wearing the suits. This hands-on training was critical for familiarizing the astronauts with the equipment they would rely on during their missions.

This preparation reflected the broader effort of Project Mercury to address every aspect of human spaceflight, from the technology and engineering of the spacecraft to the physical and mental readiness of the astronauts. The pressure suit training at B. F. Goodrich was a vital component of this comprehensive preparation, contributing to the overall success and safety of the Mercury missions.

From November 8 to December 5, 1959, the tentative design and layout of the Mercury Control Center were finalized. This facility would become the nerve center for monitoring and managing the Mercury missions, playing a critical role in ensuring their success.

The Mercury Control Center was meticulously designed to include advanced technology and systems for real-time tracking and communication with the spacecraft. Key features included trend charts to monitor the astronauts' conditions, providing vital health and safety data throughout their missions. These charts were essential for assessing the astronauts' well-being and responding promptly to any potential issues.

Mercury Control Center, Cape Canaveral, 1963

In addition to health monitoring, the control center was equipped with world map displays that tracked the spacecraft's trajectory. These maps allowed mission controllers to follow the spacecraft's position and flight path accurately, ensuring continuous oversight and the ability to make necessary adjustments.

Completing the Mercury Control Center's design marked a significant milestone in Project Mercury. It reflected the extensive planning and collaboration among engineers, scientists, and mission planners to create a facility capable of supporting the complexities of human spaceflight. The center would later become the hub for managing the Mercury missions, providing the critical infrastructure to monitor, communicate with, and support the astronauts as they ventured into space.

This development period underscored the dedication and foresight required to pioneer human space exploration. By establishing a state-of-the-art control center, NASA ensured that it had the tools and capabilities to achieve its ambitious goals, ultimately leading to the successful manned missions that defined the early years of space exploration.

On November 10, 1959, personnel from the Space Task Group visited McDonnell Aircraft Corporation to oversee the molding of the first production-type couch for the Mercury spacecraft. This visit was part of the ongoing effort to ensure that all spacecraft components met the rigorous standards for manned spaceflight.

The production-type couch was a crucial element of the Mercury spacecraft's design. It was specifically engineered to support and protect the astronaut during the intense launch, re-entry, and landing phases. The couch needed to provide both comfort and safety, absorbing and distributing the forces experienced during these critical moments.

During their visit, the Space Task Group personnel closely monitored the molding process, ensuring that the materials and techniques were of the highest quality. Their oversight was vital in verifying that the couch would meet the necessary specifications and performance criteria.

This hands-on involvement by the Space Task Group highlighted the collaborative effort between NASA and its contractors. It demonstrated the meticulous attention to detail and quality control that was essential for Project Mercury's success. The development of the production-type couch was a significant step forward, contributing to the overall safety and reliability of the Mercury missions.

The visit to McDonnell on November 10, 1959, exemplified the rigorous testing and evaluation processes that characterized the early days of human space exploration. Each component, including the astronaut's couch, was subjected to thorough scrutiny to ensure it could withstand the demands of spaceflight, ultimately helping to pave the way for America's first manned missions to space.

On November 12, 1959, NASA Administrator T. Keith Glennan and Deputy Secretary of Defense Thomas Gates signed a pivotal agreement addressing the reimbursement of costs incurred by NASA or the Department of Defense (DoD) in support of Project Mercury. This agreement formalized the financial arrangements between the two entities, ensuring that resources were efficiently allocated and expenses were transparently managed.

Project Mercury, as America's first manned space program, required extensive collaboration between NASA and the DoD. The program relied on both organizations' expertise, facilities, and personnel to achieve its ambitious goals. This included using military test ranges, tracking stations, and other critical infrastructure essential for the success of the missions.

The agreement signed by Glennan and Gates was instrumental in defining Project Mercury's financial responsibilities and cost-sharing mechanisms. It ensured that any expenses incurred by NASA or the DoD supporting the program would be reimbursed appropriately, fostering a cooperative relationship between the civilian space agency and the military.

This financial arrangement was crucial for maintaining Project Mercury's momentum. It provided a clear framework for managing the considerable expenses associated with the development, testing, and execution of space missions. By formalizing this agreement, NASA and the DoD could focus on their collaborative efforts without the hindrance of financial uncertainties.

The signing of this agreement on November 12, 1959, exemplified the strategic partnership and shared commitment to advancing America's space capabilities. It laid the groundwork for effective resource management and interagency cooperation, which were essential for the success of Project Mercury and subsequent space exploration endeavors.

From November 16 to 20, 1959, astronauts donned Mercury pressure suits and underwent training at the Navy Aircrew Equipment Laboratory in Philadelphia. During this period, they were familiarized with the anticipated reentry heat pulse, a crucial aspect of their preparation for the intense conditions they would face upon returning to Earth's atmosphere.

The reentry phase of a space mission subjects astronauts and their spacecraft to extreme temperatures due to the friction between the spacecraft and the atmosphere. Understanding and being prepared for this "heat pulse" was vital for the astronauts' safety and the mission's success. The Mercury pressure suits, designed to protect them under these conditions, were tested and evaluated for their effectiveness in such high-stress scenarios.

At the Navy Aircrew Equipment Laboratory, astronauts experienced simulations that mimicked the reentry environment. These simulations helped them understand the thermal loads and allowed them to practice procedures for maintaining safety and operational efficiency during reentry. The hands-on training provided valuable insights into how the pressure suits would perform, ensuring the astronauts were well-prepared for the real mission.

This training period underscored the meticulous preparation and rigorous testing that characterized Project Mercury. By exposing the astronauts to the expected reentry conditions in a controlled environment, NASA aimed to minimize risks and enhance the crew's readiness.

On November 20, 1959, during the fifth Mercury Coordination Meeting, the Army Ballistic Missile Agency (ABMA) proposed installing an open-circuit television system for the upcoming Mercury-Redstone second and third flights. This system was designed to observe and relay crucial data on separating the launch vehicle and the spacecraft.

The installation of the television system represented a significant technological advancement for the Mercury program. By providing real-time visual data, the system would allow engineers and mission controllers to monitor the separation process closely. This capability was essential for identifying and addressing any issues that could arise during this critical phase of the mission.

The proposal underscored the importance of thorough observation and data collection in the early stages of space exploration. Visual monitoring of the separation sequence would enhance understanding the dynamics involved, contributing to the refinement and improvement of future launch systems.

Implementing the open-circuit television system demonstrated the collaborative efforts between different agencies and the continuous drive to leverage new technologies to enhance mission safety and success. The ABMA's initiative highlighted the proactive approach taken to mitigate risks and ensure the reliability of the Mercury-Redstone flights.

This proposal marked another step forward in the meticulous planning and execution of Project Mercury, emphasizing the commitment to innovation and excellence that characterized NASA's early human spaceflight endeavors. By integrating advanced monitoring systems, NASA aimed to pave the way for safer and more successful missions, ultimately leading to the historic achievements of human space exploration.

On November 27, 1959, the Arnold Engineering Development Center (AEDC) conducted a critical test of the Grand Central solid-fuel rocket motor, a key component of the Mercury spacecraft's escape system. The primary objectives of this test were to verify the motor's ignition at high altitudes and to determine the combustion chamber pressure-time curve.

The Grand Central solid-fuel rocket motor was designed to provide the necessary thrust to propel the Mercury spacecraft away from the launch vehicle in an emergency. Ensuring the reliable performance of the escape system was paramount for the safety of the astronauts.

During the test, the AEDC simulated the conditions the rocket motor would encounter at high altitudes. This involved analyzing the motor's ignition characteristics and measuring the pressure changes within the combustion chamber over time. The data gathered from these measurements were crucial for understanding how the motor would perform under actual flight conditions.

The successful testing of the Grand Central solid-fuel rocket motor was a significant milestone for Project Mercury. It provided valuable insights into the reliability and effectiveness of the escape system, enhancing the mission's overall safety. This test also underscored the importance of rigorous testing and validation processes in developing spaceflight technology.

By verifying the performance of the escape system's rocket motor, NASA and its partners, demonstrated their commitment to astronaut safety and mission success.

Throughout November 1959, significant progress was made in the development of Project Mercury as the first manned development system tests were completed at the AiResearch Manufacturing Division of the Garrett Corporation. These tests were conducted in an altitude chamber and were designed to confirm the functionality of the life support system under simulated high-altitude conditions.

During these tests, a McDonnell subject clad in a Mercury-type pressure suit was exposed to conditions mimicking the vacuum of space. The primary objective was to assess the performance and reliability of the life support system, which was crucial for maintaining the astronaut's safety and well-being during spaceflight.

The altitude chamber provided a controlled environment where engineers could meticulously monitor and measure the system's response to extreme space conditions. Preliminary data from these tests indicated that the life support system performed satisfactorily, meeting the requirements to support a human in space.

These tests represented a critical milestone in the Mercury program, validating the design and operation of the life support system. Their successful completion demonstrated that the system could reliably provide the necessary pressure, oxygen, and thermal regulation to protect the astronaut in the harsh space environment.

Ten developmental full-pressure suits were delivered between November 1959 and January 1960 for use in Mercury training and development programs. These suits were crucial for ensuring astronaut safety and functionality during space missions. However, initial testing revealed issues with stretching and mobility, which posed significant challenges for the astronauts' ability to perform necessary tasks while suited.

The pressure suits were designed to provide a life-supporting environment, protecting astronauts from the vacuum of space and maintaining appropriate pressure, oxygen levels, and thermal regulation. The problems identified during early testing highlighted the complexities involved in creating a protective and flexible suit for practical use.

To address these issues, engineers and designers explored several potential solutions. One approach was to undersize the suits, which could help reduce excessive stretching and improve the fit. Another focus was on enhancing the insulation materials used in the suits to manage the astronauts' body temperature better and increase overall comfort and mobility.

These iterative improvements were part of the ongoing effort to refine and perfect the pressure suits for the Mercury program. Each modification brought the suits closer to meeting the stringent requirements for spaceflight. The dedication to solving these problems underscored the meticulous attention to detail and commitment to astronaut safety that defined Project Mercury.

The delivery and subsequent testing of the full-pressure suits between November 1959 and January 1960 were critical steps in the development process. By rigorously evaluating the suits and implementing necessary changes, NASA ensured that the astronauts would be well-equipped to handle the demanding conditions of space. These efforts were essential for the success of the Mercury mission and the broader goals of human space exploration.

Little Joe 1A

Spacecraft No.: Boilerplate
Launch Date: November 4, 1959
Duration: 8 minutes 11 seconds
Purpose: Test of launch escape system during flight with boilerplate capsule.
Result: Partial success

Launched on November 4, 1959, Little Joe 1A was a pivotal mission to test the launch escape system under flight conditions using a boilerplate capsule. Lasting 8 minutes and 11 seconds, the mission achieved partial success, marking significant progress in developing crucial safety systems for manned spaceflight.

The primary objective of Little Joe 1A was to validate the functionality of the launch escape system, designed to rapidly separate the spacecraft from the launch vehicle in case of an emergency. This test provided invaluable data on the system's performance during actual flight scenarios, highlighting areas for further refinement and improvement.

The mission's partial success underscored the collaborative efforts and engineering ingenuity of NASA and its partners. It contributed essential insights into enhancing the safety and reliability of future manned missions, laying the groundwork for subsequent advancements in space exploration technology.

Little Joe 1A demonstrated NASA's commitment to rigorous testing and continuous improvement, which was essential for overcoming the challenges of space exploration. Its achievements paved the way for future missions and reinforced the agency's progress toward realizing ambitious goals in human spaceflight.

Little Joe 2 (LJ-2)

Spacecraft No.: Boilerplate
Launch Date: December 4, 1959
Duration: 11 minutes 6 seconds
Purpose: Escape system test with primate at high altitude.
Result: Success

Launched from Wallops Island on December 4, 1959, Little Joe 2 (LJ-2) was a crucial milestone in Project Mercury. This mission was designed to comprehensively test the Mercury spacecraft's escape system and evaluate its performance under high-altitude conditions, using a primate as a passenger.

The primary objectives of LJ-2 included assessing the escape tower's movements during a high-altitude abort, studying spacecraft entry dynamics without a control system, testing the operation of the drogue parachute, and evaluating recovery procedures. The abort sequence was initiated successfully at an altitude of 96,000 feet and a speed of Mach 5.5.

The mission's success was underscored by the safe recovery of the primate passenger, named "Sam," demonstrating the reliability and effectiveness of the escape system. Within two hours of liftoff, all mission goals were achieved, providing critical data that advanced the development of the Mercury program.

Little Joe 2's accomplishments highlighted NASA's commitment to rigorous testing and safety in human spaceflight, paving the way for future missions and reinforcing confidence in the Mercury spacecraft's capabilities under challenging conditions.

Throughout 1959, Project Mercury made significant strides in various areas, including spacecraft testing, astronaut training, and system development. The successful completion of numerous tests and the refinement of hardware and procedures brought the program increasingly closer to its goal of manned spaceflight. Each achievement during this period was a crucial step forward, contributing to the overall readiness and reliability of the mission components.

On December 7, 1959, the Space Task Group made a significant decision in the development of Project Mercury by selecting Tenney Engineering Corporation to construct the Mercury altitude test chamber in Hangar S at Cape Canaveral. This vertical cylindrical chamber, measuring 12 feet in diameter and 14 feet in height, was specifically designed to simulate the shallow pressures equivalent to an altitude of 225,000 feet.

The Mercury altitude test chamber was critical equipment for ensuring the spacecraft's readiness for space travel. Creating a near-vacuum environment allowed engineers to conduct partial functional checks of the spacecraft systems under conditions that closely mimicked those in space. This testing was essential to verify that the spacecraft's systems could withstand the harsh conditions of space travel, including the extreme vacuum and temperature variations.

The construction of this test chamber represented a key step in the meticulous process of validating and verifying the Mercury spacecraft's performance. By testing the spacecraft in a controlled near-vacuum environment, engineers could identify and address potential issues before actual missions, thereby enhancing the spacecraft's safety and reliability.

The decision to build the altitude test chamber at Cape Canaveral underscored the importance of rigorous testing and quality assurance in the early days of human spaceflight. This facility would play a vital role in the preparation and success of the Mercury missions, ensuring that every system functioned correctly under the demanding conditions of space.

On December 8, 1959, at the Arnold Engineering Development Center (AEDC), two Thiokol retrorockets for the Mercury spacecraft underwent a series of tests. The primary objective of these tests was to evaluate the retrorockets' ignition characteristics, a crucial aspect to ensure their reliable performance during space missions.

Retrorockets play a vital role in spaceflight by providing the necessary thrust to slow down the spacecraft, enabling it to safely re-enter the Earth's atmosphere. These rockets needed to function flawlessly for the Mercury missions, as any failure could jeopardize the mission and the astronaut's safety.

The tests conducted at AEDC were designed to simulate the conditions under which the retrorockets would ignite and operate. Engineers closely monitored the ignition process, assessing ignition timing, thrust generation, and overall performance. The data collected from these tests was critical for verifying that the retrorockets met the stringent requirements for manned spaceflight.

During these tests, the retrorockets' successful ignition and operation provided confidence in their design and functionality. This was a key step in the rigorous testing regime that characterized Project Mercury, which aimed to minimize risks and ensure mission success.

On December 22, 1959, the Redstone launch vehicle designated for the first Mercury-Redstone mission (MR-1) was installed on the interim test stand at the Army Ballistic Missile Agency (ABMA) for static testing. This event marked a significant milestone in Project Mercury, as the Redstone rocket played a crucial role in the initial phases of America's efforts to send humans into space.

Static testing is a critical phase in rocket development, allowing engineers to thoroughly evaluate the vehicle's performance while it remains secured to the ground. During these tests, the Redstone rocket's engines were fired to verify their thrust output, stability, and overall functionality under controlled conditions. This process ensured that the rocket would perform reliably during an actual launch.

The Redstone rocket, originally developed as a military ballistic missile, was adapted for space exploration. It provided the necessary thrust to propel the Mercury spacecraft into suborbital flight, as a stepping stone towards achieving orbital missions. The successful integration and testing of the Redstone launch vehicle were essential for the subsequent manned missions planned under Project Mercury.

The installation and static testing of the MR-1 launch vehicle at ABMA underscored the collaborative efforts between NASA and the military in advancing the United States' space capabilities. Each successful test brought NASA closer to the historic goal of manned spaceflight, demonstrating the reliability and readiness of the launch system.

This milestone in December 1959 highlighted the meticulous preparation and rigorous testing processes that were fundamental to Project Mercury's success. The progress made with the Redstone launch vehicle was a testament to the dedication and technical expertise of the teams involved, setting the stage for the groundbreaking achievements in human space exploration that were soon to follow.

On December 31, 1959, the reliability and qualification tests for the thrust cut-off sensor fabricated by the Donner Scientific Company were completed and accepted. These sensors were crucial for the Mercury spacecraft, ensuring precise control over the rocket's thrust during launch and flight.

The thrust cut-off sensors were similar to those used in Lockheed's Polaris missile program. These sensors had already demonstrated their effectiveness and reliability through extensive environmental evaluations, providing a strong foundation of confidence in their performance for Project Mercury.

The completion of these tests marked a significant achievement in ensuring the reliability of the Mercury spacecraft's systems. The sensors were designed to detect when the rocket had reached the desired velocity and altitude, triggering the cut-off of the engine thrust at the appropriate moment. This precise control was vital for achieving the correct trajectory and ensuring the safety of the mission.

In 1959, NASA embarked on an ambitious journey with Project Mercury, a pivotal initiative in America's space exploration agenda. Rigorous testing was paramount, ensuring that every component, from sensors to spacecraft, could withstand the harsh realities of space travel.

The project's financial commitment was substantial: NASA allocated $22,830,000 to the Air Force Ballistic Missile Division for Atlas launch vehicles, $16,060,000 to the Army Ordnance Missile Command for Redstone launch vehicles, and $49,407,540 to McDonnell Aircraft Corporation for the Mercury spacecraft. McDonnell, tasked with spacecraft production, poured over 942,818 man-hours into engineering, 190,731 into tooling, and 373,232 into production, demonstrating an unprecedented dedication to achieving NASA's spacefaring goals.

By December of that pivotal year, the Mercury astronauts had transitioned from theoretical studies to practical engineering. They underwent intensive training in egress procedures, techniques to stabilize spacecraft movements, and adaptation to weightless environments. This training included rigorous sessions at Edwards Air Force Base, where they practiced essential tasks like eating, drinking, and conducting psychomotor tests in simulated weightlessness aboard F-100 aircraft.

Meanwhile, advancements in ground control were also progressing rapidly. The Space Task Group approved monitoring facilities proposed by the Stromberg-Carlson Division, crucial for the Mercury Control Center at Cape Canaveral and Bermuda. Bell Aircraft Corporation, responsible for developing the spacecraft's reaction control system, commenced flight rating tests on the automatic subsystem, marking significant strides toward ensuring the spacecraft's operational readiness.

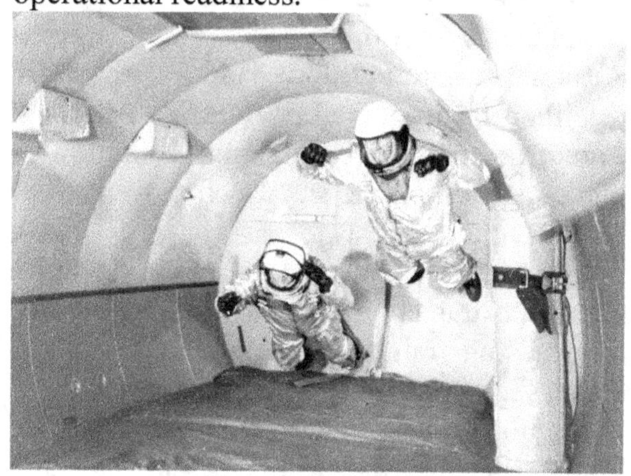

Weightlessness simulation in a C-131

Recounting the Progress

As the sun dipped below the horizon on a crisp December evening in 1959, Cape Canaveral was alive with a fervent energy. Engineers and technicians moved purposefully among the softly glowing control panels and computer screens. In Hangar S, the towering altitude test chamber, a marvel crafted by Tenney Engineering Corporation, loomed large. Soon, it would replicate the near-vacuum conditions of space, stress-testing every system aboard the Mercury spacecraft.

At the Arnold Engineering Development Center, the thunderous roar of Thiokol retrorockets reverberated through the air. These tests, meticulously designed to assess ignition characteristics, filled the facility with the unmistakable scent of burnt propellant. Each successful ignition brought engineers closer to realizing the dream of launching humans beyond Earth's atmosphere.

Across the way, at the Army Ballistic Missile Agency, the Redstone rocket stood proudly on its test stand under the bright hangar lights. Static testing of this sturdy vehicle marked a crucial milestone, as it would serve as the workhorse for the initial Mercury missions.

Meanwhile, the pace never slackened in the bustling offices of the Space Task Group. Detailed financial reports and exhaustive man-hour logs flowed in, underscoring the monumental scale of the endeavor. McDonnell Aircraft Corporation alone had invested nearly a million man-hours in engineering efforts, a testament to their unwavering commitment and pursuit of excellence, which drove Project Mercury forward.

As the year ended, America's pioneering astronauts transitioned from classroom studies to hands-on training. At Edwards Air Force Base, aboard F-100 aircraft, they experienced the exhilarating sensation of weightlessness. Amidst the roar of jet engines, laughter and shouts of excitement echoed through the cabins. Each maneuver and drill brought them closer to their ultimate goal: venturing into the uncharted realm of space.

In control rooms, laboratories, test stands, and hangars, the men and women of Project Mercury toiled tirelessly. The determination in their eyes matched the steady glow of control panel lights. They weren't just constructing rockets and spacecraft and forging humanity's path to the stars. As 1959 ended, there was a palpable sense of accomplishment and anticipation. The dawn of a new decade promised even greater challenges and triumphs as humanity stood poised to take its first audacious steps beyond the confines of Earth.

On January 6, 1960, the atmosphere within NASA's Space Task Group crackled excitedly despite the early January chill. The approval of the Project Mercury data reduction plan marked a significant milestone. Engineers and scientists, immersed in their laboratories, meticulously fine-tuned every aspect of the "Semi-Automatic Data Reduction" study. Submitted to NASA Headquarters on December 21, 1959, this initiative aimed to streamline data processing—a crucial element for the success of upcoming missions.

Five days later, on January 11, a palpable sense of achievement filled the halls as NASA sealed a contract worth $33,058,690 with the Western Electric Company. This agreement was pivotal for constructing and engineering the Mercury tracking network, essential for monitoring spacecraft journeys through the vast expanse of space. At Langley Research Center, detailed discussions and comprehensive charts underscored the magnitude of this undertaking.

By January 15, NASA Headquarters and the Space Task Group had approved the "Overall Plan for Department of Defense Support for Project Mercury Operations." This comprehensive document delineated the intricate coordination required between NASA and the Department of Defense, ensuring meticulous planning and execution of every recovery operation. The Navy's detailed "Operation Plan COMDESFLOTFOUR No. 1-60" laid out precise recovery procedures for both Mercury-Redstone and Mercury-Atlas missions, instilling a sense of order into this ambitious endeavor.

January 18, 1960, marked a pivotal moment in Project Mercury. Walter C. Williams, a key figure in the program, proposed the establishment of the Mercury-Redstone Coordination Committee. His initiative aimed to streamline the coordination of Mercury-Redstone flight tests, emphasizing the need for an organized approach to managing many tasks. Williams' formal letters underscored the importance of efficient teamwork.

Around the same time, Williams advocated for forming a standing coordination body for the Mercury-Atlas flight test working group. This group would unite experts from organizations like Convair Astronautics and McDonnell Aircraft Corporation, fostering collaborative success and ensuring seamless operations.

On January 19, engineers traveled to London to evaluate potential vendors for critical components such as SARAH beacon batteries and hydrogen-peroxide systems. Detailed reports compiled from these evaluations reflected NASA's rigorous standards in selecting partners crucial to Project Mercury's success. Workshops buzzed with activity, filled with the sharp scent of metal and machinery as assessments were conducted.

On January 21, fiscal planning took center stage as NASA's leadership gathered to draft funding estimates for fiscal year 1962. Plans included procuring 15 Atlas launch vehicles and 26 Mercury spacecraft, underscoring the agency's commitment to expanding its capabilities in manned space flight. Amid these discussions, the Little Joe 1-B (LJ-1B) successfully launched from Wallops Island, carrying a rhesus monkey named "Miss Sam." The spacecraft achieved a peak altitude of 9.3 miles and a maximum speed of 2,021.6 miles per hour before being recovered by a Marine helicopter, marking another triumph for Project Mercury.

January 25 witnessed a significant milestone as the Space Task Group at Langley received McDonnell's first production-type Mercury spacecraft. This structural shell, although initially devoid of most internal systems required for manned flight, was swiftly instrumented and designated for the Mercury-Atlas 1 (MA-1) mission. Engineers eagerly began transforming the shell, envisioning its role in future space missions.

By January 31, six chimpanzees, thoroughly trained and prepared for Mercury-Redstone or Mercury-Atlas missions, exemplified the intensity of the program's training regimen. Additional chimpanzees were en route from Africa to join the rigorous preparation efforts. Specifications for training remote-site flight controllers and Mercury control center operations personnel were sent to the Western Electric team. This comprehensive training program, comprising off-range and on-range components, ensured that every controller was impeccably prepared for their pivotal roles.

Throughout January, NASA continued to advance its fundamental communication requirements for Project Mercury, collaborating closely with Western Electric. By February, the interim proposal from Western Electric addressing these requirements was approved. Meanwhile, qualification tests on pilot and instrument viewing cameras and the Mercury spacecraft periscope were successfully completed. Each milestone and test brought the agency closer to achieving its goal of manned space flight.

Little Joe 1B

Spacecraft No.: Boilerplate
Launch Date: January 21, 1960
Duration: 8 minutes 35 seconds
Purpose: Maximum-q abort and escape test with primate using a boilerplate capsule.
Result: Success

Little Joe 1B at launch with Miss Sam, 1960

Launched on January 21, 1960, Little Joe 1B was a critical test in Project Mercury, specifically designed to evaluate the escape system's performance under maximum dynamic pressure conditions (max-q). The mission utilized a boilerplate capsule with a primate onboard.

Little Joe 1B aimed to simulate an abort scenario during the phase of flight where aerodynamic forces on the spacecraft are at their peak, known as max-q. This test was essential to ensure that the escape system could safely separate the spacecraft from the launch vehicle and protect the occupant in a high-stress environment.

The successful outcome of Little Joe 1B further validated the reliability and effectiveness of the Mercury spacecraft's escape system. This achievement provided crucial data and confidence in the system's ability to protect astronauts during critical phases of their mission.

Completing the study on "External and Internal Noise of Space Capsules" on February 1, 1960, marked a crucial step in understanding the acoustic environments of missiles and space vehicles. Despite advancements, NASA officials noted that pilots' internal noise levels remained uncomfortably high, highlighting the need for further data from actual production-model spacecraft.

Mercury-Redstone 1: launch escape system lift-off after 4'' launch, 1960

By February 5, a significant decision was made to use beryllium shingles for heat protection on the cylindrical section of the Mercury spacecraft. This choice was pivotal in ensuring the spacecraft's safety during re-entry into Earth's atmosphere. Concurrently, Colonel George M. Knauf began developing a rigorous medical-monitor training program, thoroughly preparing monitors for space medical problems to support upcoming Mercury flights.

As these months unfolded, the palpable excitement and tension within NASA were evident. Every meeting, test, and approval brought them closer to their audacious goal of sending a human into space. The corridors reverberated with the incessant clatter of typewriters, the murmur of intense discussions, and the distant roar of test launches—a testament to an era of relentless progress and ambition.

On February 8, the Army Ballistic Missile Agency initiated tests for the mission abort sensing program, crucial for integrating into the Mercury-Redstone phase. The laboratory buzzed with activity as engineers meticulously monitored data, ensuring the systems would perform flawlessly during critical mission abort scenarios.

By February 11, the Space Task Group formalized the responsibilities of the Mercury launch coordination office, following Walter C. Williams' earlier proposal. This office was charged with integrating Department of Defense support, managing launch activities, compiling launch support requirements, and representing Mercury at pivotal meetings—highlighting the meticulous planning and sense of purpose required for such a complex operation.

On February 12, as Project Mercury geared up for intensive operational activity, an operations coordination group was established at the Atlantic Missile Range, led by Christopher C. Kraft, Jr. Spirited discussions in the meeting room underscored the critical role of coordination in ensuring mission success.

On February 15, qualification tests for the Mercury spacecraft battery and landing system were completed, ensuring these components were ready for the demands of space flight and safe return. Simultaneously, Mercury remote-site flight controllers began their training with lectures covering facilities, network systems, and operations, while aeromedical staff personnel received specialized instruction to support flight teams.

On February 22, tests for the Mercury spacecraft's automatic stabilization and control system were successfully concluded, demonstrating the precision engineering ensuring spacecraft stability during flight. The same month, establishing a Project Mercury tracking site in Australia was approved, expanding the global tracking network crucial for mission success.

By February 27, the design approval and reliability tests for Mercury command receivers were completed, a critical step in ensuring seamless communication with the spacecraft. Meanwhile, on February 29, NASA Headquarters approved acquiring an analog computing facility costing $424,000. This facility promised to establish and verify Mercury system requirements and support future programs like manned circumlunar vehicles, signaling NASA's commitment to advancing space exploration capabilities.

During February, astronauts received specialized instruction in star recognition and celestial navigation at the Morehead Planetarium in Chapel Hill, North Carolina. In the tranquil setting of the planetarium, they gained practical experience using motorized trainers to correct spacecraft yaw drifts—a stark contrast to the high-energy environments they typically navigated. Dr. James Balten's guidance through the celestial sphere prepared them for the challenges of navigating the vast reaches of space.

Two Spanish firms signed agreements to provide communications support at the Grand Canary Island Mercury tracking site in February-April, highlighting the international collaboration essential to Project Mercury's success.

From March 7-10, the Wright Air Development Center hosted an intensive indoctrination program in free-floating during weightless flight. Using a modified C-131B aircraft, astronauts experienced 90 parabolas of weightlessness. This training allowed them to simulate space conditions and practice with tools and moving weights in a state of microgravity, which is crucial for preparing for their upcoming missions.

On March 9, position titles for Project Mercury operational flights were formally issued. These titles outline roles for 15 significant positions in the Mercury Control Center, 15 in the blockhouse, and two at the launch pad. This structured approach ensured clarity in responsibilities and duties for each role involved in mission operations.

March 11 saw the successful launch of Pioneer V, a mission to explore the space between Earth and Venus. The mission began transmitting valuable data on solar flare effects, particle energies, and magnetic phenomena, setting new communication records and providing critical insights into space conditions.

The Space Task Group published the recovery requirements for the Mercury-Atlas 1 (MA-1) flight test on March 16. The requirements detail procedures for safely recovering the spacecraft after its mission, ensuring comprehensive planning for all possible contingencies.

On March 19, the United States and Spain announced an agreement regarding the Project Mercury tracking station in the Canary Islands, further solidifying international cooperation in the space program.

From March 28 to April 1, astronauts underwent rigorous open-water egress training in the Gulf of Mexico near Pensacola, Florida. They practiced escaping from the spacecraft and boarding life rafts in challenging conditions, including up to 10-foot swells, preparing them thoroughly for emergencies during actual missions.

On March 29, NASA Headquarters streamlined operations by eliminating the spacecraft prelaunch facility at Huntsville, Alabama. Instead, spacecraft designated for Mercury-Redstone missions would be transported directly from McDonnell to Cape Canaveral, significantly accelerating the launch schedule.

Qualification tests for the escape tower rocket began in March and were completed by July 1960. These tests, conducted under simulated high-altitude conditions at the Lewis Research Center, confirmed that the escape rocket motors met operational requirements, ensuring the safety of astronauts during launches.

Throughout March and April, the Mercury-Atlas working panels underwent reorganization into four groups: coordination, flight test, trajectory analysis, and change control. This restructuring aimed to enhance oversight and collaboration among NASA, McDonnell, the Air Force Ballistic Missile Division, Space Technology Laboratory, and Convair-Astronautics, crucial for effectively managing the complexities of the Mercury-Atlas missions.

Mercury-Redstone 5

Pilot: Glenn (likely)
Planned Launch: March 1960
Cancellation Date: August 1961
Details: Initially planned as a suborbital flight, Mercury-Redstone 5 aimed to test the spacecraft and its systems further. The mission, likely piloted by Glenn, was scheduled for launch in March 1960. However, it was canceled in August 1961. This decision came as NASA shifted its focus towards orbital flights following the successful completion of earlier Mercury missions.

In April 1960, pivotal developments in America's burgeoning space program unfolded relentlessly and meticulously.

On April 1st, a landmark moment arrived as the first McDonnell production spacecraft reached NASA's Wallops Island facility. This spacecraft, a testament to human ingenuity, awaited rigorous testing in preparation for its role in advancing manned space exploration.

Just days later, on April 5th, the Space Task Group at Ames Research Center initiated preliminary plans to adapt the Mercury spacecraft for controlled reentry—a crucial step toward what would evolve into Project Gemini. This collaborative effort set ambitious goals for spacecraft technology by the end of the month, laying the groundwork for future missions.

Meanwhile, on April 7th, at Langley Research Center, engineers subjected nine Mercury heat shield models to critical ablation tests. These tests, conducted in the subsonic arc tunnel, were essential to ensuring the spacecraft's ability to withstand the searing temperatures of reentry—an imperative for astronaut safety.

April 8th marked another significant achievement with the completion of an altitude facility chamber in Hangar S at Cape Canaveral. This facility, designed to simulate the harsh conditions of space, represented a quantum leap in spacecraft checkout and astronaut training capabilities. Acceptance tests scheduled for completion by July 11th underscored NASA's commitment to meticulous preparation for manned missions.

As April progressed, NASA's interdisciplinary team continued to push boundaries. On April 15th, engineers began rigorous qualification tests on the Mercury spacecraft's posigrade rocket motors, subjecting them to intense vibration spectra. The successful completion of these tests alleviated concerns and paved the way for further advancements in propulsion reliability.

Simultaneously, at McDonnell, engineers finalized the fabrication of the manned environmental-control-system training spacecraft by April 18th. This milestone underscored ongoing preparations for human spaceflight, as meticulous testing ensured the system's readiness for space challenges.

Recovery Operations

In 1960, the recovery operation plans devised by NASA for the Mercury missions were so extensive that they prompted significant logistical considerations from the Navy. The scale of these plans was such that the Navy expressed concerns that executing them might necessitate deploying the entire Atlantic Fleet. This remark underscored the magnitude of the operation and hinted at the substantial resources and coordination required.

Furthermore, the Navy's estimation hinted at the potential costs involved, suggesting that the expenses associated with the recovery operations alone might approach or even exceed the entire budget allocated for the Mercury program itself. This reflection of the sheer scale and complexity of ensuring the safe recovery of astronauts and spacecraft during Mercury missions highlights the comprehensive planning and collaborative efforts between NASA and the military in pioneering manned spaceflight during that era.

Safety and Environmental Controls

The April 21, 1960 incident underscored the critical necessity of using pure oxygen in cabin and spacesuit systems for manned space missions. During a test flight conducted by McDonnell Aircraft, test pilot G.B. North experienced severe injuries when nitrogen-rich air leaked into his spacesuit.

This accident highlighted the inherent dangers of using mixed gases in a high-altitude or space environment. Unlike pure oxygen, which is crucial for sustaining human life in the vacuum of space, nitrogen can pose significant risks at reduced pressures by causing decompression sickness, also known as "the bends." North's injuries were a stark reminder of the importance of maintaining strict protocols and using pure oxygen environments to safeguard astronauts against such hazards.

As a result of incidents like this, NASA and its contractors emphasized using pure oxygen systems in spacecraft cabins and spacesuits, contributing to the safety protocols that became standard during the Mercury, Gemini, and Apollo programs.

Telemetry

Data automatically sent from the spacecraft to the ground is called telemetry. Telemetry was the automated process of collecting and transmitting data from spacecraft, aircraft, or other remote sources to ground stations or control centers. This data typically includes information about the spacecraft's status, performance, environmental conditions, and other critical parameters. Telemetry systems are essential for monitoring and controlling missions in real time, providing vital information to engineers and mission controllers to ensure the safety and success of space missions.

On April 26th, technicians meticulously tested the maximum altitude sensor fabricated by Donner Scientific Company in a modest workshop. The sensor's successful trials represented another critical step in enhancing spacecraft instrumentation and reliability.

Scientific breakthroughs also marked the month, highlighted by Explorer XI's orbital flight on April 27th. Equipped with gamma-ray detectors, the satellite detected directional fluxes of gamma radiation, challenging prevailing cosmological theories. This discovery, analyzed in dimly lit control rooms, ignited scientific discourse and expanded humanity's understanding of the universe.

Amidst these technical achievements, international cooperation was paramount. By April 29th, agreements for overseas Mercury tracking stations were finalized, from Bermuda to Australia, symbolizing a global commitment to Project Mercury's success.

As the month drew to a close, preparations intensified at Patrick Air Force Base in Florida, where Building 575 underwent refurbishment for NASA's Mercury operations. Scheduled for occupancy by July 1960, this facility promised enhanced launch, network, and data coordination capabilities, embodying the excitement and anticipation within the Space Task Group.

Through meticulous planning, rigorous testing, and international collaboration, NASA's engineers, scientists, and astronauts propelled humanity closer to the stars. Each milestone in April 1960 advanced technological capabilities and set the stage for historic achievements in manned space exploration.

Mercury-Redstone 6

Pilot: N/A
Planned Launch: April 1960
Cancellation Date: July 1961
Details: Originally planned as another suborbital flight, Mercury-Redstone 6 aimed to continue testing the spacecraft's capabilities. However, the mission was ultimately deemed unnecessary and canceled in July 1961. This decision was part of NASA's strategic shift towards prioritizing more ambitious orbital missions in the Mercury program.

In May 1960, as the United States' space program surged forward, significant milestones marked the path to manned space exploration.

On May 9th, at Wallops Island, McDonnell launched its first production spacecraft equipped with an escape rocket in a pivotal beach-abort test. This test evaluated the spacecraft's escape, parachute, and landing systems in an off-the-pad abort scenario. The flawless execution of every sequence affirmed the spacecraft's readiness, eliciting cheers from engineers and scientists alike.

Three days later, on May 12th, the Space Task Group established a crucial field office at McDonnell's St. Louis plant. Led by W. H. Gray, the office housed a team of technical experts tasked with overseeing production and quality assurance. Amidst the aroma of ink and paper, they meticulously reviewed blueprints and reports, ensuring stringent standards were met to support ongoing spacecraft development.

The spacecraft used in the beach-abort test returned to McDonnell's plant on May 14th for an integrity assessment. This meticulous scrutiny aimed to identify any potential issues that could impact future missions, reinforcing NASA's commitment to safety and reliability.

Meanwhile, on May 15th, engineers completed qualification tests for the Mercury spacecraft's explosive egress hatch. The sharp crack of fired explosive bolts marked the hatch's swift ejection in simulations, meeting rigorous safety and performance criteria. Engineers nodded with satisfaction at the hatch's flawless performance in control rooms aglow with monitor lights.

May 23rd marked a significant milestone with the delivery of Spacecraft No. 4 to Cape Canaveral for the first Mercury-Atlas mission (MA-1). Carefully prepared and instrumented by the Space Task Group and Langley Research Center, this spacecraft embodied the culmination of extensive research and development efforts. Its arrival under the Florida sun symbolized NASA's readiness for the upcoming mission, set to push boundaries in human spaceflight.

Throughout May, training initiatives intensified to prepare for manned missions. Thirty physicians the Department of Defense selected began a rigorous indoctrination program at Cape Canaveral. From medical aspects of missile operations to detailed spacecraft systems and astronaut health monitoring, these sessions equipped the medical team for their crucial role in upcoming missions.

Simultaneously, production commenced on the Mercury pressure suit tailored for manned spaceflight. Refined based on astronaut feedback and rigorous testing, these suits offered enhanced mobility and compatibility with spacecraft systems. Their development represented a pivotal advancement in astronaut safety and operational efficiency.

In parallel efforts, McDonnell delivered flight-pressurized couches for the animal phase of Mercury flight tests, addressing minor sealing issues to ensure optimal performance. Additionally, two Procedures Trainers crucial for astronaut training were delivered: the first to Langley Field on May 4th and the second to Cape Canaveral on July 5th. These trainers, later relocated to Houston as part of the Manned Spacecraft Center, were vital in preparing astronauts for space challenges.

As May transitioned into June, these milestones underscored NASA's relentless pursuit of manned space exploration, setting the stage for historic achievements.

Beach Abort

Spacecraft No.: 1
Launch Date: May 9, 1960
Duration: 1 minute 31 seconds
Purpose: Test of the off-the-pad abort system.
Result: Success
The Beach Abort test verified the spacecraft's ability to abort safely from the launch pad, demonstrating a critical safety feature.

Mercury-Redstone 7

Pilot: N/A
Planned Launch: May 1960
Cancellation Date: Not listed
Details: Initially planned as a suborbital flight, Mercury-Redstone 7 was ultimately canceled as NASA's priorities within the Mercury program shifted towards more advanced orbital missions.

In June 1960, as Project Mercury continued to advance, pivotal developments underscored America's preparations for manned spaceflight.

On June 2nd, the Space Task Group's considerations regarding meteoroid damage to the Mercury spacecraft during orbital flight concluded that the risk was minimal, even during meteor showers. However, missions were advised to avoid periods of forecasted meteor showers to mitigate any potential risk. This meticulous attention to detail reflected the group's commitment to ensuring the spacecraft's safety amidst space challenges.

Financially, on June 3rd, the funding for Contract NAS 5-59, supporting the Mercury spacecraft, reached an impressive $75,565,196. This substantial investment underscored the ambitious scope of Project Mercury and the significant financial commitment required to pioneer human spaceflight.

June 9th marked another milestone as the United States Weather Bureau allocated $50,000 for fiscal year 1961 to support Project Mercury. Their responsibilities included crucial tasks such as weather forecasting for launches and recoveries, climatological studies along the launch trajectory, and environmental assessments of specified areas. Detailed studies ensured mission planning accounted for wind velocity, visibility, and cloud coverage along the Atlantic Missile Range.

A highlight on June 18th was the delivery of the Atlas launch vehicle 50-D for the first Mercury-Atlas mission (MA-1). The sleek, silver rocket symbolized a significant leap forward in capabilities, poised to integrate with the spacecraft for its upcoming mission.

On June 20th, rigorous testing continued with the Mercury spacecraft's horizon scanner using a sandblasting technique to assess transmissibility. Simultaneously, manned tests of the Mercury environmental control system validated its effectiveness under postlanding conditions, ensuring astronaut safety and mission success.

As part of the Mercury spacecraft reliability program, June 27th saw one production spacecraft withdrawn for extensive testing under vacuum, heat, and vibration conditions, designated "Project Orbit." This rigorous testing regimen aimed to uphold the highest reliability and safety standards for manned missions.

June concluded with the delivery of Spacecraft No. 2 to the Marshall Space Flight Center in Huntsville, Alabama, on June 30th for compatibility tests with the Redstone launch vehicle. Successful integration tests paved the way for its shipment to Cape Canaveral on July 23rd, marking a significant step in readiness for upcoming launches.

Throughout June, McDonnell contributed significantly by delivering a flight-monitoring trailer to the Space Task Group. This trailer played a crucial role in real-time telemetry read-outs during Mercury-Redstone flights, enhancing mission monitoring and data collection capabilities.

Moreover, Project Mercury's unique recovery capabilities were poised to conduct various scientific and bioscience experiments during missions. These included an ultraviolet camera and bio-specimens, carefully integrated to complement primary mission objectives while advancing scientific knowledge in space.

In this dynamic environment of technological innovation, meticulous planning, and international collaboration, Project Mercury advanced steadily toward its historic achievements in manned space exploration.

Mercury-Redstone 8

Pilot: N/A
Planned Launch: June 1960

Cancellation Date: Not listed
Details: The last of the planned Mercury-Redstone suborbital flights, Mercury-Redstone 8 was also canceled as the program moved towards orbital flights.

In July 1960, pivotal developments marked Project Mercury's ambitious strides as America forged ahead in the space race. The month began with establishing a meticulous reporting plan for the Mercury-Atlas and Mercury-Redstone missions, emphasizing comprehensive documentation of mission progress and outcomes.

On July 7th, the inaugural meeting of the Mercury Network Coordination Committee convened at Cape Canaveral. This gathering charted operational procedures and ensured readiness across the sprawling network, underscoring the critical need for seamless coordination in the nascent days of space exploration.

Just two days later, Major General Leighton I. Davis assumed the role of Department of Defense representative for Project Mercury, succeeding Major General Donald N. Yates. Davis's appointment injected fresh perspectives into defense support strategies, aligning them with the evolving needs of the ambitious space program.

Mid-month, astronauts underwent rigorous desert survival training at Nevada's Stead Air Force Base despite the slim probability of such a landing scenario. The comprehensive five-and-a-half-day course equipped them with vital skills and simulated conditions using Mercury spacecraft equipment, preparing them for remote contingencies.

By July 14th, the formidable Project Mercury team boasted 543 dedicated personnel, with 419 stationed at the Space Task Group and 124 at Langley Research Center. This robust workforce underscored the immense human capital committed to the project's success.

Later in the month, Mercury spacecraft No. 2 arrived at Cape Canaveral, marking another milestone in preparation for upcoming test flights. Concurrently, recovery forces conducted exercises at the launch site, demonstrating their adeptness in simulated spacecraft recoveries off Cape Canaveral's coast.

However, July 29th saw a setback with the launch of Mercury-Atlas (MA-1), intended to validate spacecraft structure and Atlas abort systems. Just 59 seconds into the flight, a structural failure led to mission termination, with the spacecraft lost upon impact with the ocean. Despite this setback, recovery efforts salvaged crucial components, facilitating thorough post-mortem analysis crucial for subsequent missions like Mercury-Atlas 2 (MA-2).

Throughout the month, critical infrastructure progressed, culminating in completing the mobile pad egress tower, colloquially known as the cherry picker. This vehicle, delivered later in October, promised safe and efficient astronaut egress from spacecraft on the launch pad, symbolizing ongoing advancements in aerospace technology.

July 1960 encapsulated both triumphs and challenges, underscoring America's determined pursuit of space exploration amidst technical breakthroughs and operational refinements.

Mercury-Atlas 1

Spacecraft No.: 4
Launch Date: July 29, 1960
Duration: 3 minutes 18 seconds
Purpose: Test of spacecraft/Atlas combination.
Result: Failure

This mission tested the spacecraft/Atlas combination but failed, leading to essential modifications.

In August 1960, Project Mercury surged forward with meticulous planning and technical advancements, marking significant progress toward America's space ambitions.

On August 1st, the Marshall Space Flight Center published the "Final Standard Trajectory for MR-1," providing precise flight path guidance crucial for the upcoming Mercury-Redstone mission. This document ensured that every aspect of the mission's trajectory was meticulously planned and executed.

Three days later, Redstone launch vehicle No. 1 arrived at Cape Canaveral, underscoring the increasing readiness for the MR-1 mission. This milestone highlighted the complex logistics and engineering feats necessary to support manned spaceflight.

Mid-month, the Wright Air Development Center requested technical data from Project Mercury, particularly on human factors, for potential application in the X-20 Dyna Soar program. This request showcased the broader impact of Mercury's research beyond immediate space missions, influencing future aerospace endeavors.

On August 11th, a joint committee comprising representatives from NASA, McDonnell, Ballistic Missile Division, Space Technology Laboratories, and Convair convened to address the malfunction of the Mercury-Atlas 1 (MA-1) mission. Led by James A. Chamberlin, the committee meticulously analyzed flight data and conducted dynamic load tests, vibration studies, and wind tunnel experiments to resolve issues before the upcoming MA-2 mission. This collaborative effort underscored the program's commitment to continuous improvement and safety.

Meanwhile, the qualification test program for the Mercury spacecraft's landing system concluded successfully with 56 engineering models airdropped over 18 months. These rigorous tests validated the system's reliability under various conditions, ensuring astronaut safety during critical descent and landing phases.

Adjustments to the Weather Bureau's fund estimates by mid-August assured continued financial support for Project Mercury's meteorological needs, crucial for mission planning and astronaut safety.

Throughout August, astronauts actively refined spacecraft design during engineering inspections, advocating for changes to enhance operational efficiency. This collaborative effort between engineers and astronauts set standardized procedures for future spacecraft designs, ensuring rigorous standards for human spaceflight.

Towards the end of the month, a comprehensive review at NASA Headquarters evaluated the effectiveness of coordination among organizations involved in Mercury's development and testing. This review highlighted the success of information exchange and coordination panel meetings, pivotal in maintaining project momentum and overcoming challenges.

Training milestones included successful astronaut side-hatch egress exercises, vital for emergency procedures, and briefings on the Tiros weather satellite project, which aims to enhance weather observation skills during orbital missions.

Notably, advancements in spacecraft recovery included tests with fluorescein green dye, which has been proven effective for marking spacecraft locations during recovery operations. This innovation addressed previous stability issues, ensuring prompt and efficient recovery efforts.

As August drew to a close, an extensive review of the Mercury-Atlas program was initiated following setbacks in test flights. This review identified and addressed interface issues, leading to structural modifications and enhanced instrumentation for future missions like Mercury-Atlas 2 (MA-2). This set the stage for continued progress in America's pioneering space endeavors.

In September 1960, Project Mercury intensified its preparations with meticulous planning and crucial advancements as America's pioneering space program pressed onward.

The month commenced with the Space Task Group outlining stringent specification requirements for spacecraft onboard data systems to McDonnell. These tests aimed to validate the performance of communication and instrumentation systems crucial for mission success.

By September 3rd, modifications to spacecraft telemetry systems for Mercury-Atlas 3 (MA-3) and Mercury-Atlas 4 (MA-4) missions streamlined operations, omitting aircraft telemetry requirements post-main parachute deployment.

On September 9th, McDonnell submitted comprehensive plans for spacecraft systems tests and Cape Canaveral checkout procedures for spacecraft Nos. 5 and 7. Notably, spacecraft No. 7 was slated to undergo these rigorous tests first, marking a pivotal stage in validation and readiness.

Mid-month insights from the "Flight Test Evaluation Report, Missile 50-D" detailed lessons learned from the unsuccessful Mercury-Atlas 1 (MA-1) reentry test mission, informing subsequent improvements.

On September 19th, formatting for the inaugural Mercury-Redstone postlaunch (MR-1) report was finalized, setting standards for comprehensive mission documentation post-launch.

The 20th saw the arrival of Atlas launch vehicle 67-D at Cape Canaveral, pivotal for the forthcoming Mercury-Atlas 2 (MA-2) reentry test, highlighting ongoing readiness efforts.

Innovative engineering solutions emerged on September 21st following challenges during an abort test at Wallops Island, showcasing NASA's adaptive prowess in optimizing rocket performance for future missions.

Astronauts, meanwhile, engaged in rigorous weightless training aboard a modified C-135 jet aircraft, preparing them for the physiological demands of spaceflight.

As the month concluded, Mercury spacecraft No. 3 was erected at Wallops Island for the Little Joe 5 (LJ-5) mission, symbolizing the culmination of meticulous testing and engineering efforts.

Spacecraft Designations

In the Mercury space program, spacecraft received letter designations after their numerical identifier to denote specific modifications or configurations. For example, Mercury-Redstone 2 (MR-2) and Mercury-Redstone 15 (MR-15) might have variants designated as MR-2B or MR-15B, respectively.

These letter designations were used to differentiate between different iterations of the spacecraft that underwent modifications or improvements based on lessons learned from previous missions or specific mission requirements. The modifications could include changes to systems, equipment, or configurations to enhance performance, reliability, or safety for upcoming missions in the Mercury program.

Booster Terminology

The term "booster" initially referred to the first stage of the launch vehicle stack, which provided the initial thrust to lift the spacecraft off the ground. In the context of early rocketry and space exploration, this typically referred to a single rocket stage designed to boost the payload into space.

However, with the development of more complex launch systems like the Space Shuttle, the term "booster" evolved to describe additional rocket stages or solid rocket boosters (SRBs) attached to the sides of the main launch vehicle. In the case of the Space Shuttle, these solid rocket boosters were used alongside the Shuttle's main engines to provide extra thrust during liftoff.

Unlike traditional first stages that remained integrated throughout the flight, the Space Shuttle's SRBs were jettisoned after burning out and falling away from the Shuttle stack during ascent. This modular approach allowed for greater flexibility in designing and adapting launch vehicles to specific mission requirements.

Therefore, "booster" came to encompass the initial stage of the launch vehicle and additional stages or rockets attached to enhance overall thrust and performance, reflecting advancements in rocket technology and mission design capabilities.

Mercury-Redstone 2

Testing and reworking Mercury-Redstone 2 (MR-2) at Hangar S took 110 days. Hangar S, located at the NASA Langley Research Center, played a pivotal role in spacecraft preparation and in training animals like chimpanzees for early space missions.

During the Mercury program, chimpanzees such as Ham, Enos, and others were trained extensively at Hangar S for their crucial roles in testing spacecraft systems and procedures. These animals were integral to proving the viability of manned spaceflight before human astronauts undertook similar missions.

Chimpanzee Enos with handler before Mercury-Atlas 5 flight Enos became the first chimp to orbit the earth on November 29, 1961, aboard a Mercury Atlas rocket. Although the mission plan originally called for three orbits, flight controllers were forced to terminate Enos' flight after two orbits due to a malfunctioning thruster and other technical difficulties. Enos landed in the recovery area and was picked up 75 minutes after splashdown. He was found to be in good overall condition and both he and the Mercury spacecraft performed well. His mission concluded the testing for a human orbital flight, achieved by John Glenn on February 20, 1962. Enos died at Holloman Air Force Base of a non-space-related case of dysentery 11 months after his flight.

Space pioneers Ham, who became the first great ape in space during his January 31,

1961, mission, and Enos, the only chimpanzee and third primate to orbit the Earth (November 29, 1961), were research subjects in the Project Mercury program.

Three-year-old chimpanzee, Ham, in the biopack couch for the MR-2 suborbital test flight. On January 31, 1961, a Mercury-Redstone launch from Cape Canaveral carried the chimpanzee "Ham" over 640 kilometers (400 mi) down range in an arching trajectory that reached a peak of 254 kilometers (158 mi) above the Earth. The mission was successful and Ham performed his lever-pulling task well in response to the flashing light. NASA used chimpanzees and other primates to test the Mercury capsule before launching the first American astronaut Alan Shepard in May 1961.

Chimpanzee Enos

The successful flight and recovery confirmed the soundness of the Mercury-Redstone systems. 110-day period for testing

and rework of MR-2 highlights the meticulous preparation and rigorous testing required to ensure the safety and success of each mission in NASA's early space exploration efforts. Hangar S served as a hub of activity, blending technical precision with the pioneering spirit of space exploration that characterized the Mercury program.

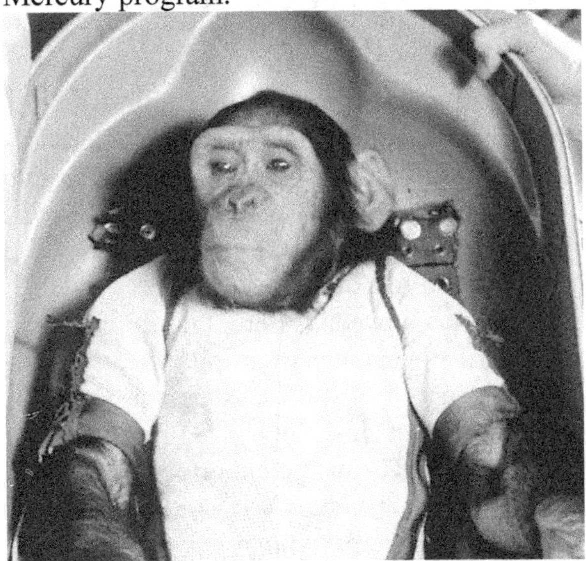

Chimpanzee Ham

Simultaneously, Mercury spacecraft No. 5 underwent rigorous booster compatibility checks at Marshall Space Flight Center, critical steps ahead of its mission with Ham the chimpanzee on Mercury-Redstone 2 (MR-2).

Throughout September, the arrival and testing of flight-type pressure suits underscored the program's commitment to astronaut safety and readiness, ensuring these suits performed flawlessly under extreme conditions.

In sum, September 1960 was marked by relentless dedication, technical precision, and collaborative innovation within Project Mercury. These efforts laid a robust foundation for subsequent missions, edging ever closer to the historic milestone of human spaceflight.

In October 1960, Project Mercury continued its rigorous preparations and advancements, solidifying its path toward manned spaceflight with meticulous planning and training.

From October 3rd to 21st, astronauts underwent intensive centrifuge training at the Aviation Medical Acceleration Laboratory. This final preparation phase for the first manned Mercury-Redstone flight focused on refining 3-axis hand-controller tasks using actual spacecraft couches and production-grade equipment. Equipped with the latest model pressure suits featuring freon coolant, astronauts practiced manual overrides in response to simulated spacecraft failures, ensuring readiness for space challenges.

On October 13th and 14th, DESFLOTFOUR personnel conducted a critical communications exercise at the Mercury Control Center's recovery room. Establishing voice and continuous-wave communications with two destroyers positioned 120 miles at sea, this exercise familiarized personnel with equipment layouts and communication protocols essential for seamless recovery operations.

October 17th marked the establishment of the Project Mercury weather support group within the Office of Meteorological Research at the United States Weather Bureau. This group was pivotal in providing accurate weather forecasts for launch and recovery operations, safeguarding missions against adverse weather conditions. Concurrently, James Carter's comprehensive study on "Crew Support Equipment," submitted from the Marshall Space Flight Center, detailed essential survival gear and tools for astronaut safety and mission success.

A significant operational shift occurred on October 18th, as the spacecraft checkout facility at Marshall Space Flight Center was relocated to Cape Canaveral. This centralization of launch preparations streamlined operational efficiency, ensuring meticulous spacecraft readiness under stringent mission rules outlined for Mercury-Redstone 1 (MR-1). Initially issued on October 18th and revised on November 1st, 1960, these rules underscored the rigorous procedures essential for mission success.

As the month ended on October 31st, Space Task Group officials presented a comprehensive update on qualification and reliability activities for Project Mercury to Dr. T. Keith Glennan, NASA Administrator. This presentation highlighted the project's significant progress and adherence to stringent standards, reaffirming confidence in the program's readiness to achieve historic milestones in manned space exploration.

October 1960 was pivotal for Project Mercury, marked by meticulous training, operational readiness enhancements, and strategic organizational developments, setting the stage for America's monumental leap into space exploration.

Little Joe assembling at Wallops Island

In November, Project Mercury continued to navigate through successes and challenges, marking significant strides toward manned spaceflight.

On November 1st, the Goddard Space Flight Center's computing and communications center became operational with two IBM 7090 computers. These systems played a crucial role in calculating spacecraft positions, predicting trajectories, and aiding in acquiring observation sites. During the Mercury-Atlas 7 mission, these computers accurately predicted the spacecraft's landing overshoot, demonstrating their essential role in mission success.

Little Joe 5

Spacecraft No.: 3
Launch Date: November 8, 1960
Duration: 2 minutes 22 seconds
Purpose: First Little Joe escape system test with a production spacecraft, at max-q.
Result: Failure

November 8th saw the launch of Little Joe 5 (LJ-5) from Wallops Island, intended to test the spacecraft under rigorous conditions. Unfortunately, a premature ignition of the escape rocket motor caused the spacecraft to remain attached to the launch vehicle until impact, resulting in its destruction. Investigations identified potential failures in spacecraft-to-adapter or escape tower clamp-ring limit switches, necessitating a reevaluation before future missions.

Little Joe 5 aimed to test the escape system with a production spacecraft under maximum dynamic pressure (max-q) conditions. However, the mission failed. Despite this setback, the mission provided valuable data that would inform future tests and improvements to the escape system.

By November 13th, system checkout tests on spacecraft No. 7 confirmed its readiness for manned missions, boosting confidence within the team as they progressed towards human spaceflight capabilities.

A pivotal meeting convened at Langley Field on November 16th to analyze the Mercury-Atlas 1 (MA-1) mission failure and ensure readiness for the upcoming Mercury-Atlas 2 (MA-2) mission. Lessons learned from MA-1 were swiftly integrated into planning and operations.

Comprehensive guidelines for mission operations were formalized on November 18th with the publication of "Standard Procedures Mercury Control Center for Flight Control and Overall Options." This document standardized procedures critical for mission success.

Mercury-Redstone 1

Spacecraft No.: 2
Launch Date: November 21, 1960
Duration: 2 seconds
Purpose: Qualification of spacecraft/Redstone combination.
Result: Failure

Mercury-Redstone 1 was intended to qualify the spacecraft and Redstone rocket combination. Unfortunately, the mission failed shortly after liftoff, marking a setback in validating this crucial launch configuration.

November concluded with a mix of setbacks and achievements in NASA's Mercury program. On November 21st, Mercury-Redstone 1 (MR-1) experienced an engine cut-off shortly after liftoff, triggering the emergency escape system. Despite the failure of the mission to qualify the spacecraft and Redstone rocket combination, the spacecraft was safely recovered with minimal damage, emphasizing the importance of robust safety systems.

Simultaneously, Phase II of the helicopter spacecraft airdrop program successfully validated spacecraft dynamics and water stability under various conditions. This achievement was crucial for refining recovery operations, demonstrating NASA's ongoing commitment to overcoming challenges in manned space exploration.

Throughout November and into December, the Mercury Control Center underwent its first operational tests during launches like Mercury-Redstone 1 (MR-1A), ensuring readiness across systems and personnel for the challenges ahead in manned spaceflight.

Erection of Redstone at Launch Complex 5

November 1960 was a pivotal period for Project Mercury, marked by technical advancements, operational refinements, and crucial lessons learned that laid the groundwork for America's imminent journey into space with human astronauts.

In the closing months of 1960, the Space Task Group and their collaborators intensified their efforts towards a monumental achievement: Project Mercury's pioneering mission into space. Their unwavering dedication propelled each meticulous step forward, laying the groundwork for future human space exploration.

On December 1, a pivotal advancement transformed the Mercury spacecraft as a 16.5-foot recovery whip antenna replaced the previous balloon-borne system. This upgrade aimed to enhance communication during recovery, crucial for ensuring astronaut safety in future missions. Simultaneously, McDonnell achieved a significant milestone by completing the first orbital timing device for spacecraft. Rigorous qualification tests swiftly followed, validating its reliability for upcoming space operations.

The following day, meticulous weight and balance parameter calculations for the Mercury-Redstone 2 (MR-2) mission were meticulously forwarded to the Marshall Space Flight Center. These precise measurements were vital in guaranteeing the spacecraft's stability and optimal performance during its critical mission phases.

Mercury-Redstone 1A

Spacecraft No.: 2
Launch Date: December 19, 1960
Duration: 15 minutes 45 seconds
Purpose: Qualification of spacecraft/Redstone combination.
Result: Success

Successfully qualifying the spacecraft/Redstone combination, this mission paved the way for manned flights.

On December 2, Redstone launch vehicle No. 3 embarked on its journey to Cape Canaveral, destined for the Mercury-Redstone 1A (MR-1A) mission. The sleek rocket, a testament to engineering excellence, symbolized months of preparation and innovation.

By December 9, excitement peaked as Spacecraft No. 7 arrived at Cape Canaveral for the highly anticipated Mercury-Redstone 3 (MR-3) mission. This mission would later make history with Alan Shepard's courageous flight, marking an undeniable leap toward human spaceflight.

However, not without setbacks. On December 14, technical challenges led to the cancellation of a satellite clock development contract with the Waltham Precision Instrument Company. Despite this setback, McDonnell's successful fabrication of an orbital timing device underscored their pivotal role in overcoming obstacles crucial to mission success.

Finally, on December 19, the MR-1A mission launched triumphantly from Cape Canaveral, repeating its earlier attempt from November 21. This mission aimed to qualify the spacecraft and flight systems for subsequent primate missions. The spacecraft soared to a peak altitude of 130.68 miles, traveling at a remarkable speed of 4,909.1 miles per hour. Fifteen minutes after landing, the recovery helicopter swiftly retrieved the capsule, affirming the efficacy of recovery procedures in real-world conditions.

The momentum continued the next day, December 20, with the delivery of Redstone launch vehicle No. 2 to Cape Canaveral. This preparation marked the next step towards the Mercury-Redstone 2 (MR-2) mission, slated to carry the chimpanzee "Ham" on an unprecedented journey into space. These successive achievements underscored America's rapid progress in the burgeoning era of manned space exploration, setting the stage for even greater triumphs ahead.

In early 1961, pivotal developments marked the rapid evolution of America's space exploration efforts under NASA's burgeoning Project Mercury.

On January 3, the Space Task Group achieved a milestone as it transitioned into a distinct NASA field element directly reporting to NASA Headquarters. Previously part of the Goddard Space Flight Center and supported by Langley Research Center, this reorganization underscored the group's escalating significance within NASA's hierarchy, coinciding with a personnel strength of 667 dedicated professionals.

Just days later, on January 16, postlaunch evaluations following the Mercury-Redstone 1A (MR-1A) mission confirmed the spacecraft's robust performance. Critical systems, including instrumentation and communications, operated flawlessly during flight, validating its readiness for forthcoming missions.

On January 20, attention turned to Wallops Island, where Spacecraft No. 14 underwent rigorous testing for the Little Joe 5A (LJ-5A) mission. Designed to simulate abort scenarios under maximum dynamic pressure, these tests subjected the spacecraft to extreme conditions, ensuring its reliability in critical moments.

Financial commitments underscored the scale of NASA's ambitions, with estimates revealing substantial investments in Project Mercury by month's end. Costs included $51,504,000 for Atlas launch vehicles, $14,918,182 for Redstone launch vehicles, and $79,245,952 allocated to Mercury spacecraft development. These figures encapsulated both processed expenditures and "Undefinitized Obligations," highlighting the fiscal commitment underpinning America's space aspirations.

January 31 marked another historic achievement with the Cape Canaveral launch of Mercury-Redstone 2 (MR-2). Carrying the chimpanzee Ham, the mission encountered challenges such as higher-than-planned thrust and premature liquid oxygen depletion. Despite these obstacles, the flight was deemed successful, affirming the spacecraft's viability. Ham's post-flight condition, showing no physiological issues, provided crucial data on the effects of space travel. Yet, his reluctance to re-enter the spacecraft after landing was a poignant reminder of the complexities and stresses inherent in space missions.

Throughout January, astronaut training intensified, focusing on comprehensive familiarization with spacecraft systems. Lectures by the Operations Division of the Space Task Group equipped astronauts with in-depth knowledge critical for their upcoming roles in manned spaceflight, underscoring NASA's meticulous preparation for the challenges ahead.

These milestones in early 1961 not only propelled Project Mercury closer to its goal of manned spaceflight and set the stage for America's continued dominance in space exploration.

NASA's Chimpanzee Missions: Validating Life Support Systems

In addition to numerous uncrewed flights, NASA conducted missions with chimpanzees to validate the Mercury spacecraft's life support systems and safety measures further. These missions were critical steps in ensuring the spacecraft could sustain human life during intense spaceflight conditions.

Chapter 9 - Chimpanzee Astronauts

The chimpanzee missions, such as Mercury-Redstone 2 (MR-2) and Mercury-Atlas 5 (MA-5), involved launching a chimpanzee aboard the Mercury spacecraft to test its systems and evaluate the spacecraft's performance with a living organism on board. Here's a brief overview of the chimpanzee missions in Project Mercury:

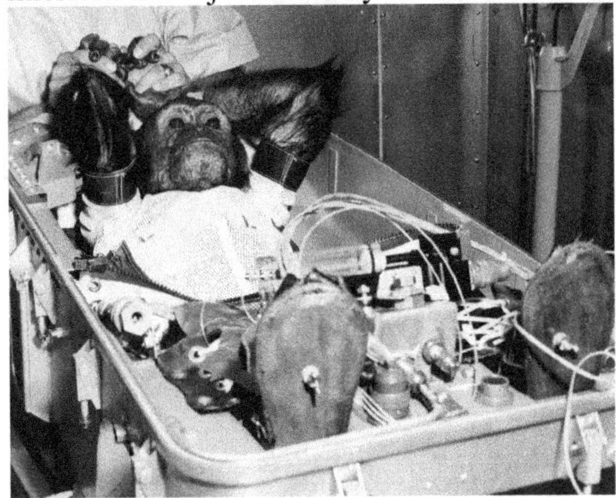

Mercury-Atlas 5: Enos, 1961

Unloading Atlas at Cape Canaveral

Mercury-Redstone 2 (MR-2): Launched on January 31, 1961, MR-2 carried a chimpanzee named Ham (an acronym for Holloman Aerospace Medical Center) into space. The mission aimed to test the Mercury spacecraft's systems and the effects of spaceflight on a living organism. Ham's successful flight demonstrated that a primate could perform tasks during spaceflight and paved the way for manned missions.

Mercury-Atlas 5 (MA-5): This mission, launched on November 29, 1961, carried another chimpanzee named Enos. Enos orbited the Earth twice, demonstrating the spacecraft's ability to support longer-duration flights and assessing the effects of extended spaceflight on living organisms.

Ham the Chimpanzee: A Pioneering Spaceflight

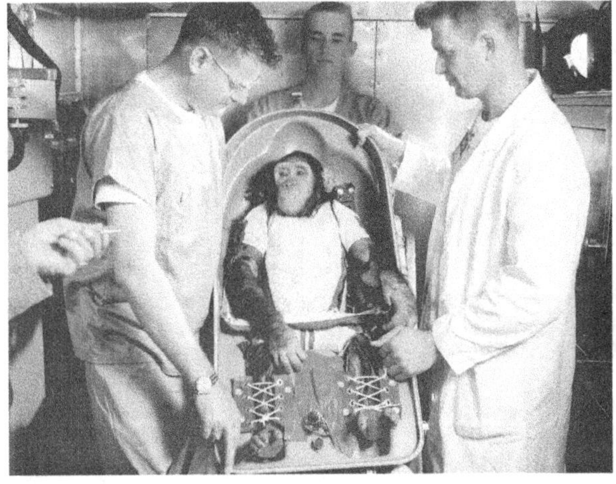

Mercury-Redstone 2: Ham, 1961

One of the most famous chimpanzee missions involved Ham, Mercury-Redstone 2: On January 31, 1961, the mission carried a chimpanzee named Ham. This 16-minute and 39-second flight was critical in validating the spacecraft's life support and environmental control systems for human spaceflight. Ham's successful mission demonstrated that a living organism could survive and function during a suborbital flight.

Mission Preparation: Ham underwent extensive training to prepare for his flight. He was taught to perform simple tasks using levers and buttons, allowing scientists to monitor his ability to function under the stress of spaceflight. This training was crucial in validating that the spacecraft's life support and environmental control systems could support a living organism capable of performing tasks.

Launch and Flight: During the flight, Ham experienced the intense g-forces of launch, weightlessness, and high reentry temperatures. The Mercury-Redstone 2 rocket propelled Ham and his spacecraft to an altitude of 157 miles and a speed of over 5,800 miles per hour.

Atlas - with spacecraft mounted - on launch pad at Launch Complex 14

In-Flight Performance: Ham's ability to perform tasks during the flight provided critical data on how spaceflight conditions affected a living organism. His successful performance demonstrated that the spacecraft's environmental systems could maintain conditions suitable for life and that an organism could function effectively in the space environment.

Recovery: After a 16-minute flight, Ham's capsule safely splashed down in the Atlantic Ocean, where a rescue team quickly recovered him. Ham emerged in good health, with only a slight bruise and dehydration from the experience. The successful recovery was a testament to the spacecraft's design and the effectiveness of NASA's recovery operations.

Ham's mission was a significant milestone in the Mercury program. His successful flight gave NASA confidence that the Mercury spacecraft was ready to support human astronauts, paving the way for the first American manned spaceflights.

Mercury-Redstone 2

Spacecraft No.: 5
Launch Date: January 31, 1961
Duration: 16 minutes 39 seconds
Purpose: Qualification of spacecraft with a chimpanzee named Ham.
Result: Success

This flight, which carried a chimpanzee named Ham, qualified the spacecraft for human spaceflight by demonstrating its systems under real flight conditions.

In February 1961, the Space Task Group's relentless efforts propelled Project Mercury forward, laying essential groundwork for America's quest for manned spaceflight.

On February 3, Eagle-Picher Company initiated a rigorous 13-week life-cycle test on Mercury spacecraft batteries. These tests aimed to ensure the batteries' reliability under the harsh space conditions, critical for sustaining spacecraft operations throughout missions.

February 10 marked significant milestones with the publication of mission rules for Mercury-Redstone 3 (MR-3), Alan Shepard's upcoming historic flight. This document underwent revisions on February 27 and April 28, reflecting ongoing adjustments in mission planning to optimize safety and mission success. Concurrently, discussions on hydrogen-peroxide fuel economy for spacecraft attitude control systems highlighted efforts to refine operational efficiency, guided by astronaut Virgil Grissom's insights.

James E. Webb's swearing-in as NASA Administrator on February 15 signaled a new era of leadership, poised to invigorate NASA's ambitious agenda and advance its pioneering space programs.

Practical concerns for post-landing safety were addressed on February 17, as the Space Task Group requested McDonnell to design and install a manual bilge pump in spacecraft No. 7. This innovation aimed to mitigate seawater ingress after landing, ensuring astronaut safety during recovery operations.

Mercury-Atlas 2

Spacecraft No.: 6
Launch Date: February 21, 1961
Duration: 17 minutes 56 seconds
Purpose: Qualified Mercury/Atlas interface.
Result: Success

This mission successfully qualified the Mercury/Atlas interface, ensuring the combination was ready for manned orbital flights.

Preparations intensified for the Mercury-Atlas 2 (MA-2) mission, culminating in its launch from Cape Canaveral on February 21. This mission tested reentry conditions under maximum heating effects, crucial for assessing spacecraft resilience during emergencies. The successful mission, reaching altitudes and ranges conducive to rigorous testing, validated mission objectives and readiness for future manned flights.

Meanwhile, extensive studies on spacecraft launch vehicle separation during emergencies were completed by February 23, ensuring meticulous preparation for unforeseen contingencies. This included a detailed analysis of separation distances and escape rocket performance, essential for astronaut safety.

February also saw progress with spacecraft No. 9 delivered for the Mercury-Atlas 5 (MA-5) orbital primate mission, setting the stage for chimpanzee Enos's pioneering flight into orbit. McDonnell's successful drop test of a boilerplate spacecraft further underscored technological advancements, confirming the spacecraft's structural integrity under simulated landing conditions.

Amidst these advancements, qualification tests commenced for the orbital psychomotor tester to enhance astronaut training and safety measures. Training sessions also focused on personal parachute use, crucial for the upcoming Mercury-Redstone 3 (MR-3) mission led by Alan Shepard, preparing astronauts for the challenges of spaceflight.

These developments in February 1961 pushed the boundaries of scientific exploration. They cemented America's pioneering role in the global race to conquer space, setting the stage for historic achievements in the years to come.

In March 1961, the intensive efforts of the Space Task Group propelled Project Mercury forward, marking significant advancements and preparations for America's ambitious space exploration goals.

On March 2, the Mercury-Atlas 2 (MA-2) flight data evaluation revealed lower-than-anticipated temperatures on the spacecraft's afterbody. This discovery provided crucial insights into thermal dynamics in space, guiding future mission designs.

On March 3, Atlas launch vehicle No. 100-D, earmarked for the Mercury-Atlas 3 (MA-3) mission, underwent a pivotal factory roll-out inspection at Convair-Astronautics. This milestone affirmed progress in ensuring mission readiness, a cornerstone of NASA's meticulous planning.

March 6 saw the publication of the "Detailed Test Objectives for NASA Mission MA-3," delineating mission goals and parameters. Concurrently, the third development engineering inspection for Mercury spacecraft (Nos. 12 and 15) yielded approximately 50 design improvement requests, underscoring ongoing efforts to enhance spacecraft functionality.

Preparations intensified for manned missions, as Spacecraft No. 11 was delivered to Cape Canaveral on March 7 for the Mercury-Redstone 4 (MR-4) flight with astronaut Virgil Grissom. Simultaneously,

Redstone launch vehicle No. 5 was readied for the Mercury-Redstone Booster Development (MR-BD) flight, pivotal steps toward manned suborbital missions.

On March 8, Spacecraft No. 10 passed acceptance tests and was dispatched to McDonnell's altitude test facility for orbital-flight environmental testing, ensuring it met rigorous space-condition criteria.

Atlas launch vehicle 100-D's delivery to Cape Canaveral on March 14 further advanced preparations for MA-3, signaling NASA's readiness for the next phase of orbital missions.

Strategic decisions continued to shape mission planning, as on March 16, the Space Task Group recommended weather reconnaissance missions preceding Mercury orbital launches, beginning with MA-4. This initiative aimed to optimize launch conditions, crucial for mission success.

Operational tests and evaluations persisted, highlighted by the launch of Little Joe 5A (LJ-5A) from Wallops Island on March 18. Despite an early tower firing, the spacecraft sustained only superficial damage, underscoring its structural integrity under high-pressure escape scenarios.

Little Joe 5A

Spacecraft No.: 14
Launch Date: March 18, 1961
Duration: 5 minutes 25 seconds
Purpose: Second test of escape system with a production Mercury spacecraft.
Result: Partial success

A second test of the escape system with a production spacecraft achieved partial success, contributing to the system's refinement.

Mercury-Redstone BD

Spacecraft No.: Boilerplate
Launch Date: March 24, 1961
Duration: 8 minutes 23 seconds
Purpose: Final Redstone test flight.
Result: Success

This final Redstone test flight was successful, validating the system for manned missions.

Mercury-Atlas 3

Phase III of the spacecraft airdrop program commenced on March 20, evaluating spacecraft flotation and impact resistance. These tests provided invaluable data for refining recovery procedures and spacecraft design.

Organizational developments mirrored technical advancements. On March 21, the Mercury-Atlas Missile Range Projects Office was integrated into the Space Task Group Director's office, streamlining project management and oversight.

Amidst these achievements, President John F. Kennedy reaffirmed NASA's leadership in space exploration on March 23, assuring congressional leaders of the agency's autonomy and pivotal role in advancing national space efforts.

March 1961 encapsulated a period of intensive progress and meticulous preparation for Project Mercury, underscoring NASA's commitment to pushing the boundaries of human exploration. Each milestone and lesson learned brought America closer to its historic goal of manned spaceflight, setting the stage for future triumphs in space exploration.

March 24, 1961, marked a pivotal moment at the Marshall Space Flight Center, where engineers and officials gathered to evaluate the outcomes of recent Mercury-Redstone missions. The atmosphere in the conference room was tense yet determined, illuminated by the clinical glow of fluorescent lights. Charts and graphs meticulously detailed Mercury-Redstone 1A (MR-1A) and Mercury-Redstone 2 (MR-2) performances.

This critical review session assessed the successes and challenges during these missions, providing valuable insights for future Mercury flights. Engineers and mission planners analyzed data on launch vehicle performance, spacecraft functionality, and mission objectives to refine strategies and enhance mission readiness.

The Marshall Space Flight Center discussions underscored NASA's commitment to meticulous planning and continuous improvement in its quest to achieve manned spaceflight. Whether successful or encountering setbacks, each mission contributed to advancing the technical and operational capabilities needed for human space exploration.

Dr. Wernher von Braun, a commanding figure with piercing blue eyes and a distinctive German accent, addressed the crowd with gravity. "We have identified several critical issues," he stated, his words cutting through the murmurs of the assembled team. "Jet-vane vibration, instrumentation compartment vibration, and thrust-controller system failures are paramount concerns. Resolving these is essential before proceeding with manned flights."

Engineers nodded thoughtfully, their brows furrowed as they took notes. Collaborative discussions filled the room, spawning debates and proposals. Whiteboards quickly filled with intricate equations and diagrams as solutions were brainstormed and refined. Despite the weight of the challenges, a collective determination permeated the atmosphere.

The Mercury-Redstone Booster Development test was swiftly scheduled, and its success was pivotal for certifying the launch vehicle for manned missions. A wave of relief swept through the center when the test objectives were met, confirming readiness for human spaceflight. Applause erupted, marking a significant milestone in NASA's quest to send humans beyond Earth's atmosphere.

Three days later, on March 27, 1961, NASA Headquarters buzzed with anticipation as a memo circulated among magazine and newspaper editors. Procedures for accrediting correspondents to cover the upcoming Mercury-Redstone 3 (MR-3) mission were finalized. By April 24, 350 correspondents had been accredited to witness and report on Project Mercury's first manned suborbital flight—a mission that captured global attention and anticipation.

On March 30, 1961, Cape Canaveral bustled with purpose as Redstone launch vehicle No. 7 rolled in for the MR-3 mission. The colossal rocket stood tall against the Florida skyline, its sleek form reflecting the morning sun, symbolizing the pinnacle of human engineering and resolve. A team of dedicated engineers meticulously surrounded the rocket, verifying every detail and fine-tuning its components. Their expressions mirrored a blend of intense focus and eager expectation, knowing that this mission marked a significant stride forward in the nation's ambitious journey into space exploration.

On March 31, 1961, a pivotal moment unfolded as the global Mercury tracking network officially commenced operations. Engineers from Western Electric Company ceremoniously handed over the $60 million network to NASA, marking a critical milestone in the annals of space exploration. Comprising a sprawling network of stations strategically positioned around the globe, this technological marvel was indispensable for tracking and maintaining communication with spacecraft during Project Mercury missions. The event underscored the network's vital role in the project's success and heralded a new era of collaborative efforts in advancing humanity's reach beyond Earth's boundaries.

Project Expectations

President Eisenhower and NASA Administrator T. Keith Glennan initially believed that achieving the first manned spaceflight would give the United States a significant advantage in the space race and potentially lead to a decisive end of the competition with the Soviet Union. This belief was rooted in the idea that demonstrating American technological prowess and capability in manned spaceflight would showcase superiority in scientific and engineering achievements.

However, the Soviet Union's successful launch of Yuri Gagarin aboard Vostok 1 on April 12, 1961, changed the dynamics of the space race. Gagarin became the first human to orbit the Earth, marking a major milestone for Soviet space exploration and intensifying the competition between the two superpowers.

The United States responded with renewed determination, accelerating efforts to catch up and eventually surpass Soviet achievements. President Kennedy's commitment to landing a man on the Moon by the end of the 1960s further galvanized NASA's efforts, leading to the successful Apollo program and Neil Armstrong's historic Moon landing in 1969.

Thus, while early optimism suggested that winning the race with the first manned flight could potentially end the competition, the reality proved to be a more complex and protracted rivalry, ultimately driving significant advancements in space exploration for both nations.

On April 2, 1961, a pivotal event unfolded within the confines of the Space Task Group facility as engineers conducted the first simulated orbital mission in an altitude chamber. Inside the chamber, the metallic walls reverberated with the steady hum of machinery as the spacecraft underwent rigorous testing under simulated orbital conditions. Dedicated engineers meticulously monitored data streams, their eyes fixed on screens displaying critical metrics, ensuring every system operated flawlessly. This meticulous preparation was crucial, laying the groundwork for future manned missions and reinforcing the nation's commitment to conquering the challenges of space exploration.

On April 3, 1961, responding to increasing public curiosity about Project Mercury, Robert R. Gilruth, Director of the Space Task Group, designated the Public Affairs Office as the official liaison for media inquiries. This strategic decision aimed to ensure transparency and provide timely, accurate information to the press while upholding stringent security protocols. By centralizing communications through the Public Affairs Office, Gilruth underscored NASA's commitment to sharing the achievements and challenges of Project Mercury with the world, fostering public support and understanding during this pivotal era of space exploration.

On April 4, 1961, the Aviation Medical Acceleration Laboratory buzzed with activity as astronauts John Glenn, Virgil Grissom, and Alan Shepard embarked on a rigorous refresher course with the centrifuge. Inside the chamber, the whirring of the centrifuge and the intense pressures of high G-forces subjected the astronauts to demanding physical tests. Undeterred, their unwavering focus reflected their determination to prepare meticulously for the upcoming first manned suborbital flight.

Simultaneously, Mercury spacecraft No. 14A arrived at Wallops Island for the Little Joe 5B (LJ-5B) mission. Freshly refitted following its earlier role in the Little Joe 5A mission, the spacecraft was meticulously readied for its next critical test under conditions of maximum dynamic pressure. This dual effort underscored NASA's meticulous approach to advancing both human endurance in space and the technological readiness of spacecraft, marking another significant stride towards the historic milestones of Project Mercury.

On April 12, 1961, the global space race entered a new phase as the Soviet Union stunned the world with the announcement of Major Yuri A. Gagarin's triumphant orbit of the Earth aboard Vostok 1. In a historic 108-minute flight, Gagarin became the first human to achieve this monumental feat. The reverberations of this achievement echoed worldwide, intensifying the competitive pressure on America's efforts within Project Mercury.

Within the corridors of NASA, the news sparked a renewed sense of determination tinged with the urgency of competition. The United States, spurred by this Soviet milestone, redoubled its efforts to accelerate Project Mercury, aiming to match and surpass the achievements of its Cold War rivals. This pivotal moment underscored the relentless pursuit of space exploration and marked a critical juncture in the quest to conquer the cosmos.

On April 18, 1961, the United States Weather Bureau made a crucial announcement, requesting $200,000 in funding to support Project Mercury for the fiscal year 1962. This financial support was essential to enhance weather forecasting capabilities vital to the mission's success. Accurate weather predictions were imperative as they could mean the difference between mission success and potential disaster, ensuring safe launch conditions and mission trajectories for astronauts embarking on America's pioneering ventures into space.

On April 20, 1961, the Space Task Group personnel meticulously reviewed spacecraft, mission, and launch vehicle flight safety. The atmosphere in the room buzzed with the rustling of papers and a steady hum of focused conversation as engineers meticulously scrutinized every aspect. Their thorough examination left no detail to chance, aiming to guarantee the utmost safety and reliability for upcoming missions. This rigorous evaluation underscored NASA's commitment to precision and preparedness to advance human exploration beyond Earth's confines.

On April 25, 1961, the launch of Mercury-Atlas 3 (MA-3) from Cape Canaveral unfolded amid tense anticipation. Shortly after liftoff, the rocket encountered critical issues, failing to achieve the correct heading and pitch. Promptly, the abort-sensing system triggered the escape rockets just moments before the range safety officer was compelled to destroy the malfunctioning launch vehicle. Despite this setback, the spacecraft, housing its "mechanical astronaut," continued its trajectory to a higher altitude, eventually deploying parachutes and safely landing in the Atlantic Ocean. The spacecraft sustained only superficial damage, yielding invaluable data crucial for refining future missions.

Simultaneously, on this pivotal day, President Kennedy underscored the nation's commitment to space exploration by signing legislation appointing the Vice President as the presiding officer of the National Aeronautics and Space Council. This legislative action underscored the elevated priority of space exploration within national policy, emphasizing its strategic importance and signaling America's steadfast dedication to advancing scientific discovery and technological prowess in the burgeoning era of space exploration.

On April 28, 1961, significant milestones marked advancements in America's space program. At Wallops Island, the launch of Little Joe 5B (LJ-5B) tested the Mercury escape system under extreme dynamic pressure conditions. Despite a delayed ignition of one rocket motor that altered its trajectory, the spacecraft demonstrated resilience by enduring higher-than-expected pressures. The successful test, followed by the spacecraft's helicopter recovery, underscored a major triumph for the Mercury program, validating crucial safety measures for future manned missions.

Simultaneously, a flawlessly executed simulated countdown for the MR-3 mission represented another pivotal achievement. This simulation brought the team one step closer to realizing the historic manned flight, refining procedures, and bolstering confidence as America prepared to embark on its first human venture into space. These dual accomplishments highlighted NASA's steady progress and determination to overcome challenges on the path to pioneering space exploration.

The competition between the superpowers was fierce. The Soviet Union won early by sending Yuri Gagarin into orbit on April 12, 1961. The world watched in awe as Gagarin completed a single orbit around Earth, cementing the Soviet Union's lead in the Space Race. The United States quickly followed, launching Alan Shepard on a suborbital flight on May 5, 1961. Although Shepard's flight did not reach orbit, it demonstrated America's growing capabilities in space travel.

Spacecraft No.: 8

Launch Date: April 25, 1961
Duration: 7 minutes 19 seconds
Purpose: Orbital flight with robot astronaut.
Result: Failure
An attempt to conduct an orbital flight with a robotic astronaut ended in failure, underscoring the challenges of achieving stable orbits.
Little Joe 5B

Spacecraft No.: 14

Launch Date: April 28, 1961
Duration: 5 minutes 25 seconds
Purpose: Third test of escape system with a production spacecraft.
Result: Success
The third test of the escape system with a production spacecraft was successful, bolstering confidence in the system's reliability.

On May 5, 1961, plans were set to evaluate the Mercury tracking and real-time computing system using a Scout test vehicle. NASA Headquarters gave tentative approval later in the month, ensuring the system's readiness for upcoming missions, where precision tracking and computing were critical for success.

The sun rose over Cape Canaveral, casting long shadows across the sprawling launch complex. There was a sense of electric anticipation in the air, a tension that rippled through the crowds gathered to witness history. Mercury-Redstone 3 (MR-3), or Freedom 7, stood proudly on the launch pad, symbolizing American ambition and technological prowess. The silver body of the rocket gleamed, standing tall against the blue Florida sky.

Astronaut Alan Shepard, the chosen pilot for this first manned suborbital flight, was a picture of calm determination as he was suited up. The bulky suit encased him, layers of protection against the unforgiving environment of space. His heart raced slightly, a mix of excitement and the gravity of the mission at hand.

As the countdown reached zero, a roar erupted from the Redstone booster. The ground trembled, and a plume of smoke and flame billowed out, propelling Freedom 7 skyward. Inside the capsule, Shepard felt the vibrations, the force pressing him into his seat. The rocket performed admirably despite some vibrations, and the cutoff occurred within the specified limits.

"All systems are go," Shepard's voice crackled over the communication lines. His fingers danced over the controls, switching the spacecraft to manual control. He tested the fly-by-wire and manual proportional modes, and each response was precise and smooth. The attitude control system hummed in response, the thrusters firing as needed with minimal fuel leaks.

The spacecraft soared to a peak altitude of 116.5 statute miles, a breathtaking view of the curvature of the Earth unfolding before Shepard's eyes. For approximately five minutes, he experienced weightlessness, floating within the confines of the small capsule. Then, as the spacecraft descended, the g-forces returned with a vengeance. Shepard felt his body press hard against the seat, experiencing up to 12 g's upon reentry.

The Atlantic Ocean came into view, and Freedom 7's parachutes deployed as planned. Helicopters tracked the descent, their rotors chopping through the air as they closed in on the splashdown site. Recovery teams were ready, and within two minutes of impact, they contacted Shepard. The capsule was hoisted from the water, and Shepard emerged unscathed and triumphant. The mission was a success, a flawless testament to human ingenuity and courage.

On May 8, 1961, the White House's East Room glittered with the flash of cameras as President John F. Kennedy presented NASA's Distinguished Service Medal to Alan Shepard. The ceremony was a moment of national pride, attended by dignitaries and reporters alike. President Kennedy's words echoed the bravery of Shepard's pioneering flight into space, marking a historic milestone in America's journey into the cosmos.

Three days later, on May 11, 1961, Cape Canaveral hummed with purpose as engineers meticulously prepared Mercury spacecraft 8A for its role in the upcoming Mercury-Atlas 4 (MA-4) mission. Technicians fine-tuned every detail, ensuring the spacecraft was primed for its pivotal unmanned flight—a critical step toward manned orbital missions. The atmosphere crackled with anticipation, underscoring America's ambitious strides in space exploration.

Meanwhile, on May 13, 1961, NASA submitted its legislative program to the 87th Congress, seeking new authorities crucial for its operations. These included property leasing rights, patent acquisitions, and contractor indemnity against hazardous risks—essential measures to streamline operations and foster innovation.

Just days later, on May 17, 1961, an Atlas investigation board convened to analyze the Mercury-Atlas 3 (MA-3) mission's failure. Delving into test data, the board identified three likely causes for the launch vehicle's malfunction, signaling a somber but determined effort to rectify issues and ensure future success.

Simultaneously, on May 19, 1961, NASA Headquarters and the Space Task Group initiated a comprehensive review of Mercury's

advancements. This effort aimed to identify breakthrough technologies that could benefit other government agencies and American industries, promoting a culture of collaboration and continuous improvement.

From May 23-24, 1961, engineers at McDonnell conducted the fourth development engineering inspection on Mercury spacecraft No. 18. The meticulous examination yielded 45 requests for modifications, ensuring each component met rigorous standards for human spaceflight.

On May 25, 1961, President Kennedy addressed Congress with a visionary pledge: to land a man on the moon and return him safely to Earth. This ambitious declaration underscored America's commitment to space exploration, and it was accompanied by a request for $611 million in appropriations for NASA and the Department of Defense.

The first "Peaceful Uses of Space" conference convened in Tulsa, Oklahoma, from May 26-27, 1961, gathering scientists, engineers, and policymakers. Robert R. Gilruth highlighted Project Mercury's progress, emphasizing its potential to advance mankind's understanding of space.

Meanwhile, the Mercury spacecraft Freedom 7 (MR-3) captivated over 650,000 visitors at the Paris International Air Show from May 26 to June 4, 1961. Displayed as a testament to technological prowess, it showcased Alan Shepard's historic flight and the marvels of engineering that made it possible.

Lastly, from May 29 to June 30, 1961, astronauts underwent rigorous centrifuge training at the Aviation Medical Acceleration Laboratory. Enduring intense G-forces simulated by the centrifuge, they prepared for the challenges of Mercury-Atlas orbital missions, demonstrating unwavering determination in the face of daunting conditions.

On June 1, 1961, as preparations intensified for the Mercury-Atlas 4 (MA-4) mission, stringent prelaunch mission rules were published. These guidelines meticulously outlined procedures and protocols essential for ensuring a flawless launch, underscoring the meticulous planning required for every aspect of the mission.

A significant milestone followed on June 6, 1961, in Washington, where a joint conference sponsored by NASA, the National Institute of Health, and the National Academy of Sciences reported on the biomedical findings from Alan Shepard's historic flight. These findings were pivotal in understanding the physiological impacts of space travel, providing crucial insights for future missions.

Published on June 8, 1961, recovery requirements for the MA-4 mission detailed protocols for safely retrieving the spacecraft post-flight, emphasizing safety measures for the spacecraft and potential occupants.

On June 12, 1961, Cape Canaveral received Redstone launch vehicle No. 8, marking progress toward the Mercury-Redstone 4 (MR-4) suborbital flight mission— a pivotal step in advancing human space exploration.

Simultaneously, on June 13, 1961, the Space Task Group forwarded details of the Mercury-Scout instrumentation system to NASA Headquarters. This mission aimed to evaluate the operational effectiveness of the Mercury global tracking network, critical for ensuring mission success.

From June 13-25, 1961, the Freedom 7 spacecraft was showcased at the Rassegna International Electronic and Nuclear Fair in Rome, Italy, attracting approximately 750,000 visitors. This display symbolized technological achievement and international collaboration in the realm of space exploration.

On June 16, 1961, an Ad Hoc Task Group reported to NASA on the key challenges and decisions necessary for a manned lunar landing mission, including intensive studies on the direct ascent method—a pivotal precursor to President Kennedy's visionary lunar exploration goals.

Meanwhile, from June 21 to July 15, 1961, chimpanzees underwent rigorous training at the University of Southern California as part of the Mercury-Atlas animal program. This training included exposure to noise, vibration, centrifuge runs, and weightlessness aboard a C-131 aircraft, preparing them for their crucial role in space research.

On June 22, 1961, the Navy was forwarded recovery requirements for the Mercury-Redstone 4 (MR-4) mission. The Redstone booster was erected on Pad 5 at Cape Canaveral, a visible testament to ongoing preparations for manned spaceflight.

On June 24, 1961, the spacecraft designated for the second manned suborbital Mercury flight underwent modifications, including replacing viewports with observation windows and installing an improved manual control system, which enhanced functionality and safety.

A seaworthiness test conducted east of Wallops Island on June 28, 1961, confirmed spacecraft No. 5's satisfactory flotation characteristics in varying sea conditions, affirming its resilience for maritime operations.

Concurrent discussions between the Space Task Group and Goddard Space Flight Center personnel focused on tracking network enhancements for extended-range Mercury missions, ensuring robust support for Project Mercury's ambitious objectives.

Finally, on June 29-30, 1961, Convair conducted a meticulous factory roll-out inspection of Atlas launch vehicle 88-D, slated for the MA-4 mission. This thorough examination ensured the vehicle met rigorous specifications, highlighting the collaborative effort and dedication of the 794-strong Space Task Group personnel in propelling humanity further into space.

Throughout 1961, as Project Mercury advanced, Robert R. Gilruth and James E.

Webb implemented a tradition allowing each astronaut to name their spacecraft. This personalization added a poignant touch to the missions, reflecting the spirit of exploration and the quest for human freedom in space. For the inaugural flight, the spacecraft was christened "Freedom 7" by the astronaut, symbolizing the aspirations of mankind beyond Earth's bounds. The name was officially registered with the Federal Communications Commission, underscoring the spacecraft's integration into the national space exploration effort.

The journey towards manned spaceflight during this period was marked by challenges, triumphant achievements, and the steadfast dedication of countless individuals committed to expanding the frontiers of human potential. Each milestone, from technical advancements to personal touches like spacecraft names, contributed to the profound narrative of America's pioneering efforts in space exploration.

Mission Profile - Suborbital Missions

The suborbital missions of Project Mercury were a pivotal step in America's journey to explore the cosmos. These missions utilized the formidable Redstone rocket to propel the Mercury capsule on a brief yet significant voyage. The rocket ignited with a thunderous roar, its liquid-fueled engine burning alcohol and liquid oxygen to produce an impressive 75,000 pounds-force (330 kN) of thrust. Though insufficient for achieving orbit, this thrust was more than adequate for the suborbital flights that formed the foundation of early human space exploration.

At liftoff, the Redstone rocket stood tall at 83 feet (25 meters), with the capsule and escape system attached. It carried the Mercury capsule upwards for the first two minutes and thirty seconds, reaching 32 nautical miles (59 km). At this point, the booster separated from the capsule, which continued to ascend on a ballistic trajectory. The launch escape system, having served its purpose, was jettisoned shortly after.

The retrorockets were fired as the capsule approached its apogee—the highest point in its flight. These rockets were tested during suborbital missions, although they were not needed for reentry since orbital speed was not achieved. The capsule then began its descent, eventually landing in the Atlantic Ocean. The entire suborbital mission, from liftoff to splashdown, lasted approximately 15 minutes. During this brief yet intense period, the capsule reached an apogee altitude of 102 to 103 nautical miles (189 to 191 km) and traveled a downrange distance of 262 nautical miles (485 km).

Throughout the ballistic flight, the astronaut inside experienced weightlessness, a sensation that began with the booster's separation and lasted until reentry. As the capsule reentered the atmosphere, atmospheric drag gradually slowed its descent. The drogue parachute deployed around 12 minutes into the flight, followed by the main parachute two minutes later, ensuring a safe splashdown at the 15-minute mark.

The Mercury-Redstone Launch Vehicle itself was an engineering marvel. Derived from the German V-2 rocket, seized and repurposed by the U.S. Army after World War II, the Redstone was modified extensively for space missions. Engineers removed the warhead and added a specially designed collar to support the spacecraft. They also incorporated additional materials to dampen the intense vibrations experienced during launch, ensuring the safety and stability of the spacecraft and its human cargo.

North American Aviation produced the rocket motor, which included fins capable of altering the rocket's direction during flight. These fins could direct airflow or thrust—or both simultaneously—providing crucial control over the rocket's trajectory. A critical safety feature of the Mercury-Redstone was its automatic abort sensing system. This system could activate the launch escape mechanism if any anomalies were detected during the flight, offering vital protection for the astronauts onboard.

While the Jupiter rocket was considered for intermediate suborbital flights, the proposal was ultimately abandoned. Developed by Wernher von Braun's team at the Redstone Arsenal, the Jupiter was deemed too expensive to modify for manned missions. Instead, NASA opted for the Atlas rocket for later missions, a decision driven by economic practicality.

Visualizing a Mercury-Redstone launch conjures a scene of intense preparation and anticipation. With its sleek and towering presence, the rocket stands ready on the launch pad, surrounded by a hive of activity. Engineers and technicians perform their final checks, ensuring every component is primed for the mission. The astronaut awaits the countdown inside the capsule, encapsulated in a small yet robust spacecraft.

Countdown and Launch Procedures

The countdown procedure for Mercury and subsequent space missions typically involved several stages of control and monitoring:

Blockhouse Control: The initial phases of the countdown were managed from the blockhouse located near the launch pad. This facility housed personnel responsible for overseeing the final preparations of the spacecraft, ensuring that all systems were ready for launch.

Transfer to Mission Control: Approximately two minutes before launch, control of the countdown was transferred to the Mission Control Center. This transition allowed for continuous monitoring and coordination of the launch sequence from a centralized location.

Final Countdown: Mission controllers in Mission Control conducted the countdown in the final moments leading up to launch. This included giving the astronaut, who would be

inside the spacecraft preparing for liftoff, the final 10-second countdown.

Inside Control Center at Cape Canaveral (Mercury-Atlas 8)

Broadcasting: The final countdown, particularly the last 10 seconds, was often broadcast live on television and radio, allowing the public to witness the critical moments before launch. This broadcast helped to build anticipation and awareness of the space missions among the general public.

This procedure ensured that all aspects of the countdown and launch were carefully monitored and managed, with clear communication channels established between ground control and the astronaut. The transfer of countdown control from the blockhouse to Mission Control marked a pivotal moment in the launch sequence, transitioning from local oversight to centralized command as the spacecraft prepared to embark on its mission into space.

As the clock ticks down, anticipation builds. At zero, the powerful liquid-fueled engine ignites, shaking the ground and filling the air with the roar of thrust. The rocket begins its ascent, a blazing trail marking its path through the sky—a bold step in humanity's journey into space. This moment, marked by meticulous engineering and brave exploration, signifies the dawn of a new era in space exploration that would eventually lead to humanity's footprints on the Moon.

Orbital Missions

The countdown for orbital flights in Project Mercury was a meticulously choreographed process that began 6.5 hours before launch (T - 390 min) and continued until the spacecraft achieved orbital insertion (T + 5 min). Preparations for these groundbreaking missions started a month in advance, with the selection of the primary and backup astronauts. These astronauts trained rigorously together, honing their skills and preparing for the mission's demands.

Three days before the scheduled launch, the astronaut adhered to a special diet designed to minimize the need for defecation during the flight, a practical yet crucial aspect of mission preparation. The astronaut's routine began on launch day with a hearty steak breakfast. Following this, medical sensors were applied to monitor vital signs, and the astronaut donned the iconic pressure suit. Breathing pure oxygen, the astronaut acclimatized to the spacecraft's atmosphere, reducing the risk of decompression sickness.

The astronaut arrived at the launch pad two hours before launch, a scene bustling with anticipation and activity. Ascending the launch tower via elevator, the astronaut entered the spacecraft, where final preparations occurred. Once securely fastened inside the Mercury capsule, the hatch was bolted shut, the launch area was evacuated, and the mobile tower was rolled back. With these steps completed, the launch vehicle was fueled with liquid oxygen, a critical final preparation for liftoff.

For orbital missions, Project Mercury relied on the Atlas LV-3B, a man-rated adaptation of the Atlas D originally developed by Convair for the Air Force as the United States' first operational intercontinental ballistic missile (ICBM) in the mid-1950s. This rocket was pivotal in America's effort to send astronauts into orbit, marking a significant leap in space exploration capabilities.

The Atlas rocket was a distinctive "one-and-one-half-stage" vehicle powered by kerosene and liquid oxygen (LOX). Standing 67 feet (20 meters) high on its own, the combined height of the Atlas-Mercury space vehicle reached an impressive 95 feet (29 meters) at launch. This towering structure was a marvel of engineering, designed to withstand space travel's immense forces and complexities.

The rocket's first stage featured a booster skirt equipped with two powerful liquid-fuel engines, providing the initial thrust necessary to lift the vehicle off the ground. The second stage, the sustainer stage, contained a single engine that continued to power the rocket after the first stage completed its task. Notably, both stages fired simultaneously at liftoff, with the thrust from the second-stage sustainer engine passing through an opening in the first stage.

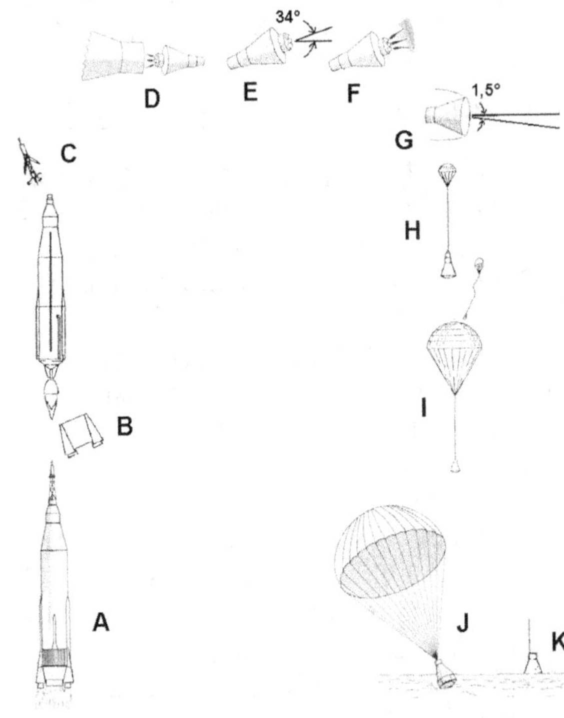

Launch and reentry profiles: A-C: launch; D: orbital insertion; E-K: reentry and landing

As the rocket ascended, the first stage separated, allowing the sustainer stage to continue its journey alone. This stage was guided by gyroscope-controlled thrusters, ensuring precise navigation as the vehicle approached orbit. To further refine its trajectory and maintain stability, smaller vernier rockets were attached to the sides of the Atlas, providing fine-tuned maneuvering capabilities essential for the delicate orbital insertion process.

Sensory Experience

Imagining the launch of the Atlas LV-3B evokes a scene of intense preparation and coordination. The rocket, gleaming under the sunlight, stands poised on the launch pad, surrounded by the meticulous work of engineers and scientists—the best minds of their generation. Every detail is scrutinized, ensuring all systems are go. Inside the Mercury capsule, the astronaut is securely fastened, a mix of anticipation and focus etched on their face, acutely aware of the mission's historic significance.

Mercury crewed launches

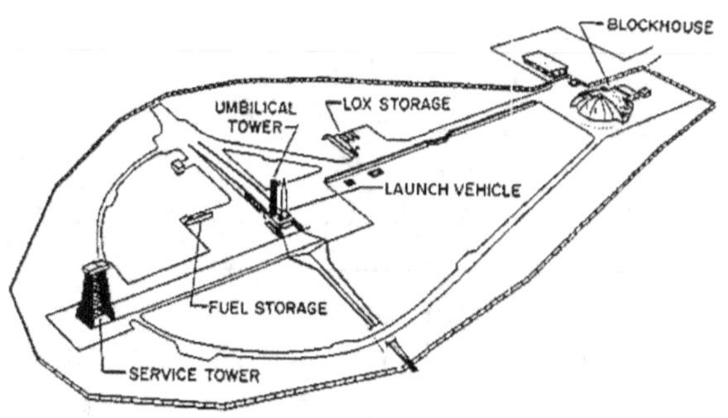

Launch Complex 14 just before launch (service tower rolled aside). Preparations for launch were made in the blockhouse.

The engines ignite as the countdown reaches zero, releasing a thunderous roar reverberating across the launch site. The ground trembles as the Atlas-Mercury space vehicle rises, propelled by a column of fire and smoke. The intricate ballet of stage separation and sustainer ignition unfolds flawlessly, a testament to cutting-edge technology and human ingenuity. This moment signifies not just a step forward in space exploration but a monumental leap in human achievement, showcasing the prowess of Project Mercury and the Atlas LV-3B in the relentless pursuit of the stars.

Inside the Mercury spacecraft, the atmosphere is tense and electrifying. Every sound is amplified in the cramped quarters—the hum of the environmental control system, the clicking of switches, and the faint creaking of the metal structure as it endures the launch forces. Strapped into the custom-fitted seat, the astronaut feels the intense vibrations and g-forces pressing them firmly back as the rocket roars to life and climbs through the atmosphere.

As the spacecraft ascends, the view outside the small portholes transitions from the deep blue of Earth's atmosphere to the inky blackness of space, speckled with stars. The astronaut's heart pounds with a mix of anticipation and awe, knowing they are among the first humans to experience such a journey. The intense forces and sensations during launch and reentry are both thrilling and harrowing, culminating in the relief and triumph of a successful splashdown.

The experience of being an astronaut aboard the Atlas LV-3B is unparalleled. The powerful rocket engines generate a deafening roar and a bone-rattling vibration that permeates the entire spacecraft. As the Atlas rocket lifts off, the astronaut feels the immense g-forces pressing them back into their seat, a testament to the raw power required to break free from Earth's gravitational pull.

During the ascent, the spacecraft's environmental control system works tirelessly to maintain a stable and habitable environment within the Mercury capsule. The astronaut hears the steady hum of the system, a constant reminder of the life-sustaining technology surrounding them. Each switch click and adjustment made by mission control are crucial elements of a meticulously choreographed ballet between human and machine.

As the rocket climbs higher, transitioning from the familiar blue sky to the vast, inky blackness of space is breathtaking and humbling. The small portholes offer glimpses of the cosmos, a stark contrast to the confined interior of the Mercury capsule. Stars, previously twinkling faintly in the night sky, now shine steadily against the dark void.

The journey is not without its challenges. The intense forces during launch test the limits of the astronaut and the spacecraft. Reentry brings its own set of trials, as the capsule faces the searing heat and immense pressure of returning to Earth's atmosphere. Yet, through it all, the astronaut remains focused and resolute, driven by the knowledge that they are part of a historic endeavor.

Finally, as the capsule descends toward the ocean, the parachutes deploy, slowing the spacecraft's descent. The tension of the mission culminates in the splashdown, where the astronaut's relief and triumph are palpable. Having endured the rigors of space travel and returned safely, they contribute to a pivotal chapter in human exploration. This moment, marked by meticulous engineering and brave exploration, signifies the dawn of a new era in space exploration that would eventually lead to humanity's footprints on the Moon.

Launch and Reentry Profiles

A-C: Launch

A: Four seconds before liftoff, the Atlas rocket engines ignited, their roar reverberating across the launch site. The vehicle, held to the ground by sturdy clamps, built up sufficient thrust before being released, propelling the rocket skyward with a tremendous force.

B: At 2 minutes and 10 seconds into the flight, the two outboard booster engines shut down and were jettisoned, leaving the center sustainer engine to continue powering the ascent. This separation was critical, shedding unnecessary weight and allowing the spacecraft to accelerate further into the upper atmosphere.

C: Following the booster jettison, the jettisoned rocket activated to separate the launch escape system from the spacecraft, streamlining the vehicle's journey into space.

Orbital Insertion

As the space vehicle ascended, it gradually shifted to a horizontal attitude. At an altitude of 87 nautical miles (161 km), the sustainer engine shut down, and the spacecraft was inserted into orbit precisely 5 minutes and 10 seconds after liftoff. Pointing east to leverage Earth's rotational speed, the postgrad rockets fired for a second to separate the spacecraft from the launch vehicle, setting it on a stable orbital path.

Orbital Insertion

The direction of insertion for Mercury missions and subsequent manned spaceflights typically aimed east and slightly northward. This trajectory was chosen strategically to optimize several key factors:

Tracking Network: By launching eastward from Cape Canaveral in Florida, the spacecraft could take advantage of the extensive tracking network available along the eastern seaboard of the United States and across the Atlantic Ocean. This network included ground-based radar stations and ships equipped with telemetry equipment, crucial for monitoring the spacecraft's trajectory and vital signs during its flight.

Orbital Dynamics: Launching eastward into a slightly inclined orbit northward helped achieve orbital parameters that could facilitate a safe trajectory and potential abort scenarios while allowing for flexibility in the mission profile.

Recovery Zones: The trajectory also ensured the spacecraft could safely splash down in designated recovery zones during an abort or mission termination. For Mercury missions, this often meant aiming for a landing in the North Atlantic Ocean, where recovery forces were stationed and ready to retrieve the astronauts and their spacecraft.

Overall, the east-northeast trajectory was carefully selected to maximize the efficiency of the mission's operations, from launch through to recovery, ensuring the safety and success of manned spaceflights during the early years of space exploration.

Sustainer Stage

After the launch of Friendship 7, which was the Mercury-Atlas 6 mission, parts of the sustainer stage of the Atlas rocket were indeed found in South Africa. The sustainer stage was a crucial component of the Atlas rocket, responsible for providing the main propulsion during the initial stages of the launch.

During the ascent phase of the mission, once the sustainer stage had expended its fuel and completed its role in propelling the spacecraft towards orbit, it would separate from the rest of the rocket. This separation was typically planned over the ocean to minimize risk to populated areas. However, in some cases, residual debris from the rocket stages could fall into various parts of the world.

In the case of Friendship 7, debris from the sustainer stage was discovered in South Africa after the launch. This occurrence highlights the global reach of space missions and the potential for rocket debris to land in unexpected locations despite careful planning and tracking efforts.

Attitude Control System

The Mercury spacecraft's Attitude Control System (ASCS) utilized small hydrogen peroxide thrusters to maintain the desired orientation and counteract any unintended drift. These thrusters provided precise control over the spacecraft's orientation in the vacuum of space.

During missions, especially those of longer durations where fuel conservation was crucial, there were periods when the spacecraft was intentionally allowed to drift. This strategy aimed to conserve fuel by minimizing the use of attitude control thrusters during non-critical phases of the mission. By strategically allowing the spacecraft to drift, mission planners could optimize fuel usage while maintaining operational readiness for critical maneuvers such as reentry orbital adjustments.

This approach balanced the need for precise control with conserving resources during extended missions. It showcased the careful planning and operational strategies employed by NASA to ensure the success and longevity of Mercury missions and beyond in human spaceflight.

Recovery Measures

After the success of the first orbital flight in the Mercury program, several initial recovery measures, such as radar chaff and a SOFAR bomb, were indeed deemed unnecessary for subsequent missions. Here's a breakdown of these recovery measures and their adjustments:

Radar Chaff: Radar chaff, which consists of small, reflective metal strips deployed to confuse radar systems, was initially considered for use during Mercury missions to help track and locate the spacecraft during reentry and descent. However, after the first orbital flight demonstrated successful tracking and recovery capabilities without chaff, it was determined that this measure was unnecessary and could be omitted from future missions.

SOFAR Bomb: The SOFAR (Sound Fixing and Ranging) bomb, designed to emit a sound pulse that could be detected acoustically to aid in locating the spacecraft upon splashdown, was also reconsidered after the first orbital flight. The success of recovery operations and the reliability of tracking systems rendered the SOFAR bomb unnecessary for subsequent missions, as other methods proved sufficient for pinpointing the spacecraft's location in the ocean.

Flotation Collar: A flotation collar, which helps stabilize the spacecraft and prevent it from capsizing upon water landing, was initially developed but not ready for use during early suborbital missions. This device became more critical for orbital missions where the spacecraft would land in open water after completing multiple orbits around the Earth. Its introduction was prioritized for subsequent orbital missions to enhance the safety and stability of the recovery process.

These adjustments reflect the iterative nature of space mission planning and operations, where lessons learned from each mission are applied to improve equipment readiness, streamline recovery procedures, and enhance overall mission success and astronaut safety in subsequent flights.

Capsule Exit

In the Mercury spacecraft, astronauts could indeed exit the capsule through the nose cylinder in emergencies. This feature was designed as a contingency method for astronaut egress in case the main hatch or other standard exit points were unavailable or compromised.

Scott Carpenter became the only Mercury astronaut to utilize this method during his flight on Aurora 7 in 1962. After splashdown in the Atlantic Ocean, Carpenter experienced difficulties with the main hatch's automatic release mechanism. As a result, he manually initiated the release of the nose cylinder and egressed through this alternate exit point. This allowed him to successfully exit the spacecraft and await recovery by the nearby USS Intrepid.

Carpenter's use of the nose cylinder exit highlighted the importance of contingency planning and the versatility of the Mercury spacecraft design in ensuring astronaut safety during potentially critical situations. This feature underscored NASA's commitment to astronaut safety and mission success throughout the early stages of human spaceflight.

Launch and Control Personnel

In the Mercury program, T.J. O'Malley was the Launch Conductor for John Glenn's historic flight aboard Friendship 7. Friendship 7 was launched from Launch Complex 14 (LC-14) at Cape Canaveral Air Force Station, Florida. O'Malley was crucial in overseeing the final countdown and launch preparations for this mission, ensuring all systems were ready for Glenn's orbital flight on February 20, 1962.

Reactions to Delays

The repeated delays in the launch of Friendship 7, which carried John Glenn on his historic orbital flight, did indeed cause frustration among many involved in the Mercury program. Senator John F. Kennedy famously compared it to a "Rube Goldberg device on top of a plumber's nightmare" during a congressional hearing in 1961. This metaphor highlighted the complexities and challenges in preparing for the first American orbital mission.

The delays were primarily due to technical issues and weather conditions. NASA and its contractors worked meticulously to ensure the spacecraft and launch vehicle were ready for the demanding task of orbiting the Earth. Despite these setbacks, John Glenn launched successfully on February 20, 1962, becoming the first American to orbit the Earth and a national hero.

Kennedy's colorful criticism underscored the pressure and scrutiny faced by NASA during the early days of manned spaceflight, emphasizing the high stakes of the space race and the importance of overcoming technical hurdles to achieve ambitious goals in space exploration.

Calvin D. Fowler, as the Site Manager and Launch Conductor at Complex 14, handled overseeing the launches of Scott Carpenter (Aurora 7), Wally Schirra (Sigma 7), and Gordon Cooper (Faith 7). Fowler's role involved managing LC-14's operational readiness, coordinating with mission control, and ensuring the safe and successful execution of these Mercury missions.

Both O'Malley and Fowler played pivotal roles in the Mercury program, contributing to the success of the manned spaceflights during this pioneering era of American space exploration. Their expertise and leadership were instrumental in preparing and executing these historic missions.

E-K: Reentry and Landing

E: To initiate reentry, the spacecraft fired its retrorockets at an angle of 34° downward from its flight path, beginning the descent back to Earth.

F: The retrorockets fired in sequence for 10 seconds each, slowing the spacecraft and altering its trajectory for reentry.

G: As the spacecraft plunged through the atmosphere, the astronaut experienced up to 8 g (11–12 g on suborbital missions), with the heat shield temperature soaring to 3,000 °F (1,600 °C). This intense heating caused a two-minute radio blackout due to the ionization of the surrounding air, a tense period of silence in mission control.

H: At 21,000 feet (6,400 meters), a small drogue parachute deployed, stabilizing the spacecraft's descent.

I: The main parachute deployed at 10,000 feet (3,000 meters). It initially opened with a narrow aperture, gradually expanding to reduce strain and ensure a controlled descent.

J: Just before splashdown, the landing bag inflated, cushioning the impact as the spacecraft hit the water.

K: Upon landing, the parachutes were released, and an antenna was raised to send out signals for recovery ships and helicopters. Green marker dye spread in the water made the location more visible from the air.

Recovery

Frogmen from helicopters swiftly inflated a collar around the capsule to keep it upright in the water. The recovery helicopter hooked onto the spacecraft, and the astronaut blew the escape hatch to exit. The astronaut was then hoisted aboard the helicopter, bringing the astronaut and the spacecraft to the recovery ship.

Sensory Experience

Inside the Mercury spacecraft, the astronaut experienced a symphony of intense sensations. During launch, the deafening roar of the engines and the powerful vibrations pressed the astronaut firmly into the custom-fitted seat. Through the small portholes, the view transitioned from the vibrant blue of Earth's atmosphere to the stark, inky blackness of space. After booster separation, the astronaut felt the surreal sensation of weightlessness, a sharp contrast to the high g-forces encountered during reentry. The temperature around the heat shield rose dramatically, adding to the physical strain and tension, especially during the brief radio blackout. The deployment of parachutes brought a sense of relief, and the gentle landing in the ocean marked the end of a harrowing yet triumphant journey. This blend of sensations underscored the historic nature of the mission, highlighting the challenges and achievements of early human spaceflight.

Ground Control

Inside the Control Center

The nerve center for Project Mercury was the Mercury Control Center at Cape Canaveral, Florida. This hub of activity was where mission control personnel meticulously managed each flight, their operations dominated by a large control board that displayed the spacecraft's position above the Earth. Inside the control center, the atmosphere was charged with tension and focus as rows of engineers and scientists closely monitored the various data feeds streaming in from the spacecraft.

Approximately 18,000 personnel supported each Mercury mission, with around 15,000 dedicated to recovery operations. The remaining personnel, scattered across the globe, were responsible for tracking the spacecraft using the World Wide Tracking Network. This network, a chain of 18 stations around the equator, was established in 1960 and adapted from satellite tracking systems. It provided crucial two-way communication between the astronaut and ground control, with each station covering a range of 700 nautical miles (1,300 km) and allowing for a pass duration of about 7 minutes.

The Mercury astronauts not on the mission took turns as Capsule Communicator, or CAPCOM, serving as the primary link between the spacecraft and mission control. Data from the spacecraft were sent to Earth and processed at the Goddard Space Center by a pair of redundant IBM 7090 computers. This data was then relayed to the Mercury Control Center, displayed on boards flanking a world map. These boards showed the spacecraft's position, ground track, and potential emergency landing sites within the next 30 minutes.

Additional computer support for Mercury's ground control came from an IBM 709 system at Cape Canaveral, which calculated mid-launch abort scenarios, and another IBM 709 in Bermuda, serving as a backup for Goddard's 7090 machines. The Burroughs-GE system provided radio guidance for the Atlas rocket during launch.

The World Wide Tracking Network was vital to the Mercury missions, ensuring continuous monitoring and communication with the spacecraft. This network continued to serve subsequent space programs until it was replaced by a satellite relay system in the 1980s. In 1965, Mission Control's operations were moved from Cape Canaveral to Houston, Texas, marking a new chapter in space mission management.

Sensory Experience

Inside the Mercury Control Center, the air was thick with anticipation. Engineers and scientists, seated at their consoles, were surrounded by monitors and control panels. Each screen and gauge provided real-time data on the spacecraft's status, from its speed and trajectory to the astronaut's vital signs. The constant hum of machinery and the occasional crackle of radio communications added to the intensity of the environment.

As the mission progressed, the control room buzzed with activity. Technicians adjusted dials and toggled switches while supervisors issued commands and verified calculations. The large control board at the front of the room displayed the spacecraft's position against a world map, its path illuminated by a moving light that traced its orbit around the Earth.

The CAPCOM communicated directly with the astronaut throughout the mission, their calm and steady voice providing a crucial link between the spacecraft and mission control. Each transmission was followed by a flurry of data analysis as the team worked to ensure the mission proceeded smoothly.

The mission's culmination brought a mixture of relief and elation. The spacecraft's successful splashdown was met with cheers and applause as the tension in the control room gave way to celebration. The astronaut's safe return marked the end of a successful mission and a significant milestone in the journey of human space exploration.

Tracking Network

The World Wide Tracking Network was crucial to Project Mercury's success. With its 18 strategically placed stations around the equator, this network ensured continuous communication and data relay for the spacecraft. Each station covered a range of 700 nautical miles (1,300 km), allowing for a pass duration of about 7 minutes, thus providing near-constant monitoring and communication with the Mercury capsule.

The robust nature of this network was vital for the mission's operations. It facilitated the real-time relay of telemetry data and voice communications between the astronaut and mission control. It was essential for tracking the spacecraft's position, monitoring its systems, and ensuring the astronaut's safety. This system's reliability and effectiveness were not only pivotal for the success of Project Mercury but also set the foundation for future space programs. The World Wide Tracking Network continued supporting subsequent missions until it was eventually replaced by more advanced satellite relay systems in the 1980s, demonstrating its enduring legacy in space exploration.

Flights and Landing Sites

Project Mercury achieved six successful crewed missions, each meticulously planned and executed, marking significant milestones in the early days of human space exploration.

Freedom 7 (MR-3): Launched on May 5, 1961, with astronaut Alan Shepard aboard, Freedom 7 was the first American manned spaceflight. This historic mission achieved a suborbital trajectory, propelling Shepard into space and making him the first American to experience the weightlessness of space. The mission lasted about 15 minutes, reaching an apogee of 116.5 miles (187.5 km) and a maximum speed of 5,134 mph (8,260 km/h). It concluded with a successful splashdown in the Atlantic Ocean, where recovery forces promptly retrieved both Shepard and the capsule.

On July 1, 1961, operational control of the Mercury global network transferred to the Goddard Space Flight Center. This ensured robust management during active missions while maintaining readiness under the Space Task Group during critical phases.

Ten days later, on July 11, 1961, Walter C. Williams, Project Mercury Operations Officer, finalized key personnel assignments for the upcoming Mercury-Redstone 4 (MR-4) manned suborbital flight. These appointments spanned critical roles across various locations, including the Mercury Control Center, Atlantic Missile Range Central Control, and support vessels, highlighting the intricate coordination required for mission success.

Mission rules for MR-4 were issued on July 13, 1961, alongside static testing of the Redstone launch vehicle for MR-6 at the Marshall Space Flight Center. A comprehensive safety review ensued, scrutinizing spacecraft and launch vehicle operations in preparation for the impending mission.

Atlas launch vehicle 88-D arrived at Cape Canaveral on July 15, 1961, swiftly prepared for its role in the Mercury-Atlas 4 (MA-4) unmanned orbital flight, marking another stride forward in America's space exploration efforts.

Anticipation peaked on July 18-19, 1961, with two unsuccessful launch attempts of MR-4 due to adverse weather conditions. Finally, on July 21, 1961, Virgil Grissom piloted Mercury-Redstone 4, nicknamed Liberty Bell 7, into space. The mission proceeded smoothly until an unexpected malfunction during recovery caused the spacecraft to flood and sink. Grissom egressed safely but lost Liberty Bell 7 in the process. Despite this setback, the mission was hailed as a success, leading the Space Task Group to conclude the Mercury-Redstone suborbital phase.

The following day, July 22, 1961, Virgil Grissom received the NASA Distinguished Service Medal from Administrator James Webb at a press conference, recognizing his courage and the mission's achievements.

A pivotal briefing between McDonnell and NASA officials on July 27-28, 1961 introduced Project Gemini, outlining plans for an advanced two-man spacecraft capable of longer missions—a significant step toward future space exploration endeavors.

By July 31, 1961, the astronaut centrifuge training program at the Aviation Medical Acceleration Laboratory focused intensively on preparing for Mercury-Atlas orbital missions. This rigorous training regimen simulated the harsh conditions of space, ensuring astronauts were fully physically and mentally prepared for the challenges ahead.

Liberty Bell 7 (MR-4): Piloted by Gus Grissom, Liberty Bell 7 took to the skies on July 21, 1961, achieving another successful suborbital flight. The mission mirrored Shepard's in many respects, reaching an apogee of 118.26 miles (190.29 km) and a maximum speed of 5,310 mph (8,547 km/h). However, it faced a near-disaster when the capsule's hatch blew open after splashdown, causing the spacecraft to flood and sink. Despite losing the capsule, Grissom was safely recovered by rescue forces.

August 1-3, 1961, saw rigorous testing of the operational Mercury spacecraft's seaworthiness under challenging sea conditions off Wallops Island. Over a 33-hour period, the spacecraft endured ground swells ranging from 5 to 15 feet and wave heights of 2 to 10 feet, demonstrating robust flotation capabilities crucial for safe recovery missions.

On August 6, 1961, the USSR achieved a significant milestone in the space race by launching Vostok II with Major Gherman S. Titov aboard. Titov's orbital flight, slightly heavier than its predecessor Vostok I, captivated the world as his journey was broadcast live on Radio Moscow, highlighting Soviet advances in human space exploration.

NASA's preparations for the unmanned Mercury-Atlas 4 (MA-4) orbital flight intensified with the publication of retrofire-from-orbit mission rules on August 9, 1961. Concurrently, key personnel assignments were finalized, ensuring coordinated efforts for the upcoming critical test of the Mercury program.

Mercury spacecraft No. 15 arrived at Cape Canaveral on August 13, 1961, but was promptly returned to McDonnell for reconfiguration to support the ambitious orbital-manned 1-day mission. This redesign process resulted in the spacecraft's reissue as No. 15A (later 15B), delivered on November 16, 1962, ready for its revised mission objectives.

A pivotal announcement came on August 18, 1961, as NASA completed all objectives for the Mercury-Redstone suborbital phase. This marked a significant achievement for Project Mercury, leading to the cessation of further Mercury-Redstone flights as the focus shifted toward more ambitious orbital missions.

From August 22 to September 12, 1961, extensive safety reviews were conducted for the unmanned Mercury-Atlas 4 (MA-4) orbital flight, ensuring meticulous planning and readiness across mission, spacecraft, and launch vehicle components.

The planned Mercury-Atlas 4 (MA-4) unmanned orbital flight faced a postponement on August 24, 1961, prompting the team to refine preparations for the crucial mission test.

Meanwhile, Explorer XIII, tasked with measuring micrometeoroid impacts on spaceflight, encountered setbacks on August 25, 1961, necessitating further research and development in this critical area of space exploration.

Mercury spacecraft No. 13 arrived at Cape Canaveral on August 27, 1961, designated for the historic first manned Mercury-Atlas orbital flight (MA-6), piloted by John Glenn. Extensive test and checkout procedures commenced immediately to ensure the spacecraft's readiness for its groundbreaking mission.

A thorough investigation into the premature activation of the Mercury-Redstone 4 (MR-4) explosive egress hatch was initiated on August 30, 1961. Despite rigorous testing under severe conditions, no repeat incidents occurred. As a precaution, McDonnell was tasked with designing a mechanical-type hatch, accompanied by revised operational procedures to enhance safety during recovery operations.

In August of 1961, NASA embarked on a pivotal mission to establish a permanent manned spacecraft center, setting the stage for a new era in American space exploration. A dedicated site selection team led by John F. Parsons meticulously evaluated locations across the United States. Their criteria encompassed advanced scientific facilities, robust infrastructure, favorable climate conditions, and supportive community amenities, all crucial for advancing the nation's ambitious space endeavors.

During this period, NASA faced significant technical challenges and milestones. From August 5 to October 12, 1961, rigorous environmental tests were conducted on spacecraft components, particularly the explosive egress hatch, following issues encountered during the Mercury-Redstone 4 mission. These tests aimed to enhance reliability and safety under various operational conditions.

Meanwhile, three rocket sled tests were executed at the Naval Ordnance Test Station in China Lake, California, between September 5 and 14, 1961. These tests focused on refining the separation dynamics between launch vehicles and spacecraft clamp rings, crucial for ensuring flawless mission executions. Each run incorporated minor adjustments, culminating in perfected separation mechanisms by the third test.

A seminal report outlined technological advancements resulting from the Mercury development program on September 8, 1961. Innovations included impact force attenuation through crushable honeycomb structures, adaptable couch configurations, and improved stability in abort attitudes with modified tower clamp rings. Further enhancements encompassed thrust chamber and pressure transducer improvements and innovations like destabilization flaps and landing bag designs to optimize spacecraft reentry and recovery.

Mercury-Atlas 4

Spacecraft No.: 8
Launch Date: September 13, 1961
Duration: 1 hour 49 minutes 20 seconds
Purpose: Test of environmental control system with robot astronaut in orbit.
Result: Success

Successfully testing the environmental control system with a robot astronaut, this mission was a key step towards manned orbital flights. September 13, 1961, marked a historic milestone with the Mercury-Atlas 4 (MA-4) launch from Cape Canaveral. Equipped with specialized instrumentation, including vibration and noise sensors and a mechanical crewman simulator, this mission achieved the first-ever Earth orbit for a Mercury spacecraft. Following a successful orbit and precise retrograde rocket maneuvers, the spacecraft splashed down in the Atlantic Ocean, east of Bermuda, and was swiftly recovered by the USS Decatur. MA-4 demonstrated the readiness of the Atlas booster, spacecraft systems, and global tracking network for future manned orbital flights.

Amidst these technical feats, significant administrative decisions shaped NASA's future. On September 19, 1961, NASA Administrator James Webb announced the establishment of a new center for manned spaceflight at a sprawling 1,000-acre site near Houston, Texas. Donated by Rice University in Harris County, this site would become pivotal as NASA's operations expanded.

As organizational changes took shape on September 24, 1961, NASA restructured key leadership roles to streamline decision-making and program oversight. D. Brainerd Holmes assumed the helm of the Office of Manned Space Flight. At the same time, Robert R.

Gilruth was appointed Director of the Manned Spacecraft Center in Houston, underscoring NASA's commitment to centralizing and advancing its manned space programs.

These pivotal events throughout August and September 1961 underscored NASA's technical prowess and laid the groundwork for Houston's emergence as a hub for American space exploration, solidifying the nation's commitment to pushing the boundaries of human achievement in space.

In October 1961, NASA's progress in space exploration reached new heights with significant milestones and preparations for upcoming missions. The month began with the factory roll-out inspection of Atlas booster No. 93-D at Convair on October 1st, designated for the pivotal Mercury-Atlas 5 (MA-5) mission. This booster, crucial for advancing orbital capabilities, underwent meticulous scrutiny to ensure readiness.

By October 9th, Atlas booster No. 93-D arrived at Cape Canaveral, marking a crucial step towards the upcoming MA-5 orbital flight mission. The delivery underscored NASA's logistical precision and commitment to timeline adherence in its ambitious spaceflight agenda.

On October 13th, NASA Headquarters approved expansive construction projects for a permanent manned spacecraft center at Clear Lake, southeast of Houston, Texas. This facility would serve as a cornerstone for NASA's operations, encompassing essential infrastructure such as an auditorium, project management offices, technical laboratories, and support facilities. This approval highlighted NASA's strategic investment in establishing a robust operational base to support ongoing and future missions.

As preparations intensified, NASA published the Mercury-Atlas 5 (MA-5) data acquisition plan on October 20th, delineating comprehensive strategies for gathering mission-critical data during the upcoming orbital flight. This meticulous planning was essential for maximizing mission success and scientific returns.

On October 23rd, in a poignant moment, NASA presented Freedom 7, the Mercury-Redstone 3 (MR-3) spacecraft, to the National Air Museum of the Smithsonian Institution, commemorating America's pioneering steps into space.

The month concluded with pivotal decisions and operational tests. On October 25th, NASA Headquarters officially greenlit the Mercury extended range or 1-day mission program, expanding the scope of Mercury missions and advancing capabilities for longer-duration spaceflights.

Practical preparations continued with ship retrieval tests conducted on October 26-27, aimed at refining procedures for recovering manned Mercury spacecraft from sea landings. These tests validated critical recovery protocols without encountering significant challenges, further enhancing mission safety.

On October 29th, NASA announced plans for a Mercury-Scout launch to verify the readiness of the global Mercury Tracking network for upcoming orbital flights, underscoring NASA's commitment to ensuring robust mission support infrastructure worldwide.

Throughout October 1961, NASA's relentless pursuit of space exploration milestones was evident as preparations for the MA-5 mission and future endeavors advanced. The delivery of Mercury spacecraft No. 12 to Cape Canaveral as a backup for the MA-8 mission, with considerations for extended mission configurations, underscored NASA's proactive approach in expanding human spaceflight capabilities. These efforts set the stage for continued advancements in America's space program, marking a transformative period in the history of space exploration.

In October 1961, NASA Headquarters decided to extend the Mercury program by adding one-day missions after the three- and six-orbit flights. Additionally, during the same year, follow-on manned space programs, later known as Projects Gemini and Apollo, began to take shape. These events were unusual, representing a program expansion to a higher level of difficulty before the basic objectives of Project Mercury—launching and safely returning a man from Earth orbit—had been attained. The confidence in Project Mercury, initially guided by the Space Task Group and later by the Manned Spacecraft Center, was bolstered by the highly successful suborbital flights of Alan Shepard and Virgil Grissom and the orbital flight of the "mechanical astronaut."

In October 1961, a pivotal decision emerged from NASA Headquarters: Project Mercury would expand to include one-day missions, following the planned three and six-orbit flights. Concurrently, new manned space programs began to take shape, later becoming known as Projects Gemini and Apollo. This announcement, made even before Project Mercury had achieved its primary goal of launching and safely returning a man from Earth's orbit, underscored NASA's profound confidence in its burgeoning space venture.

This confidence stemmed from the successes of astronauts Alan Shepard and Virgil Grissom's suborbital flights and the orbital journey of the "mechanical astronaut." These milestones demonstrated that the Space Task Group, later succeeded by the Manned Spacecraft Center, had built a solid foundation for the program.

The answer lay in the program's concurrency of effort. Project Mercury pursued simultaneous development across all facets, unlike other major national initiatives. From the outset, work was underway on spacecraft components, launch vehicle adaptations, worldwide tracking network readiness, astronaut selection and training, and ground support equipment development. No detail was deemed too minor for the scientists and engineers making critical decisions about humanity's leap into orbit.

Every organization with relevant technical expertise or capability was tapped for assistance, facilities, or equipment. The team was a vast network of dedicated individuals and institutions working towards a common goal. Their combined efforts propelled the United States into a new era of space exploration, setting the stage for the monumental achievements of Projects Gemini and Apollo.

The reliability program for Mercury hardware was exhaustive, ensuring no corners were cut. This meticulous approach refutes the idea of a "crash program"; "accelerated" aptly describes the effort. Despite the American public's eager anticipation, managers resisted rushing the program, maintaining a steadfast commitment to precision and safety.

Several catalysts led to the approval of Project Mercury and directly contributed to its success. Post-World War II, experimental missile tests at White Sands, New Mexico, reached altitudes beyond the sensible atmosphere. Simultaneously, rocket aircraft research aimed to break the sound barrier. Throughout the early to mid-fifties, the National Advisory Committee for Aeronautics (NACA) and industry scientists tackled the thermal barrier to solve the reentry problem for ballistic missiles. These research endeavors naturally progressed towards the challenges of manned space flight.

Interest in a national space program grew, bolstered by planning and research for the artificial Earth satellite program during the International Geophysical Year. The launch of Sputnik I in 1957 galvanized public support for a manned space flight project. At the same time, the Atlas launch vehicle reached a developmental milestone and was recognized as capable of lifting the payload required for manned orbital missions.

Project Mercury's journey can be divided into three major phases: conception, research and development, and operation. Each phase demonstrated the meticulous attention to detail from NASA personnel, other government agencies, and American industry, all united in ensuring the success of America's first manned step into space.

The excitement and urgency of the era were palpable as scientists, engineers, and astronauts worked tirelessly to overcome each new challenge. Laboratories buzzed with activity, launch sites were abuzz with preparations, and control rooms were filled with the focused intensity of those monitoring every critical second. Each successful test and launch brought the nation closer to realizing its dream of manned space exploration, inspiring awe and pride across the country.

This was a time of extraordinary innovation and collaboration, setting the stage for the incredible achievements that would follow in the years to come.

Chapter 10 - NASA Looking to the Future

Mercury-Scout 1

Spacecraft No.: -
Launch Date: November 1, 1961
Duration: 44 seconds
Purpose: Special satellite to test Mercury tracking network.

Result: Failure

An attempt to launch a specialized satellite for testing the tracking network failed shortly after liftoff, highlighting the importance of robust booster systems.

In November 1961, NASA's relentless pursuit of manned spaceflight milestones continued with significant achievements and preparations leading up to historic missions. The month began with a setback on November 1st, as the launch attempt of Mercury-Scout 1 (MS-1) failed shortly after liftoff. The launch vehicle experienced erratic motions and high aerodynamic loads, leading to its destruction by safety protocols. Despite this setback, the successful missions of Mercury-Atlas 4 (MA-4) and the impending Mercury-Atlas 5 (MA-5) confirmed the readiness of the Mercury global network for orbital missions.

Simultaneously, on November 1st, NASA's Space Task Group was redesignated as the Manned Spacecraft Center, solidifying its focus on manned space missions under the leadership of Robert R. Gilruth. This reorganization underscored NASA's commitment to advancing human spaceflight capabilities.

On November 15th, Mercury spacecraft No. 18 was delivered to Cape Canaveral, earmarked for the upcoming Mercury-Atlas 7 (MA-7) mission. This mission, set to be piloted by astronaut Scott Carpenter, aimed further to expand America's capabilities in manned orbital flight.

Critical milestones continued with the factory roll-out inspection of Atlas launch vehicle 109-D on November 19th, designated for the groundbreaking Mercury-Atlas 6 (MA-6) mission. This mission, slated to be the first manned orbital spaceflight, marked a pivotal moment in NASA's history.

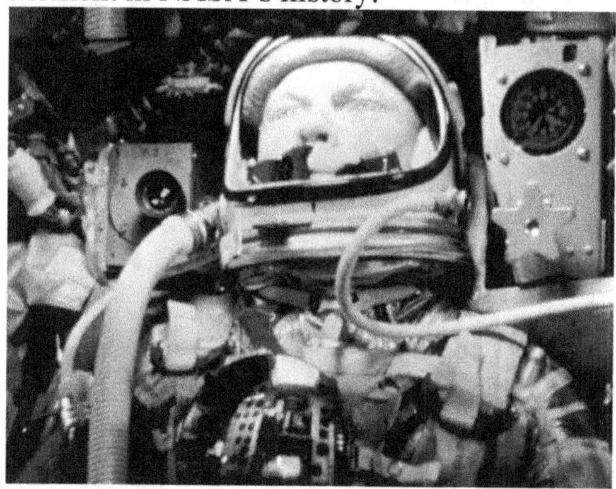

John Glenn in orbit, 1962 (Mercury-Atlas 6)

As preparations intensified, on November 29th, Mercury astronauts were assigned as spacecraft communicators for the MA-5 mission at tracking stations across the Mercury global network. The MA-5 mission, launched from Cape Canaveral with the chimpanzee Enos aboard, aimed for three orbits but returned after two due to technical issues. Despite challenges with a roll reaction jet and electrical system overheating, the mission was deemed highly successful, validating the Mercury spacecraft's readiness for manned orbital flight. Enos's performance in space validated the feasibility of human spaceflight under rigorous conditions, highlighting NASA's progress in biological studies essential for human space exploration.

Looking forward, astronaut John Glenn was selected as the pilot for the first manned orbital flight, with Scott Carpenter serving as the backup. Intensive training commenced immediately to prepare Glenn and Carpenter for their historic mission, while other astronauts focused on engineering and operational roles within the newly restructured Manned Spacecraft Center.

November 30th saw the delivery of Atlas launch vehicle 109-D to Cape Canaveral for the MA-6 mission, marking another crucial step toward achieving America's first manned orbital spaceflight. These achievements in November 1961 set the stage for America's pioneering endeavors in space, positioning NASA on the brink of achieving unprecedented milestones in human space exploration.

On November 29, 1961, NASA launched Mercury-Atlas 5, a pivotal mission that carried a chimpanzee named Enos into orbit. This mission was a critical step in demonstrating the spacecraft's ability to support life during an orbital mission, ultimately paving the way for John Glenn's historic flight.

Mission Objectives and Preparation

Mercury-Atlas 5 aimed validate the environmental control system in orbit.

Ensure the spacecraft could sustain a living organism during an extended orbital mission.

Gather data on the effects of spaceflight on a biological payload, providing a precursor to human spaceflight.

Enos, the chimpanzee chosen for this mission, underwent extensive training to prepare for the flight. He was trained to perform specific tasks during the mission, which would help scientists assess his physical and psychological responses to the space environment.

The Flight

The mission lasted 3 hours and 20 minutes, during which Enos completed two orbits around the Earth. The spacecraft experienced various conditions that future human astronauts would encounter, including exposure to microgravity and the stresses of launch and reentry.

Launch: Mercury-Atlas 5 was launched from Cape Canaveral, Florida, atop an Atlas rocket. The launch was smooth, and the spacecraft successfully entered orbit.

Orbital Operations: Enos performed tasks using levers and buttons while in orbit, allowing scientists to monitor his responses and the functionality of the spacecraft's systems. The environmental control system was tested extensively, ensuring it maintained a livable environment for Enos.

Reentry and Recovery: After completing the mission's objectives, the spacecraft reentered Earth's atmosphere. The heat shield performed as expected, protecting Enos and the spacecraft from the intense heat generated during reentry. The capsule splashed down in the Atlantic Ocean, where recovery teams quickly retrieved Enos and the spacecraft.

Results and Impact

The mission was a success, achieving its primary objectives:

Environmental Control System: The flight validated the spacecraft's environmental control system, proving it could maintain suitable conditions for life during an orbital mission.

Life Support: Enos completed the mission tasks and demonstrated that the spacecraft could support a living organism in space, an essential precursor to human spaceflight.

Data Collection: The data collected from Enos's physiological responses provided valuable insights into the effects of spaceflight on biological organisms, informing the preparations for human missions.

Mercury-Atlas 5

Spacecraft No.: 9

Launch Date: November 29, 1961

Duration: 3 hours 20 minutes 59 seconds

Purpose: Test of the environmental control system in orbit with a chimpanzee named Enos.

Result: Success

This mission, which carried a chimpanzee named Enos, successfully tested the environmental control system in orbit and provided critical data for upcoming manned missions. Mercury-Atlas 5's success was a crucial milestone in the Mercury program, directly contributing to the confidence and readiness for John Glenn's orbital flight on Mercury-Atlas 6. Enos's mission underscored the importance of rigorous testing and validation, ensuring that the spacecraft systems could fully support human astronauts during their journeys into space.

Sensory and Technical Experience

These uncrewed and chimpanzee missions were filled with technical challenges and meticulous planning. Engineers and scientists monitored every aspect of the flights from the control center. The launches were accompanied by the roar of the engines and the vibrations felt by everyone nearby. The data collected during these missions were crucial in understanding the behavior of the spacecraft and the effects of spaceflight on living organisms.

Legacy

The successful flight of Mercury-Atlas 5 and the safe return of Enos marked a significant achievement in NASA's journey toward manned spaceflight. This mission demonstrated the technical capabilities of the Mercury spacecraft and highlighted the critical role of biological testing in ensuring the safety and success of human missions.

Impact on Human Spaceflight

The success of Ham's mission had far-reaching implications for NASA and the future of space exploration. It confirmed that the Mercury spacecraft's life support systems could sustain human life and that the launch, flight, and recovery procedures were sound. This critical validation was necessary before risking human lives on similar missions.

Ham's flight demonstrated the importance of thorough testing with biological payloads, providing valuable insights that helped refine spacecraft systems and procedures. This approach ensured that when the first American astronauts ventured into space, they did so with the highest possible assurance of safety and mission success.

The chimpanzee missions, particularly Ham's flight, remain a testament to NASA's meticulous and step-by-step approach to developing human spaceflight capabilities. By proving the viability of the Mercury spacecraft with living organisms, NASA laid the groundwork for the successful manned missions that followed.

The Mercury spacecraft's rigorous uncrewed and chimpanzee testing provided NASA with invaluable data. This allowed engineers to refine the spacecraft's design and address any issues before manned missions. These early tests were foundational in ensuring the success and safety of America's first human spaceflights.

The lessons learned from these tests laid the groundwork for future space programs, proving the importance of thorough testing and validation in developing spaceflight technology.

Canceled Missions: Shifting Priorities and Changing Times

The Evolution of Project Mercury: Strategic Decisions and Canceled Missions

During Project Mercury, the landscape of human spaceflight was rapidly evolving. As the United States and the Soviet Union vied for dominance in space, strategic decisions, and technological advancements led to the cancellation of several planned missions. Initially, there were ambitious plans for multiple suborbital flights involving additional astronauts. However, these plans were scaled back and ultimately canceled, particularly after the successful flight of Soviet cosmonaut Gherman Titov demonstrated the capability for extended human spaceflight.

In December 1961, NASA continued to forge ahead with ambitious plans and preparations for pioneering milestones in manned spaceflight. On December 6th, NASA Headquarters announced the scheduling of the first Mercury manned orbital flight for early 1962. This decision, based on the success of the Mercury-Atlas 5 (MA-5) mission, affirmed the readiness of spacecraft systems, launch vehicles, and the global tracking network for human spaceflight. In a historic joint ceremony, astronauts Alan Shepard and Virgil Grissom were awarded the first Astronaut Wings, recognizing their pivotal roles in advancing space exploration.

The scope of NASA's ambitions expanded on December 7th with Robert R. Gilruth, Director of the Manned Spacecraft Center, announcing plans to develop a two-man Mercury spacecraft. This initiative, officially designated Project Gemini on January 3, 1962, aimed to enhance capabilities and prepare for more complex missions beyond Mercury.

On December 11th, the Army Corps of Engineers awarded Brown and Root, Incorporated a significant contract to design major portions of the permanent facilities at the Manned Spacecraft Center. This marked a critical step in establishing the infrastructure needed to support ongoing and future space missions.

Practical preparations continued with spacecraft egress exercises conducted from December 11-13 near Langley Field, focusing on astronaut training for manned orbital missions and helicopter recovery operations. These exercises validated procedures for both top and side hatch egress from the spacecraft, ensuring emergency readiness.

Technical evaluations proceeded with spacecraft ultimate pressure tests conducted on December 12th, reaching 20 pounds per square inch. Subsequent inspections confirmed the spacecraft's structural integrity, crucial for withstanding the rigors of spaceflight.

On December 14th, Walter C. Williams addressed the University of Houston, highlighting the Mercury spacecraft's role as a pioneering testbed for developing orbital flight techniques and hardware, setting the stage for future ambitious space programs.

From December 14-18, two Mercury spacecraft solid-bottom water drop tests were conducted, evaluating structural resilience and heat shield performance upon impact with water. These tests confirmed the spacecraft's robust design and readiness for reentry.

Further technical evaluations on December 18th included spacecraft external pressure tests, verifying structural integrity under pressures up to 15 pounds per square inch with minimal bulkhead deflection.

On December 29th, NASA announced key appointments critical to its mission success, including Dr. Joseph F. Shea as Deputy Director for Systems Engineering at NASA Headquarters and Dr. Arthur Rudolph as Assistant Director of Systems Engineering. Dr. Rudolph's role as liaison between the Marshall Space Flight Center and the Manned Spacecraft Center aimed to streamline vehicle development and operational integration.

As the year drew to a close, the Manned Spacecraft Center's personnel reached 1,152, underscoring the growth and organizational depth required for NASA's expanding missions. The establishment of a Management Council, chaired by D. Brainerd Holmes and comprising senior officials from NASA Headquarters and the Marshall Space Flight Center, aimed to ensure efficient progress and coordination across manned spaceflight programs.

December 1961 marked a significant advancement and preparation for NASA, setting the stage for America's historic achievements in manned orbital flight and paving the way for future explorations beyond Earth's orbit.

Mercury 13

The Mercury 13 program, also known as the "Women in Space Program," was an unofficial program conducted independently of NASA. It involved thirteen American women who underwent rigorous testing at the Lovelace Clinic in New Mexico in 1961.

These tests were designed to assess their suitability for spaceflight, and they underwent many of the same physical and psychological evaluations as the male Mercury astronauts.

The women who participated in the Mercury 13 program performed well in these tests, often exceeding the expectations set by the criteria established for the male astronauts. However, despite their qualifications and capabilities, the program was ultimately canceled before any of the women could complete their training. NASA did not officially sanction the Mercury 13 program, and at the time, the agency did not consider women as candidates for astronaut selection.

The cancellation of the Mercury 13 program disappointed the women involved and advocates for equal opportunities in space exploration. Nevertheless, the program played a significant role in demonstrating that women could meet the rigorous demands of spaceflight, paving the way for future opportunities for women in the U.S. space program.

The question of whether women could qualify for the astronaut program was raised during Project Mercury. Although no women met the qualifications then, the media coined the term "Mercury 13" for thirteen American women who successfully underwent the same tests as the male astronauts. These tests were conducted by NASA physician William Randolph Lovelace, but the privately funded program was quickly canceled, and the women never completed the training. It wasn't until 1978 that women qualified for the Space Shuttle program. The thirteen women were:

Jerrie Cobb - A highly experienced pilot who advocated women's inclusion in the space program.

Janet Dietrich
Marion Dietrich
Myrtle Cagle
Jane Hart
Jean Hixson
Rhea Hurrle (Woltman)
Gene Nora Stumbough (Jessen)
Irene Leverton
Sarah Gorelick (Ratley)
Bernice Steadman
Jerri Sloan (Truhill)

Wally Funk became the oldest woman to fly to space in 2021 when she joined Blue Origin's New Shepard crew.

These women demonstrated exceptional skill and determination, and although they did not fly during the Mercury program, their efforts paved the way for future generations. It wasn't until 1978 that women were officially included in NASA's astronaut program with the Space Shuttle, a significant milestone in the history of space exploration.

In January 1962, NASA's preparations for pivotal milestones in manned space exploration continued with focused determination and meticulous planning:

On January 1st, a survey at the Manned Spacecraft Center revealed that most personnel intended to relocate from Langley Field to Houston, Texas, solidifying the Center's transition plans.

On January 3rd at Lynnhaven Roads Anchorage near Norfolk, Virginia, exercises validated the feasibility of using an auxiliary flotation collar in recovery operations. This successful test led to its adoption, enhancing spacecraft recovery capabilities. Concurrently, flight controllers for the Mercury-Atlas 6 (MA-6) mission received final briefings before deployment to remote tracking sites, underscoring the thorough preparations for the upcoming mission.

The organizational framework of the Manned Spacecraft Center's Mercury Project Office was finalized by January 15th. Under Kenneth Kleinknecht's leadership as Project Mercury's manager, key divisions, including the Project Manager's Office, Project Engineering, and Engineering Operations, were established. This structured approach ensured coordinated efforts toward mission success.

Training from January 15-17th at Pensacola Naval Air Station prepared recovery area swimmers for the MA-6 mission, encompassing deployment procedures of the auxiliary flotation collar and helicopter operations. This rigorous training underscored NASA's commitment to efficient and effective recovery operations.

Spacecraft 16's delivery to Cape Canaveral on January 16th marked a significant step for the third manned orbital flight, Mercury-Atlas 8 (MA-8), reinforcing ongoing preparations for future missions.

Director Robert R. Gilruth received the prestigious Louis W. Hill Space Transportation Award on January 23rd from the Institute of Aerospace Sciences. The award recognizes his exemplary leadership in advancing spacecraft technology for manned spaceflight.

Anticipation heightened on January 27th when weather conditions forced a postponement of the MA-6 mission just 29 minutes before launch. This delay underscored the complexities and stringent safety protocols of spaceflight.

Technical challenges persisted as the MA-6 mission faced another postponement on January 30th due to launch vehicle issues, prompting intensive efforts by engineers to resolve the setbacks.

Throughout January, preparations for future missions intensified with identifying potential recovery areas and technical enhancements, including using submarine cables for high-speed data transmission and rigorous parachute drop tests at El Centro, California, under Project Reef. These initiatives aimed to optimize mission readiness and safety for upcoming extended-duration flights.

In January 1962, meticulous planning, rigorous training, and technical advancements culminated in pivotal events shaping America's future in manned space exploration. This marked a period of intense preparation leading up to a historic milestone in February 1962.

The stage was set with NASA's announcement on February 1st, scheduling the Mercury-Atlas 6 (MA-6) mission for February 13th, following repairs to the Atlas launch vehicle's fuel tank leak. However, adverse weather conditions on February 14th postponed the mission, highlighting the stringent safety protocols of spaceflight. Undeterred, Project Mercury Operations Director Walter C. Williams announced a new launch date of February 20th, capitalizing on improved weather forecasts.

On February 20th, 1962, the eagerly anticipated MA-6 mission launched successfully from Cape Canaveral, carrying astronaut John Glenn aboard the Friendship 7 spacecraft. Glenn's journey included three orbits of the Earth, lasting 4 hours, 55 minutes, and 23 seconds, captivating an estimated 60 million viewers worldwide via live television broadcast. During the flight, Glenn encountered challenges, including a yaw attitude control jet issue and a false heat shield alarm, necessitating careful reentry precautions. Nevertheless, the spacecraft splashed down in the Atlantic Ocean and was swiftly recovered by the USS Noa, achieving Project Mercury's primary goal of manned orbital flight.

The mission's impact was global; a metal fragment from the Atlas booster landed in South Africa on February 21st, underscoring its international significance. Two days later, on February 23rd, President John F. Kennedy honored John Glenn and Robert R. Gilruth with the NASA Distinguished Service Medal at Cape Canaveral, recognizing their pivotal roles in the mission's success.

Looking forward, progress continued with the February 25th factory roll-out inspection of the Mercury-Atlas 7 (MA-7) mission's Atlas launch vehicle at Convair. This milestone signaled ongoing advancements and preparations for future missions.

The nation celebrated John Glenn's historic achievement on February 26th, as Washington, D.C., declared "John Glenn Day." The festivities included receptions, a parade, and Glenn addressing a joint session of Congress, reflecting America's pride and excitement over his groundbreaking flight.

These events of early 1962 underscored the meticulous planning, technical innovations, and collaborative efforts that defined Project Mercury's success. They solidified America's leadership in the space race, laying a firm foundation for the remarkable achievements and advancements that would follow in exploring space.

Friendship 7 (MA-6): On February 20, 1962, John Glenn made history with Friendship 7, becoming the first American to orbit the Earth. Glenn completed three orbits during his nearly five-hour flight, experiencing the awe-inspiring views of Earth from space. The mission, which reached an altitude of 162 statute miles (261 km) and a maximum speed of 17,544 mph (28,260 km/h), concluded with a safe splashdown in the Atlantic Ocean. This milestone flight cemented Glenn's place in history and demonstrated the feasibility of orbital spaceflight for Americans.

The defining moment for Project Mercury came on February 20, 1962, with John Glenn's historic flight. The world held its breath as Glenn orbited the Earth three times in the cramped confines of the Mercury capsule. His calm yet tinged with excitement voice cracked over the radio, providing updates from space. The view from Glenn's capsule was breathtaking, the curvature of the Earth set against the vast, dark expanse of space.

On March 1, 1962, New York City became the stage for "John Glenn Day," where an estimated 4 million people gathered to honor the astronaut's monumental achievement. Mayor Robert Wagner presented John Glenn and Robert R. Gilruth with the city's Medal of Honor, reflecting the profound impact of Glenn's successful orbital mission on public morale.

Simultaneously, McDonnell submitted Mercury Report No. 8140, detailing the status of component qualification tests in their "Contractor Furnished Equipment Status Report." This meticulous oversight ensured the ongoing reliability and safety of mission equipment crucial for future spaceflights.

The following day, March 2, 1962, John Glenn and his fellow Mercury astronauts were guests at the United Nations, underscoring the mission's global significance. Glenn, acting as spokesman, participated in an informal reception hosted by Acting Secretary General U Thant, emphasizing international collaboration in space exploration.

On March 4-5, 1962, Scott Carpenter and Walter Schirra, designated for the upcoming Mercury-Atlas 7 (MA-7) mission, underwent rigorous water-egress exercises. These drills included multiple side-hatch egresses and helicopter pickups, preparing the astronauts for emergency scenarios they might encounter during their mission.

March 6, 1962, marked another milestone as Atlas launch vehicle 107-D arrived at Cape Canaveral for the MA-7 mission, signifying continued progress in preparations for manned spaceflights.

The scientific community celebrated on March 7, 1962, when the first Orbiting Solar Observatory (OSO) successfully conducted thirteen experiments in space. The data collected on solar radiation, cosmic dust effects, and spacecraft surface materials' thermal properties provided valuable insights crucial for future manned missions.

John Glenn's stature as an American hero was further cemented on March 9, 1962, when he became the third astronaut to receive Astronaut Wings in a ceremony at the Pentagon, honoring his pioneering contributions to space exploration.

Amidst these achievements, on March 12, 1962, the Manned Spacecraft Center's transition from Langley Field to Houston, Texas, ensured continuity in Mercury's operational activities, preventing disruptions during the move.

NASA Headquarters made a critical announcement on March 15, 1962, revealing Scott Carpenter as the new pilot for the MA-7 mission, replacing Donald Slayton due to a minor health concern. This decision highlighted NASA's rigorous health standards for astronauts.

Spacecraft 19's delivery to Cape Canaveral in its manned orbital configuration on March 20, 1962, was a significant step, although the mission was later canceled following Walter Schirra's successful six-orbit flight.

On March 22, 1962, personnel from the Manned Spacecraft Center briefed the Chief of Naval Operations on recovery requirements for the MA-7 mission and subsequent Mercury flights, ensuring thorough preparation for all operational contingencies.

Throughout March, the experimental implementation of the PERT reporting system marked a pivotal advancement in project management and efficiency. This system, fully operational by April 26, 1962, with the issuance of the first PERT report on the Mercury mission schedule and cost analysis, exemplified NASA's commitment to meticulous planning and execution in the evolving realm of space exploration.

On April 6, 1962, NASA hosted a significant symposium in Washington to analyze the findings from John Glenn's pioneering three-orbit flight aboard Mercury-Atlas 6 (MA-6). Central to discussions was the phenomenon of "fireflies," or luminous particles observed during the mission, which sparked scientific inquiry into their origin and implications.

Three days later, on April 9, 1962, John Glenn received the prestigious Hubbard Medal from the National Geographic Society. This honor, bestowed sparingly since 1906, placed Glenn in esteemed company alongside legendary figures like Admiral Robert A. Peary, Charles A. Lindbergh, Roald Amundsen, and Admiral Richard E. Byrd, recognizing his exceptional contributions to exploration and science.

April 15, 1962, marked Scott Carpenter and Walter Schirra's intensive water exercise training. The training focused on life raft boarding procedures and survival pack utilization for the upcoming Mercury-Atlas 7 (MA-7) manned orbital mission. These rigorous drills ensured their preparedness for emergency scenarios during the mission.

On April 19, 1962, NASA announced that Friendship 7, the spacecraft from John Glenn's MA-6 mission, would embark on a global tour under the auspices of the United States Information Agency. Dubbed the "fourth orbit of Friendship 7," this tour spanned 20 international stops across all continents. It aimed not only to showcase American technological prowess but also to promote international goodwill and highlight the achievements of Project Mercury.

By April 30, 1962, NASA's focus extended to astronaut nutrition and recovery procedures for the upcoming MA-7 mission. Testing included 27 bite-size food samples for suitability in space, ensuring astronauts had nutritious and manageable dietary options. Meanwhile, swimmer training commenced for MA-7's recovery area, incorporating film study, briefings, and practical exercises in deploying flotation equipment and helicopter egress techniques.

Throughout April, NASA developed an advanced pressure suit and helmet for the Mercury extended-range missions. Objectives included enhancing comfort, ventilation, mobility, and incorporating features like an electrically heated helmet visor with light attenuation capabilities and a robust mechanical visor seal mechanism. These innovations aimed to improve astronaut safety and performance during extended-duration missions.

These events underscored Project Mercury's meticulous preparation, technological innovation, and collaborative spirit, paving the way for historic achievements in space exploration and solidifying America's leadership in the global space race.

From May 5, 1961, to May 1962, Project Mercury exemplified meticulous preparation and evaluation, demonstrating a steadfast commitment to achieving a successful manned orbital flight. These efforts underscored the rigorous processes and collaborative spirit that defined the project's operational phase, culminating in the historic success of John Glenn's orbital flight and laying the groundwork for future advancements in space exploration.

On May 1, 1962, a gas analysis laboratory was established in Hangar S at Cape Canaveral. This facility played a crucial role in analyzing gases used in the Mercury spacecraft, ensuring the integrity and safety of the spacecraft's atmosphere, vital for the astronauts' health and the success of their missions.

May 4, 1962, saw significant developments with the issuance of a memorandum outlining proposed experiments for inclusion in future Mercury manned orbital flights. This initiative aligned with Walter C. Williams' vision of using the spacecraft as a test bed for more ambitious space projects. On the same day, Scott Carpenter, slated as the primary pilot for the Mercury-Atlas 7 (MA-7) mission, completed a simulated exercise of the MA-7 mission, enhancing his readiness and familiarity with the spacecraft's systems.

However, delays affected the MA-7 mission schedule. On May 7, 1962, NASA announced a postponement due to checkout issues with the Atlas launch vehicle, part of ongoing efforts to ensure all systems were functioning perfectly before launch. Subsequent postponements on May 17 and May 19 were caused by modifications to altitude-sensing instrumentation and irregularities in the Atlas flight control system, respectively, highlighting the rigorous safety protocols and attention to detail essential for mission success.

Despite these challenges, on May 24, 1962, the Mercury-Atlas 7 (MA-7) mission, piloted by Scott Carpenter aboard spacecraft Aurora 7, successfully launched into Earth orbit. Carpenter's three-orbit flight achieved all mission objectives despite technical issues with the pitch horizon scanner and concerns over fuel consumption. Notably, Carpenter improvised by allowing the spacecraft to drift to conserve fuel. The mission lasted 4 hours and 56 minutes, with Carpenter landing in the Atlantic Ocean northeast of Puerto Rico and being recovered after nearly three hours in the water.

The success of MA-7 was recognized on May 27, 1962, when Scott Carpenter and Walter C. Williams received the NASA Distinguished Service Medal for their contributions to the mission's success, presented by NASA Administrator James Webb at Cape Canaveral.

Further advancements followed in May 1962, including removing retrorocket heater blankets from the spacecraft and distributing Mercury-type survival kits to the Air Force's X-20 Dyna Soar program and the Navy, showcasing the broader application of Mercury program technologies.

Additionally, technical reports such as Technical Report No. 138 documented the valuable data obtained from Project Mercury's ballistic and orbital chimpanzee flights, further contributing to scientific understanding and mission planning.

Adjustments like removing leg support on the Mercury couch throughout these developments aimed to enhance astronaut comfort and operational efficiency, reflecting ongoing efforts to optimize mission capabilities.

In summary, the events of May 1962 underscored the meticulous preparations, collaborative efforts, and commitment to excellence that defined Project Mercury's operational phase. These efforts secured historic milestones in space exploration and set a solid foundation for future missions and advancements in human spaceflight.

Aurora 7 (MA-7): Scott Carpenter followed in Glenn's footsteps on May 24, 1962, with the Aurora 7 mission. Carpenter orbited the Earth three times, conducting scientific experiments and observing the planet's features from space. However, his reentry was off-target due to a delay in the retrorocket firing, resulting in a splashdown 250 miles off course in the Atlantic. Despite the off-target landing, Carpenter was safely recovered, and the mission provided valuable data for future flights.

PART III (B): Operational Phase of Project Mercury

June 1962 through June 12, 1963

On June 13, 1962, the Manned Spacecraft Center unveiled a pioneering experiment to understand meteoroid impacts in space. Proposed as part of the Mercury spacecraft mission, aluminum sheets would be deployed and exposed to the harsh meteoroid environment for two weeks before being safely retracted for reentry and recovery. This initiative sought to gather crucial data that could enhance spacecraft durability and astronaut safety.

Just twelve days later, on June 25, 1962, Scott Carpenter was awarded Astronaut Wings, marking him as the fourth individual from Project Mercury to receive this prestigious recognition. His achievement underscored his invaluable contributions to the ambitious Mercury program and highlighted the bravery and dedication of America's early astronauts.

The following day, Project Reef concluded its pivotal tests, focusing on the Mercury spacecraft's ringsail parachute. Designed to support increased weight for extended missions, the successful tests validated the parachute's reliability, ensuring critical safety measures for future space missions.

June 27, 1962, marked another significant milestone with NASA's Mercury-Atlas 8 (MA-8) mission announcement. Planned for up to six orbits, this manned orbital mission was a pivotal advancement, with Walter Schirra chosen as the primary pilot and Gordon Cooper as backup. This decision underscored NASA's commitment to expanding the boundaries of human space exploration during the Mercury era.

Simultaneously, NASA's Office of Advanced Research and Technology appointed Dr. Eugene B. Konecci as Director of Biotechnology and Human Research. Tasked with developing life support systems and enhancing human performance in space, Dr. Konecci's leadership heralded new advancements in astronaut safety and mission success.

On June 28, 1962, the Manned Spacecraft Center collaborated with Langley Research Center to conduct critical acoustic tests on ablation materials. These materials were intended to provide lightweight heat protection for future Apollo-class vehicles, laying the groundwork for NASA's ambitious lunar exploration goals.

Engineering efforts completed the spacecraft reaction control system reserve fuel tank by June 29, 1962. This enhancement bolstered the Mercury spacecraft's capabilities for extended-range missions, facilitating longer-duration flights crucial to NASA's evolving space exploration objectives.

As June drew to a close, the Manned Spacecraft Center's workforce expanded to 1,802 personnel, underscoring the program's increasing scale and complexity. This growth reflected America's steadfast commitment to leading the world in space exploration, laying the foundation for future achievements beyond Earth's atmosphere.

On July 1, 1962, NASA's Manned Spacecraft Center completed its move from Langley Field to Houston, Texas, marking a pivotal shift in the operational base for America's manned spaceflight endeavors. This relocation to Houston solidified the city as a hub for NASA's ambitious missions and laid the groundwork for future milestones in space exploration.

A week later, on July 8, 1962, Operation Dominic achieved a historic milestone with the successful detonation of a megaton-plus hydrogen device over Johnston Island in the Pacific. This high-altitude thermonuclear explosion, observed from Wake Island to New Zealand, illuminated the Pacific sky and contributed crucial data on the effects of nuclear blasts on the Van Allen radiation belts. Subsequent satellite observations revealed the creation of an artificial radiation belt, furthering scientific understanding of Earth's upper atmosphere.

The following day, July 9, 1962, NASA scientists identified a phenomenon known as "airglow," previously reported by astronauts Glenn and Carpenter. Carpenter measured this luminous layer using a photometer, which spans about 2 degrees across the night sky. Airglow, primarily caused by chemical reactions in the upper atmosphere, contributes significantly to the natural illumination of Earth's night sky.

On July 11, 1962, NASA officials decided to adopt Lunar Orbit Rendezvous (LOR) as the primary mission mode for Project Apollo. This strategic choice, informed by extensive study and analysis over a year, enabled NASA to accelerate planning, research and development, procurement, and testing programs for the ambitious lunar exploration missions ahead.

On July 13, 1962, tests were conducted with a subject wearing a Mercury pressure suit equipped with a B-70 (Valkyrie) harness in a modified Mercury spacecraft couch. Initial results suggested advantages over the existing Mercury harness, prompting plans for further evaluation in spacecraft tests to enhance astronaut safety and comfort during missions.

President John F. Kennedy made a notable announcement on July 21, 1962, honoring Robert R. Gilruth, Director of the Manned Spacecraft Center, with the President's Award for Distinguished Federal Civilian Service. This prestigious accolade recognized Gilruth's leadership in achieving the milestone of manned orbital flight, marking a significant achievement in American aerospace history.

By July 27, 1962, the Mercury-Atlas 8 (MA-8) mission moved closer to launch readiness with the inspection and acceptance of the Atlas launch vehicle No. 113-D at Convair. This milestone underscored the meticulous preparation and progress toward America's next manned orbital mission, solidifying NASA's commitment to pushing the boundaries of space exploration.

On August 6, 1962, the historic Friendship 7 spacecraft, famously flown by John Glenn during the Mercury-Atlas 6 (MA-6) manned orbital mission, was prominently displayed at the Century 21 Exhibition in Seattle, Washington. This public exhibition celebrated Glenn's groundbreaking flight and showcased America's achievements in space exploration. Subsequently, on February 20, 1963, the spacecraft was formally transferred to the National Air Museum of the Smithsonian Institution, where it remains a cherished artifact of American space history.

Two days later, on August 8, 1962, Spacecraft 9 (later redesignated as 9A) was integrated into the Project Orbit program, preparing for the Mercury extended-range or 1-day mission. Actual testing commenced in September 1962, marking significant progress towards NASA's goals of extended-duration space missions. Concurrently, Atlas launch vehicle 113-D was delivered to Cape Canaveral in readiness for the upcoming

Mercury-Atlas 8 (MA-8) manned orbital mission, highlighting NASA's ongoing preparations for manned spaceflight endeavors.

On August 10, 1962, NASA announced the appointment of Dr. Robert L. Barre as a Scientist for Social, economic, and Political Studies in the Office of Plans and Program Evaluation. Dr. Barre's role focused on assessing the broader implications of NASA's activities and shaping the agency's long-term strategies through comprehensive social, economic, and political analysis.

August 11, 1962, marked the completion of a critical spacecraft reaction control system test. This test provided essential data on thermal management and thruster performance for the Mercury extended-range or one-day mission spacecraft. This testing phase was crucial in refining the spacecraft's operational capabilities for extended missions in space.

From August 11-12, 1962, the Soviet Union achieved a significant milestone in space exploration with the concurrent launches of VOSTOK III and VOSTOK IV. Piloted by Major Andrian G. Nikolayev and Lt. Colonel Pavel R. Popovich respectively, these missions demonstrated near-rendezvous capability in orbit, showcasing Soviet advancements in manned spaceflight technology.

Beginning on August 15, 1962, Navy swimmers designated for the Mercury-Atlas 8 (MA-8) recovery area underwent refresher training at Pensacola, Florida. This training regimen included practical exercises such as installing auxiliary flotation collars on spacecraft and briefings on assisting astronauts during spacecraft egress, ensuring readiness for mission recovery operations.

On August 21, 1962, a technical conference focusing on the Mercury-Atlas 7 (MA-7) manned orbital mission (Carpenter flight) convened at the Rice Hotel in Houston, Texas. This conference provided a platform for detailed discussions and analysis of mission technical aspects, fostering collaboration among key stakeholders in mission planning and execution.

Throughout August and September 1962, negotiations with McDonnell Corporation culminated in spacecraft configuration changes necessary to support the Mercury 1-day manned orbital mission. A June 7, 1962 design engineering inspection outlined these modifications, ensuring spacecraft readiness and compatibility with extended-duration space missions.

These milestones and developments underscored NASA's rapid progress and strategic advancements during a pivotal era in space exploration, laying the groundwork for future missions beyond Earth's atmosphere.

On September 7, 1962, a joint study involving the Atomic Energy Commission, the Department of Defense, and NASA addressed concerns about the artificial radiation belt generated by Operation Dominic and its potential impact on the upcoming Mercury-Atlas 8 (MA-8) manned orbital mission. The study predicted that astronaut Walter M. Schirra would be exposed to approximately 500 roentgen radiation outside the spacecraft during his six-orbit flight. However, rigorous shielding measures integrated into the vehicle structures and flight suit were projected to reduce this exposure to around 8 roentgen on Schirra's skin, well within established safety limits.

The following day, September 8, 1962, Atlas launch vehicle 113-D underwent a static firing at Cape Canaveral. This test aimed to validate modifications designed to enhance engine combustion performance for the MA-8 mission, ensuring the reliability and readiness of the launch vehicle.

On September 10, 1962, NASA announced the postponement of the MA-8 mission to September 28, 1962. This delay allowed additional time for meticulous flight preparations and final checks, underscoring NASA's commitment to mission safety and success.

President John F. Kennedy visited the Manned Spacecraft Center on September 12, 1962, where he toured exhibits showcasing Mercury, Gemini, and Apollo spacecraft hardware. During his visit, NASA disclosed plans to launch a specialized satellite by year-end. This satellite aimed to gather crucial data on radiation effects for future satellite missions and further understand the environment influenced by the artificial radiation belt.

Studies conducted by the Navy Biophysics Branch on September 17, 1962, at the Navy School of Aviation Medicine in Pensacola, Florida, revealed that astronaut John Glenn had received significantly less cosmic radiation during his Mercury-Atlas 6 (MA-6) orbital flight than initially projected. This finding highlighted the effective protective measures incorporated into the MA-6 spacecraft design, ensuring astronaut safety during extended missions in space.

On September 18, 1962, Donald Slayton, one of the original Mercury Seven astronauts, was appointed Coordinator of Astronaut Activities at the Manned Spacecraft Center. This role positioned Slayton to oversee astronaut training and mission readiness, contributing to the operational efficiency of NASA's manned spaceflight programs.

September 22, 1962, marked preparations for the MA-8 mission, with Walter Schirra planning to use a special 2.5-pound hand camera onboard. This experiment aimed to test photography techniques applicable to future advanced weather satellites like Nimbus, showcasing NASA's efforts to leverage manned missions for broader scientific advancements.

Finally, on September 28, 1962, Walter Schirra conducted a 6.5-hour simulated flight aboard the MA-8 spacecraft. This exercise was supported by a global tracking network comprising 21 ground stations and ships, validating mission procedures and enhancing confidence in the upcoming manned orbital mission's operational readiness.

Sigma 7 (MA-8): On October 3, 1962, Walter Schirra's mission focused on technical evaluations and refining spacecraft systems. Sigma 7 orbited the Earth six times, setting a new U.S. duration record with a nine-hour flight. Schirra's precise control of the spacecraft and efficient use of onboard systems highlighted the advancements in spaceflight technology and operational procedures.

On October 1, 1962, as Tropical Storm "Daisy" loomed, preparations for the Mercury-Atlas 8 (MA-8) mission persisted undeterred. Just two days later, on October 3rd, the historic launch of Mercury-Atlas 8, christened Sigma 7, occurred at Cape Canaveral. Astronaut Walter Schirra commanded the spacecraft on a meticulously planned six-orbit journey. This mission was pivotal, featuring crucial modifications to enhance flight performance: a refined reaction control system and upgraded communication antennas ensured smoother operations than previous endeavors.

Schirra executed what he described as a "textbook flight," navigating through the challenges of adjusting suit temperatures while conducting precision fuel conservation studies. The mission, spanning 9 hours and 13 minutes, culminated in Sigma 7's safe landing northeast of Midway Island, near the USS Kearsarge. Live coverage via the Telstar satellite broadcasted the mission across Western Europe, underscoring global fascination with space exploration.

Following its return on October 5th, Sigma 7 underwent meticulous inspection and postflight maintenance at Hangar S, poised for a future as a permanent exhibit at Cape Canaveral. Meanwhile, pioneering medical assessments by Dr. Charles A. Berry at the Manned Spacecraft Center indicated unexpectedly low radiation exposure for Schirra, contrasting earlier concerns about high-altitude nuclear test fallout.

The significance of MA-8 reverberated through subsequent events: from Schirra's confident post-flight briefing at Rice University on October 7th, to the delivery of Spacecraft 20 for the upcoming MA-9 mission on October 9th. Meanwhile, advancements in high-frequency direction-finding systems aimed to enhance spacecraft tracking capabilities, reflecting ongoing technological strides in space exploration.

By October 15th, accolades flowed as Schirra received the NASA Distinguished Service Medal in his hometown of Oradell, New Jersey. This was followed by the award of Astronaut Wings on October 16th, solidifying his place among Mercury's elite. McDonnell's confirmation of Spacecraft 20's readiness for MA-9 on October 19th underscored the program's rapid progress.

Behind the scenes, coordination efforts led by Major General Leighton Davis and recognition ceremonies, including Outstanding Leadership Awards to key contributors like Maxime A. Faget and George B. Graves Jr., on October 25th, underscored the collaborative spirit driving America's space ambitions.

As October drew to a close, NASA's reorganization under Robert C. Seamans Jr. aimed to streamline operations, emphasizing the program's strategic evolution under dynamic leadership. These developments marked a transformative period in space exploration, setting the stage for extended missions and broader scientific horizons.

On November 1, 1962, a pivotal step in astronaut training occurred as the Mercury Procedures Trainer No. 1, rebranded as the Mercury Simulator was relocated from Langley Field to the Manned Spacecraft Center at Ellington Air Force Base in Houston, Texas. This move marked its readiness for active duty, enhancing training capabilities crucial for upcoming missions.

Chimpanzee Training and Reward Systems

Chimpanzees were trained for spaceflight missions at Hangar S, located at the Holloman Air Force Base in New Mexico. The training involved various exercises to prepare the chimps for the conditions they would experience during spaceflight, such as acceleration and confinement in a capsule.

During training exercises, chimpanzees were rewarded with banana pellets for completing tasks or responding correctly to stimuli. Conversely, mild electric shocks were sometimes used as negative reinforcement for incorrect responses or to discourage specific behaviors.

This training regimen was designed to acclimate the chimps to the spaceflight environment and condition them to respond appropriately during the missions. The data gathered from these training sessions with chimpanzees provided valuable insights into the physiological and psychological effects of space travel on living organisms, helping to pave the way for human spaceflight in the Mercury and subsequent programs.

Tragically, on November 4, 1962, Enos, the chimpanzee celebrated for his two-orbit flight aboard Mercury-Atlas 5 (MA-5) the previous year, passed away at Holloman Air Force Base in New Mexico after two months of observation and treatment. Unrelated to his space mission, his death highlighted the risks and sacrifices inherent in early space exploration endeavors.

November 13, 1962, brought significant announcements for future missions: Gordon Cooper was designated the pilot for Mercury-Atlas 9 (MA-9), slated for April 1963, with Alan Shepard named his backup. Concurrently, the B.F. Goodrich Company achieved a milestone by successfully developing and testing a specialized tinted visor for Mercury helmets, enhancing astronaut visibility and protection during missions.

Preparations intensified on November 16, 1962, with the Manned Spacecraft Center outlining recovery and network support requirements for MA-9 to the Department of Defense. Meanwhile, Mercury spacecraft 15A was delivered to Cape Canaveral in anticipation of the Mercury-Atlas 10 (MA-10) mission, marking continued progress toward extended orbital flights.

By November 28, 1962, as the countdown continued towards MA-9, the Mercury Simulator 2 underwent crucial modifications to simulate the specific conditions of a one-day orbital mission. McDonnell Aircraft Corporation reported intensive efforts in engineering and production to support Project Mercury, underscoring the collaborative effort across aerospace industries.

Additionally, lessons from MA-8 prompted a thorough review of Mercury's orbital timing device, addressing a minor delay during retrofire. These meticulous adjustments aimed to optimize mission precision and astronaut safety, reflecting NASA's commitment to continuous improvement and readiness for the challenges ahead in space exploration.

In early December 1962, preparations for the Mercury-Atlas 9 (MA-9) mission intensified with a pre-operational conference at Patrick Air Force Base, Florida. This critical meeting, held on December 3-4, focused on reviewing plans and assessing the Department of Defense's readiness to support the upcoming one-day orbital flight. Insights gleaned from the successful six-orbit MA-8 mission guided strategic planning and operational contingencies.

On December 4th, news arrived that seven individuals associated with the Manned Spacecraft Center would be honored with monetary awards for their inventive contributions crucial to Project Mercury's development. These accolades underscored the program's reliance on innovative solutions to overcome technical challenges.

Meanwhile, the Massachusetts Institute of Technology's Instrumentation Laboratory, entrusted with developing the Apollo guidance system, pursued new avenues in space navigation. They explored using Earth's sunset limb as a reference point for mid-course observations during MA-9, building on data from earlier Mercury missions like MA-7 piloted by Carpenter.

December 14th brought diplomatic strides as NASA's International Programs Office secured clearance for a survey mission to Changi Air Field in Singapore, essential for contingency recovery operations linked to MA-9. Additionally, informal assurances from the United Kingdom highlighted Aden's potential role in supporting aircraft recovery efforts.

As the month progressed, strategic decisions were made regarding telemetry and communication facilities. Woomera, Australia, a key segment of Mercury's global network, was deemed unnecessary for future flights, streamlining operational logistics.

By December 31st, comprehensive documents outlining Department of Defense support commitments for MA-9 were forwarded to operational units for review, ensuring alignment with mission requirements. Financially, the Mercury spacecraft program, managed by McDonnell Aircraft Corporation under Contract NAS 5-59, incurred significant costs of $135,764,042. Despite this, ongoing efficiencies reduced labor force requirements to approximately 325 personnel, down from a peak of over 1,600 in earlier phases of development.

Looking ahead to MA-9, slated for April 1963, experiments were categorized into three areas: space flight engineering, biomedical research, and space science. These included innovative experiments like a trailing balloon, a flashing beacon on the retro package, and radiation measurement devices aimed at advancing scientific understanding in the unique environment of orbital spaceflight.

As 1962 drew to a close, these preparations underscored the collaborative efforts and technological advancements driving America's pioneering efforts in space exploration, paving the way for future missions and scientific discoveries.

On January 3, 1963, NASA made a significant decision to potentially extend the upcoming Mercury-Atlas 9 (MA-9) mission from its planned 18 orbits to 22 orbits. This extension aimed to enhance the mission's scientific value, allowing for more comprehensive data collection during the prolonged orbital duration.

Just days later, on January 7th, final acceptance tests were completed on the Mercury space flight simulator at Ellington Field, Texas. Initially known as the procedures trainer and originally stationed at Langley Field, this simulator had been relocated to Houston. The meticulous testing involved personnel from the Manned Spacecraft Center and the Farrand Optical Company, ensuring the simulator's readiness for training astronauts in simulated spaceflight conditions.

From January 10-16, 1963, Mercury spacecraft No. 9A underwent rigorous testing at McDonnell Aircraft Corporation's facilities in preparation for Project Orbit Mission Runs 108, 108A, and 108B. Despite encountering cabin pressure issues, the tests demonstrated the spacecraft's capability to sustain a one-day mission lasting 40 hours and 30 minutes, adhering closely to the MA-9 flight plan.

Throughout mid-January, intensive modifications were undertaken on Mercury spacecraft 15A, re-designated as 15B and designated as a backup for the MA-9 mission. These modifications focused on critical components like hand controller rigging and pitch/yaw control valves, ensuring redundancy and reliability for mission success.

Meanwhile, preparations for scientific experiments continued to evolve. The Manned Spacecraft Center proposed repeating the ground light visibility experiment conducted during Walter Schirra's MA-8 flight. This experiment evaluated astronauts' ability to visually discern ground lights of known intensities from orbit, contributing valuable insights into space navigation and observation capabilities.

By January 17th, Dr. Robert C. Seamans addressed a Congressional committee, affirming expectations that the original Mercury program and the forthcoming MA-9 mission would be completed within fiscal year 1963. Contingency plans, including backup launch vehicles and spacecraft, were poised to ensure mission objectives could be met should additional test data be required.

As training intensified, the Simulations Section devised a comprehensive simulator training plan tailored for the MA-9 mission. Astronauts engaged in over 20 simulated launch and reentry scenarios, focusing on critical aspects such as spacecraft attitude control and fuel consumption. Simulated faults were inserted to prepare for a range of potential mission contingencies.

Technical investigations continued, with McDonnell Aircraft Corporation studying the thermal effects of white-painted spacecraft surfaces. Preliminary findings suggested that white-painted patches on previous missions exhibited cooler temperatures than oxidized surfaces, hinting at potential advantages for extended-duration missions. Further analysis was slated to refine these findings.

On January 26th, the Manned Spacecraft Center (later renamed Johnson Space Center) made significant announcements regarding specialized assignments for its astronaut corps, outlining roles across various future projects:

L. Gordon Cooper and Alan B. Shepard were designated for piloting phases within Project Mercury, emphasizing their roles in advancing the early manned space missions.

Virgil I. Grissom was assigned to Project Gemini, NASA's program focusing on developing capabilities for longer-duration space missions and rendezvous.

John H. Glenn was tasked with responsibilities related to Project Apollo, NASA's ambitious lunar exploration program aimed at landing humans on the Moon.

M. Scott Carpenter was assigned to lunar excursion training, highlighting his role in preparing for future lunar missions.

Walter M. Schirra received Gemini and Apollo operations and training responsibilities, emphasizing his contributions to both programs.

Donald K. Slayton assumed the role of Coordinator of Astronaut Activities, overseeing and coordinating the assignments and activities of the astronaut corps.

These assignments were tailored to leverage each astronaut's expertise and capabilities, reflecting NASA's strategic approach to deploying human resources aligned with its expanding space exploration goals. The announcement underscored NASA's commitment to preparing for increasingly complex missions and milestones in the years to come.

On January 27th, John A. Powers from the Manned Spacecraft Center hinted to Texas Associated Press editors about the ambitious plans for Gordon Cooper's upcoming Mercury-Atlas 9 (MA-9) mission. The significant detail revealed was the potential extension of the mission to up to 22 orbits, which could span approximately 34 hours in total duration. This announcement underscored NASA's commitment to maximizing the scientific returns and operational achievements from every mission opportunity in human spaceflight exploration.

Extending the MA-9 mission to such a duration would have allowed Gordon Cooper to conduct extensive scientific experiments, observations, and tests in orbit. It reflected NASA's strategic goal of pushing the boundaries of human endurance and spacecraft capabilities, gathering valuable data on human physiological responses to extended periods in space, spacecraft performance over long durations, and the effects of space radiation.

Moreover, planning for a 34-hour mission highlighted NASA's confidence in the Mercury spacecraft, the Atlas launch vehicle, and the ground support infrastructure. It demonstrated NASA's evolving capabilities in mission planning, orbital mechanics, and operational logistics necessary for longer-duration space missions.

In early February 1963, NASA's Mercury program experienced several pivotal events and intensive preparations in anticipation of the Mercury-Atlas 9 (MA-9) mission. These activities underscored NASA's meticulous and rigorous approach to advancing manned space exploration.

Key developments during this period included:

Preparations for MA-9 Mission: NASA intensified its preparations for the Mercury-Atlas 9 mission, which would be the final manned mission of the Mercury program. This mission aimed to orbit astronaut Gordon Cooper around the Earth for an extended period, marking a significant milestone in American spaceflight.

Flight Readiness Reviews: NASA conducted comprehensive flight readiness reviews to ensure that all systems, including the spacecraft, Atlas launch vehicle, ground support, and mission operations, were ready for the upcoming MA-9 mission. These reviews involved meticulously scrutinizing technical data, simulations, and contingency plans to mitigate risks.

Training and Simulation: Astronaut Gordon Cooper and the mission control teams engaged in extensive training and simulations. These sessions focused on mission procedures, emergency scenarios, spacecraft operations, and communication protocols to ensure readiness for all possible mission contingencies.

On February 1st, Kenneth S. Kleinknecht, Manager of the Mercury Project Office, made significant announcements regarding preparations for the Mercury-Atlas 9 (MA-9) mission. He confirmed that a peroxide expulsion experiment planned for the mission had been canceled. This experiment likely involved testing the behavior of peroxide in the space environment, which was intended to gather scientific data but was deemed unnecessary or unfeasible for this specific mission.

Despite canceling the peroxide experiment, Kleinknecht also affirmed that the zodiacal light experiment would proceed as planned. The zodiacal light experiment aimed to observe and study a faint glow caused by sunlight scattering off interplanetary dust in the plane of the Solar System, providing insights into cosmic dust distribution and dynamics.

In addition to these updates, modifications to the astronaut gloves were underway. These modifications were specifically focused on enhancing the astronauts' ability to operate cameras during the mission. This adjustment was crucial for the success of the mission's scientific objectives, as it aimed to improve the handling and dexterity of the astronauts while performing visual observations and experiments in space.

On February 5th, NASA's Manned Spacecraft Center encountered a setback. Electrical wiring issues were identified in the control system of the Atlas launch vehicle assigned to the Mercury-Atlas 9 (MA-9) mission, provoking a critical delay in the mission schedule.

The electrical wiring issues in the Atlas launch vehicle's control system were serious enough to warrant immediate attention. Given the complexities and potential hazards involved in launching a manned spacecraft, particularly one intended for multiple orbits like MA-9, ensuring the reliability and safety of the launch vehicle's control systems was paramount.

As a result of this setback, NASA initiated a thorough inspection and rewiring process. This delay was crucial as it allowed engineers and technicians the necessary time to address the electrical issues comprehensively. By conducting these inspections and making necessary corrections, NASA aimed to mitigate any risks associated with the control system malfunction and to ensure the overall safety and success of Gordon Cooper's upcoming mission.

Delays in space missions, especially due to technical issues like wiring problems, are common as they prioritize safety and mission integrity. For the MA-9 mission, this delay underscored NASA's commitment to rigorous testing and preparation, ensuring that all systems were functioning optimally before proceeding with the launch.

From February 5th to February 14th, personnel from NASA's Manned Spacecraft Center meticulously reviewed the status of spacecraft 15B and 20 at McDonnell's St. Louis plant. This review was crucial to assessing and enhancing the readiness of these spacecraft for upcoming missions within the Mercury program.

During this period, fifteen modifications were implemented for spacecraft 15 B. These modifications primarily focused on refining its control panel, interior setup, and external connections. The goal was to ensure that spacecraft 15B met stringent operational standards and was fully prepared to fulfill its role in future missions. These modifications were likely aimed at improving functionality, reliability, and crew comfort, factors critical for successful manned spaceflight operations.

In contrast, spacecraft 20 underwent the status review without requiring additional modifications beyond those already implemented. This spacecraft closely adhered to simulator configurations, indicating that it already met the necessary specifications and was deemed ready for its intended missions without further adjustments.

Such detailed status reviews and modification processes are standard practices in the aerospace industry, especially in NASA's manned spaceflight programs. They ensure that spacecraft are thoroughly inspected, any necessary improvements are made, and all systems function optimally before they are cleared for mission assignments. This rigorous approach underscores NASA's commitment to safety, reliability, and mission success in human space exploration endeavors during the Mercury era.

On February 7th, NASA conducted a Development Engineering Inspection for spacecraft 15B and identified 42 necessary alterations to refine its design for the upcoming MA-10 mission. This inspection and subsequent modification process highlight NASA's dedication to ensuring the reliability and performance standards necessary for manned space exploration during the Mercury program.

Development Engineering Inspections are critical milestones in the spacecraft development process. They involve comprehensive reviews by engineering teams to assess every aspect of the spacecraft's design, systems, and components. The goal is to identify any deficiencies, potential risks, or areas where improvements can be made to enhance functionality, safety, and mission success.

The identification of 42 necessary alterations in spacecraft 15 B underscores NASA's meticulous approach to refining and optimizing the spacecraft's design. These alterations likely addressed various aspects such as structural integrity, electrical systems, propulsion, instrumentation, and crew accommodations. Each modification is intended to ensure that the spacecraft meets or exceeds the rigorous standards required for its intended mission, in this case, the MA-10 mission.

By February 12th, NASA published the objectives for the MA-9 mission, which aimed to achieve nearly 22 orbits to evaluate the effects of prolonged spaceflight on astronauts. The primary objectives outlined for MA-9 included:

Assessing Astronaut Capability in Weightlessness: This objective focused on studying how astronauts adapt to the weightless environment of space over an extended duration. It aimed to gather data on physiological changes, spatial orientation, and overall performance in microgravity.

Monitoring Spacecraft Performance: MA-9 aimed to thoroughly evaluate the performance of the Mercury spacecraft systems during an extended mission. This included testing life support systems, telemetry, communication systems, and thermal control mechanisms to ensure they functioned effectively throughout the mission.

Validating the Global Tracking Network: Another critical objective was to validate the effectiveness of the global tracking network for monitoring and communicating with spacecraft during orbital missions. This network was essential for maintaining constant communication with the astronaut and tracking the spacecraft's position and health parameters throughout its orbit.

These objectives underscored NASA's strategic goals of advancing manned space exploration capabilities and understanding the physiological and operational challenges associated with extended space missions. MA-9 represented a significant step forward in preparing for future longer-duration spaceflights, contributing valuable data to enhance astronaut safety, spacecraft reliability, and mission planning for subsequent projects like Gemini and Apollo.

From February 18th to 22nd, McDonnell Aircraft Corporation reported successful results from Project Orbit Run 109, a 100-hour simulated mission aimed at validating MA-9's capability for a 1-day mission despite initial challenges. This milestone affirmed the spacecraft's readiness for extended missions.

Project Orbit Run 109 was crucial in demonstrating that the Mercury spacecraft, designated for the MA-9 mission, could function effectively and sustainably over a prolonged period similar to the planned duration of the MA-9 flight. The simulation likely subjected the spacecraft to various operational scenarios, including tests of its life support systems, thermal management, communication systems, and overall structural integrity.

The successful completion of Project Orbit Run 109 provided reassurance to NASA and McDonnell Aircraft Corporation that the Mercury spacecraft was well-prepared for the demands of the MA-9 mission. It validated the spacecraft's capability to support astronaut Gordon Cooper during his approximately 34-hour flight, confirming that all systems could operate reliably throughout the extended duration in space.

On February 20th, marking the first anniversary of John Glenn's historic MA-6 flight, Kenneth S. Kleinknecht emphasized that cumulative manned orbital time had surpassed 1,144 minutes. This milestone not only underscored humans' adaptability in space but also highlighted the reliability of spacecraft systems.

Accumulating over 1,144 minutes of manned orbital time was a testament to the success and progress made in manned space exploration through projects like Mercury. It demonstrated NASA's capability to sustain human presence in space for extended periods, paving the way for future ambitious missions such as Gemini and Apollo.

Kleinknecht's remarks also acknowledged the collaborative efforts among various stakeholders, including government agencies, industry partners, and academia. These cooperative endeavors were crucial in advancing space exploration capabilities and fostering confidence in tackling more complex missions in the future.

Another notable event occurred on February 20th, alongside commemorating the first anniversary of John Glenn's MA-6 mission: the Smithsonian Institution received Friendship 7, the spacecraft carrying Glenn during his historic orbital flight. This momentous occasion included the spacecraft itself, John Glenn's flight suit, and other symbolic artifacts from the MA-6 mission.

The transfer of Friendship 7 and its associated artifacts to the Smithsonian marked a significant milestone in space history. It honored Glenn's pivotal role as the first American to orbit the Earth and celebrated the technological achievements and scientific advancements made during the early days of human spaceflight.

By preserving Friendship 7 and Glenn's personal items, the Smithsonian Institution ensured that future generations could learn about and appreciate the courage, dedication, and scientific contributions of early astronauts like John Glenn. This event underscored the cultural and historical significance of the Mercury program and its lasting impact on space exploration.

On February 21st, following the significant milestone of transferring Friendship 7 to the Smithsonian Institution, preparations for the MA-9 mission continued with meticulous attention to detail:

Gordon Cooper and Alan Shepard, the astronauts assigned to MA-9, engaged in detailed experiment briefings. These sessions were crucial for familiarizing them with the specific scientific objectives and procedures planned for the mission.

Simultaneously, McDonnell Aircraft Corporation began preparations for ultra-high frequency (UHF) transceivers. These preparations aimed to ensure effective communication capabilities, particularly crucial during critical phases of the MA-9 mission. The UHF transceivers were designed to facilitate communication over distances of up to 20 miles, enhancing operational effectiveness and astronaut safety during the mission.

These activities underscored the rigorous planning and technical readiness efforts leading up to MA-9, highlighting NASA's commitment to conducting successful manned space missions and advancing scientific exploration objectives in space.

On February 23rd, 1963, several critical developments contributed to the ongoing preparations for the Mercury-Atlas 9 (MA-9) mission and highlighted NASA's commitment to meticulous planning and collaboration in human space exploration:

Spacecraft 20 Checkout and Adjustments: At Cape Canaveral, extended checkout procedures and hardware adjustments were underway for Spacecraft 20. These activities were necessitated by the removal of experimental hardware and testing challenges. The meticulous nature of these adjustments ensured that the spacecraft was thoroughly prepared and met all mission requirements before launch.

Air Force Mercury Tracking Chart Update: The Air Force published an updated Mercury tracking chart. This update was crucial as it accommodated missions up to 22 orbits. The ability to track the Mercury spacecraft across extended mission durations was vital for global monitoring and ensuring mission success. It reflected the collaborative efforts between NASA and the Air Force in establishing and maintaining a robust tracking and communications network necessary for manned space missions.

On March 1, 1963, Spacecraft 9A, which was being prepared for a manned 1-day mission, underwent Project Orbit Run 110. This test specifically evaluated the reaction control system (RCS), which is crucial for maneuvering and stabilizing the spacecraft in space. The test identified areas for improvement in the pressurization and fuel systems of the RCS. As a result, these systems were modified to ensure they function effectively during the upcoming mission. This rigorous testing and subsequent adjustments underscored NASA's commitment to refining spacecraft systems and ensuring their reliability for extended-duration missions like the upcoming Mercury-Atlas 9 (MA-9) flight.

NASA Headquarters published a study on March 5, 1963, focusing on ejecting an instrument package from an orbiting spacecraft. This study was critical as it aimed to determine the observation requirements necessary for future Gemini rendezvous flights. Rendezvous, particularly for the upcoming Gemini program, was a pivotal capability that NASA sought to develop for more complex space missions, including lunar missions planned under the Apollo program. Understanding the logistics and requirements for deploying and observing instrument packages from orbiting spacecraft was essential for advancing these capabilities and ensuring successful future missions.

On March 11, 1963, McDonnell reviewed the clearances between Mercury spacecraft 15B's retropack and the launch vehicle adapter during separation maneuvers, as requested by the Manned Spacecraft Center. The study confirmed that adequate safety margins were in place, ensuring that the spacecraft's retropack could safely separate from the launch vehicle adapter during mission operations. This verification was crucial for mission safety and success, as it ensured that all components would function as intended without risk of collision or interference during critical maneuvers in space.

On March 15, 1963, General Dynamics conducted the factory roll-out inspection of Atlas launch vehicle 130. This inspection had been delayed by 15 days due to flight control wiring rework. Despite the delay, the vehicle successfully passed inspection and was subsequently shipped to Cape Canaveral on March 18, 1963. This milestone marked a significant step forward in preparations for upcoming missions, underscoring the meticulous attention to detail and readiness efforts essential for successful launches during the Mercury era of space exploration.

On March 19, 1963, a slow-scan television camera system was integrated from Lear Siegler, Incorporated into the Mercury-Atlas 9 (MA-9) mission. Weighing 8 pounds, this system captured interior and exterior views during Gordon Cooper's historic flight. This technological addition aimed to provide visual documentation of various aspects of the mission, enhancing scientific observations and public engagement with the space exploration efforts of the Mercury program.

On March 22, 1963, the National Rocket Club awarded John Glenn the Robert H. Goddard Trophy in recognition of his significant contributions to missile, rocket, and space flight programs. This prestigious award highlighted Glenn's pioneering achievements and leadership in advancing aerospace technology and exploration during the early years of space exploration.

Gordon Cooper and Alan Shepard underwent runs on the Johnsville centrifuge on March 28, 1963, to review and acclimate to the acceleration profile planned for the MA-9 mission. This centrifuge testing was crucial for familiarizing the astronauts with the G-forces they would experience during various mission phases, ensuring their preparedness for the upcoming flight.

In early April 1963, Gordon Cooper and Alan Shepard embarked on a crucial mission preparation phase at the Morehead Planetarium in North Carolina. Their visit focused on several key aspects of mission training:

Reviewing the Celestial Sphere Model: They refreshed their knowledge of celestial navigation by studying the celestial sphere model, essential for accurately determining spacecraft orientation and position using celestial bodies as references.

Refining Star Navigation Techniques: Cooper and Shepard honed their skills in star navigation techniques, crucial for maintaining course and orientation during the MA-9 mission, especially in the absence of visual landmarks in space.

Participating in Simulations: They engaged in simulations that likely simulated aspects of the MA-9 mission experiments, preparing them for the tasks and procedures they would perform during the actual flight.

This intensive preparation at the Morehead Planetarium underscored NASA's commitment to meticulous training and readiness, ensuring that Cooper and Shepard were well-prepared for the challenges and scientific objectives of the upcoming MA-9 mission.

At Langley Research Center, preparations for the Mercury-Atlas 9 (MA-9) mission were in full swing. Engineers meticulously set up the tethered balloon experiment at Cape Canaveral, while astronauts Gordon Cooper and Alan Shepard immersed themselves in rigorous training sessions from April 10-11. Their training regimen covered spacecraft egress, helicopter pickups, dinghy boarding, and the deployment of survival equipment, ensuring they were ready for any scenario during the mission.

By mid-April, the focus shifted to the Manned Spacecraft Center where detailed planning for MA-9 was underway. A comprehensive flight plan was meticulously crafted, outlining the scheduled experiments and observations defining Cooper's mission. The plan even allocated an 8-hour sleep period, highlighting the meticulous minute-by-minute scheduling to maximize mission efficiency.

As launch day neared, astronauts and support teams underwent extensive briefings on recovery procedures, communication protocols, spacecraft systems, and mission protocols. On April 22, 1963, spacecraft 20 was transported to Complex 14 at Cape

Canaveral and integrated with the Atlas launch vehicle, a pivotal milestone in mission readiness.

Concurrently, advancements in spacecraft technology were unfolding. The Bendix Corporation finalized an innovative airlock design for the Mercury spacecraft, tailored to capture micrometeorites during orbital flights. This modular airlock concept would later prove indispensable, finding adaptation and refinement in the Gemini and Apollo programs, thus enhancing its scientific versatility.

Amid these technical strides, Gordon Cooper's pressure suit underwent significant upgrades. Improved helmet seals, enhanced gloves, and a more flexible torso section were integrated to enhance comfort and mobility during the extended duration of the MA-9 mission. These modifications underscored NASA's commitment to advancing astronaut safety and performance in the demanding space environment.

Gordon Cooper's Faith 7 mission, launched on May 15, 1963, marked the culmination of NASA's Mercury program. It was the longest and final Mercury mission, lasting over 34 hours and encompassing 22 orbits around Earth. Cooper's objective was clear: to push the boundaries of the Mercury spacecraft's endurance and assess the astronaut's capability to function effectively during an extended spaceflight.

During Faith 7, Cooper successfully executed experiments and observations to advance scientific understanding and operational procedures in space. His manual control of reentry, a critical maneuver, demonstrated the skill and precision of NASA's astronauts. Faith 7's safe landing in the Atlantic Ocean underscored the program's achievements in astronaut safety and mission success.

The Mercury missions collectively laid a solid foundation for human space exploration, enhancing our understanding of spaceflight's challenges and capabilities. These achievements were pivotal in preparing for the ambitious Apollo missions, ultimately culminating in astronauts landing on the Moon. Project Mercury showcased the bravery and expertise of its astronauts and highlighted the dedication and innovation of the engineers and scientists who propelled humanity toward the stars.

On May 12, 1963, Dr. Charles A. Berry, from the Aerospace Medical Operations Office, officially confirmed Gordon Cooper's exceptional physical and mental readiness for the upcoming Faith 7 mission. This endorsement marked a pivotal moment in America's space exploration history, solidifying Cooper's preparedness to embark on the longest and final Mercury flight.

Dr. Berry's assessment underscored the rigorous training and meticulous preparation Cooper and the entire NASA team underwent. It highlighted the critical role of medical science in ensuring astronaut health and performance under the extreme conditions of space. With Cooper declared fit for duty, all was set for the mission that would test the limits of both human endurance and spacecraft capabilities, further advancing America's quest to conquer the challenges of space exploration.

Ticker tape parade for Gordon Cooper in New York City, May 1963

The Press Convergence: May 12-19

In the warm mid-May of 1963, Cape Canaveral buzzed with global media activity as over 1,020 reporters, commentators, and technicians gathered on the Florida coast for a momentous occasion: NASA's Mercury-Atlas 9 (MA-9) mission, a landmark in America's pioneering Project Mercury.

The atmosphere crackled with excitement as journalists from across the United States and worldwide set up camp, their equipment humming with anticipation. An additional 130 media professionals stationed at Hawaii's NASA News Center added to the comprehensive coverage, ensuring every aspect of the mission was meticulously documented.

During these bustling days, Western Union estimated that reporters churned out approximately 600,000 written content, with 140,000 words swiftly transmitted across the Atlantic to eager European audiences. Yet, this torrent of information only scratched the surface, as numerous stories were phoned in and reported from the Pacific News Center, complemented by extensive radio and television broadcasts.

Television was pivotal in bringing the mission to life for global audiences. Viewers were treated to unprecedented glimpses inside and outside the spacecraft, with nearly two hours of meticulously planned broadcast time showcasing the astronaut's activities and the vast expanse of space. Satellite technology bridged continents, beaming live images to Europe and beyond, ensuring this monumental event was witnessed in homes far and wide.

Beyond its scientific achievements, the MA-9 mission highlighted NASA's capacity to captivate and inform a worldwide audience, setting a precedent for the era of space exploration that was yet to unfold fully.

The Delayed Launch: May 14

On May 14, 1963, Cape Canaveral brimmed with anticipation as the countdown for the Mercury-Atlas 9 mission commenced. Engineers, technicians, and astronaut Gordon Cooper readied themselves for the historic launch. However, just 60 minutes before liftoff, a critical setback halted proceeding: a malfunction in the fuel pump of the gantry's diesel engine, crucial for moving the launch structure away from the rocket.

The gantry, encasing the spacecraft and its towering Atlas rocket, stood silent amid the unexpected delay. The air was thick with the smell of fuel, blending with the tension permeating the control center. Engineers worked urgently to resolve the issue, and repairs were completed after a nerve-wracking 129-minute delay.

Yet, as optimism began to return, another blow struck, this time at the Bermuda tracking station. A malfunction in a vital computer converter responsible for key orbital decisions forced a countdown halt at T-13 minutes. Disappointment swept through the control rooms as the mission was scrubbed for the day.

In the evening, senior NASA official Walter C. Williams addressed a gathered press corps, his demeanor a mix of determination and frustration. He confirmed the repairs at Bermuda and announced the rescheduling of the mission for May 15, 1963. The room buzzed again with renewed anticipation, the prospect of a second attempt hanging in the air like a promise of redemption.

The events of May 14 vividly underscored the unpredictable nature of space exploration and the resilience required to persevere through setbacks. As the world awaited the next day's launch, the collective hope was palpable, knowing each attempt brought humanity closer to the thresholds of space and the promise of new discoveries.

Launch Day: May 15-16

In the quiet predawn of May 15, 1963, Cape Canaveral's launch complex was illuminated by soft floodlights, casting a serene glow over the sprawling site. The rhythmic hum of generators and the subdued chatter of technicians filled the air as final preparations for Mercury-Atlas 9, known as Faith 7, unfolded. At 5:33 a.m. EDT, astronaut L. Gordon Cooper, dressed in his silver pressure suit, calmly entered the compact confines of his spacecraft. Every movement was executed precisely, reflecting the meticulous training that had prepared him for this historic mission.

Lt. Colonel John A. Powers' voice resonated over the loudspeakers, reassuring NASA personnel and the gathered media that, barring unforeseen technical issues, the launch was scheduled for 8:00 a.m. EDT. Setting into his seat, Cooper took a moment for a brief nap, a testament to his ability to maintain composure under immense pressure.

The countdown progressed smoothly until T-11 minutes and 30 seconds when a minor glitch in the guidance equipment prompted a brief hold. The control room tensed momentarily, but the issue was swiftly resolved. Another pause at T-19 seconds ensured the automatic sequencing systems were in order. Hearts raced in unison with the countdown, culminating in a thunderous roar at 8:04 a.m. as the Atlas rocket ignited, propelling Faith 7 into the clear morning sky.

Visual tracking followed the spacecraft's ascent through the cloudless atmosphere for two minutes. As the booster engine cut off and the escape tower was jettisoned, the sustainer engine maintained its burn, guiding Faith 7 into orbit. At 8:09 a.m. EDT, the spacecraft achieved near-perfect insertion, hurtling through space at an astonishing speed of 17,546.6 miles per hour. The flight's trajectory reached a perigee of approximately 100.2 statute miles and an apogee of 165.9 miles.

Inside Faith 7, Cooper swiftly adjusted to the conditions, stabilizing cabin temperatures at 92 and 109 degrees Fahrenheit, ensuring comfort within manageable limits. By the second orbit, he settled into the routine, taking advantage of the conducive environment to rest briefly. Cooper's steady performance and the spacecraft's flawless operation as the mission continued, underscored a remarkable day in America's space exploration journey.

Orbits and Experiments

During the historic Mercury-Atlas 9 mission on May 15, 1963, L. Gordon Cooper achieved several groundbreaking milestones as he orbited the Earth aboard Faith 7. Cooper successfully deployed a flashing light experiment in the third orbit, marking a man-launched satellite's first operation in orbital flight. This accomplishment was a significant step forward in space exploration, demonstrating human capability to conduct experiments in space.

Cooper's dedication paid off further when he spotted the beacon on the night side of the fourth orbit, validating the experiment's functionality. However, an attempt to deploy a balloon during the sixth orbit was unsuccessful despite repeated efforts. Undeterred, Cooper pressed on with the mission, critically observing a ground light in South Africa during the same orbit—an observation that would prove invaluable for future Gemini and Apollo missions.

Faith 7 entered a programmed drifting flight phase by the eighth orbit, allowing Cooper a well-deserved rest period. Communicating with the telemetry command ship, Rose Knot Victor, off the coast of Chile, Cooper informed them of his intention to begin his rest period. He even spoke with fellow astronaut John Glenn during the ninth orbit before sleeping as Faith 7 completed its tenth orbit.

Cooper monitored suit temperatures closely throughout his rest, adjusting as needed to ensure his comfort. Despite a slight increase in temperature, he resumed his rest, waking refreshed and alert during the fourteenth orbit to establish contact with Muchea, Australia. Cooper's ability to manage the mission's operational and personal aspects highlighted his skill and resilience in the demanding space environment.

The accomplishments of the Mercury-Atlas 9 mission, including Cooper's pioneering experiments and steadfast performance, paved the way for future space exploration endeavors, shaping NASA's approach to manned spaceflight in the years to come.

The First Manual Reentry

During the nineteenth orbit of the Mercury-Atlas 9 mission on May 16, 1963, tranquility gave way to a brief moment of uncertainty when a .05g light illuminated on Faith 7's instrument panel. Typically signaling orbital decay, telemetry data indicated no such issue, attributing the light's activation to a possible water-induced relay short-circuit—an erroneous alert.

In response to the anomaly, mission control instructed Gordon Cooper to prepare for manual reentry, a departure from the automated systems used in previous missions. Cooper remained composed and methodical, manually firing the retrorockets and using

scribe marks on his observation window to maintain the correct reentry attitude. Over the command ship off Japan's coast, fellow astronaut John Glenn provided the countdown for the retrosequence and guidance on jettisoning the retropack.

As Faith 7 descended through Earth's atmosphere, the main parachute deployed flawlessly at 11,000 feet. The spacecraft gently splashed down 7,000 yards from the awaiting USS Kearsarge, marking the end of a remarkable 34-hour, 19-minute, and 49-second mission. Cooper remained inside until safely hoisted aboard the carrier, ensuring every post-splashdown procedure was meticulously followed, cementing the success of NASA's longest manned orbital mission.

Mission Success and Reflections

The Mercury-Atlas 9 mission, culminating in Gordon Cooper's triumphant return aboard Faith 7, was celebrated as an unequivocal triumph. Cooper's meticulous management of essential resources—electrical power, oxygen, and attitude control fuel—was particularly noteworthy throughout the mission.

By the fifteenth orbit, Cooper had preserved 75 percent of the primary oxygen supply, prompting light-hearted banter from mission control. Playfully dubbing him a "miser," they joked that he should "stop holding his breath." This humorous exchange highlighted Cooper's disciplined approach to resource management, ensuring that critical supplies would sustain him comfortably until reentry and splashdown.

Cooper's ability to stretch resources while maintaining mission objectives underscored his skill and preparedness. This contributed to the overall success of the Mercury-Atlas 9 mission and advanced NASA's capabilities in manned spaceflight.

Contractor Personnel at Cape Canaveral: May 15

On May 15, 1963, the launch day of Mercury-Atlas 9 (MA-9), Cape Canaveral was abuzz with activity as an influx of contractor personnel worked tirelessly to ensure the mission's success. Harold G. Collins, the Contracting Officer, meticulously recorded the manpower involved in a detailed memo dated May 17, 1963.

According to Collins' report, McDonnell allocated 251 personnel to Contract NAS 5-59, with an additional 23 individuals assigned to spacecraft 15B, designated for the upcoming MA-10 mission. Federal Electric Corporation contributed 8 individuals to support various aspects of the operation. This concerted effort transformed Cape Canaveral into a collaboration and technical expertise hub.

The culmination of these combined efforts was the triumphant launch of astronaut Gordon Cooper aboard Faith 7. This achievement underscored NASA's strides in manned space exploration and highlighted the effectiveness of coordinated teamwork between contractors and NASA personnel at Cape Canaveral.

Community Impact

Cocoa Beach, located just south of Cape Canaveral, emerged as a bustling activity center during the early days of space exploration. With each scheduled launch, tens of thousands of spectators flocked to the shores, their gaze fixed on the horizon where the heavens met the Earth. The distant roar of rockets ascending into the sky became an iconic symbol of humanity's relentless drive to reach beyond the bounds of our planet.

As the Mercury, Gemini, and later Apollo missions took flight, Cocoa Beach became a place where many people's dreams of space exploration came true. Families, tourists, and enthusiasts gathered on the sandy coastline, their excitement palpable as they witnessed history unfolding overhead. The launches were not just technical achievements; they were moments of collective awe and inspiration, uniting people from all walks of life in shared wonderment at the vastness of space and the courage of those who ventured into its depths.

The sight and sound of rockets piercing the atmosphere from nearby Cape Canaveral left an indelible mark on the Cocoa Beach community. It reminded the community that amidst the ordinary rhythms of beach life, extraordinary feats were being accomplished above, pushing the boundaries of human knowledge and ambition.

Media and Public Relations

During the era of Project Mercury, the world's fascination with space exploration reached new heights, driven by intense media coverage that turned the Mercury Seven astronauts into national icons. Reporters from around the globe meticulously documented their every move, transforming these pioneering astronauts into symbols of bravery and exploration. Among them was Yuriko "Yuri" Azuma, a daring Japanese journalist renowned for her bold attempts to secure exclusive stories and insights, contributing to the vibrant tapestry of America's early space endeavors.

Yuriko "Yuri" Azuma was a courageous and determined Japanese journalist known for covering the Mercury astronauts during the early years of space exploration. Her exploits were marked by her fearless pursuit of stories about space missions and the astronauts. Yuri Azuma gained recognition for her in-depth reporting and exclusive, reportedly seductive interviews with the Mercury astronauts, providing insights into their lives, training, and the challenges they faced as pioneers of space travel.

As a journalist, Yuri Azuma distinguished herself by breaking barriers and overcoming cultural and logistical challenges to cover the American space program. Her contributions helped bridge the gap between East and West in space reporting, offering a unique perspective to her audience in Japan and beyond.

Yuri Azuma's legacy lies in her pioneering efforts to bring the stories of the Mercury astronauts to a global audience, highlighting their bravery and the scientific advancements of the space age. Her work continues to inspire journalists and space enthusiasts alike, showcasing the importance of investigative journalism in exploring the frontiers of human achievement.

The legacy of Project Mercury endures as a testament to human ingenuity, courage, and the unwavering pursuit of knowledge. It laid the groundwork for subsequent space programs, proving that the once-unthinkable dream of space exploration was possible and within our reach. Looking back, Project Mercury's achievements continue to serve as a beacon of inspiration for future generations, encouraging them to dare to dream and aspire to explore the cosmos beyond our planet's confines.

Post-Mission Press Conference: May 19

On May 19, 1963, astronaut Gordon Cooper captivated a national audience with a televised press conference from Cocoa Beach, Florida, where he reflected on his historic mission aboard Faith 7. The atmosphere was anticipating as Cooper, displaying his characteristic calm demeanor, methodically recounted his experiences from launch to recovery.

He commended Calvin Fowler of General Dynamics for his pivotal role at the console during the Atlas rocket's launch. Cooper then shared captivating insights from his time in orbit, such as observing the atmospheric haze noted by previous astronauts and the mesmerizing "fireflies" that illuminated the space around him, echoing John Glenn's earlier descriptions.

Addressing the challenge of sleeping in space, Cooper explained his technique of anchoring his thumbs to the helmet restraint strap to prevent inadvertent activation of switches—a practical solution born of necessity in weightlessness.

Perhaps most striking were Cooper's detailed visual observations of Earth from orbit. He vividly described identifying various landmarks, including an African town where an experiment was conducted, expansive oil refineries in Perth, Australia, wisps of smoke rising from rural Asian homes, and familiar sights like Miami Beach and Clear Lake near Houston. These observations underscored the clarity and potential of space-based Earth observation missions.

Cooper candidly discussed technical challenges, notably issues with the condensate water pumping system, which required careful monitoring throughout the mission.

When queried about the possibility of a Mercury-Atlas 10 (MA-10) flight, Dr. Robert C. Seamans tempered expectations, indicating it was "quite unlikely," hinting at the evolving priorities and future directions of NASA's manned spaceflight program.

Cooper's articulate and insightful reflections highlighted the successes and challenges of his mission and underscored the pioneering spirit and scientific endeavor that defined NASA's early space exploration efforts.

Presidential Recognition: May 21

On May 21, 1963, a significant ceremony unfolded at the White House, where President John F. Kennedy bestowed the NASA Distinguished Service Medal upon astronaut Gordon Cooper. This event marked a solemn yet celebratory occasion, honoring Cooper's pivotal contributions to the success of Project Mercury and his trailblazing role in advancing human spaceflight.

The ceremony, attended by dignitaries and members of the Mercury operations team, stressed the mission's profound national importance and its place in the annals of American achievement in space.

Alongside Cooper, other key figures instrumental to Project Mercury's triumph were also recognized for their exceptional leadership. G. Merritt Preston, the Manager of Project Mercury Operations at Cape Canaveral, received a commendation for his adept management of the mission's intricate logistics. Floyd L. Thompson from the Langley Research Center, Kenneth S. Kleinknecht, Manager of the Mercury Project Office, and Christopher C. Kraft, Director of the Flight Operations Division at the Manned Spacecraft Center, were honored for their indispensable roles in planning and executing the mission.

Major General Leighton I. Davis, Commander of the Air Force Missile Test Center, was also acknowledged for his crucial support and coordination. This highlighted the collaborative efforts across military and civilian sectors underpinning the success of America's early space endeavors.

The ceremony at the White House underscored the achievements of individuals and the collective effort and national resolve that propelled America to the forefront of space exploration during Project Mercury's pioneering days.

Presidential Insight and Technical Clarifications

On May 22, during a routine press conference, President John F. Kennedy responded to a question concerning the possibility of another Mercury mission with measured consideration. His reply, covered by the New York Times on May 23, highlighted the administration's stance of entrusting NASA with the decision-making process regarding future space missions. President Kennedy emphasized that the determination for any subsequent Mercury missions would ultimately lie within NASA's purview.

His statement underscored the administration's confidence in NASA's scientific and technical expertise, affirming their role as the authority in charting the course of America's space exploration efforts. This approach reflected a commitment to advancing space science and exploration underpinned by rigorous planning and evaluation by the nation's premier space agency.

Clarifying Consumable Usage

Two days later, on May 24, William M. Bland, Deputy Manager of the Mercury Project Office, addressed the Aerospace Writers' Association Convention in Dallas, Texas. He aimed to correct a prevalent misconception about the Mercury spacecraft and its consumables. Bland emphasized that despite the challenges of space travel, essential resources such as electrical power, coolant water, oxygen, and carbon dioxide absorption capacity were meticulously managed with significant safety margins.

He highlighted a notable example: astronaut Walter Schirra had a surplus of 9 hours of primary oxygen supply after his mission. This surplus illustrated the careful planning and robust execution of the Mercury program. Bland's presentation aimed to assure the aerospace community and the public of the program's capability to manage resources effectively throughout its missions, reinforcing confidence in NASA's technical and operational competence in human spaceflight during the pioneering Mercury era.

Comprehensive Defense Support

The Department of Defense (DoD) was pivotal in supporting the Mercury-Atlas 9 (MA-9) mission, a landmark in space exploration detailed in a comprehensive report submitted on May 29. This report underscored the meticulous preparations and substantial resources mobilized to ensure the mission's success and the safety of its astronaut, Gordon Cooper.

Central to this effort was the provision of the Atlas launch vehicle and the strategic positioning of assets for tracking telemetry, and recovery operations. The Air Force Coastal Sentry Quebec, stationed south of Japan, was critical in monitoring and supporting retrofire maneuvers during crucial orbits. Meanwhile, the Atlantic Missile Range telemetry command ship, Rose Knot Victor, commanded operations from strategic positions in the southeast Pacific. The Atlantic Missile Range's C-band radar ship, Twin Falls Victory, was instrumental in tracking reentry from Cape Canaveral to Bermuda. Simultaneously, the Navy's Pacific Missile Range offered similar vital services through its Range Tracker.

The DoD's extensive support network also included fixed island stations and aircraft from multiple branches, ensuring comprehensive coverage. Task Force 140, under Rear Admiral Harold G. Bowen, was positioned in the Atlantic Ocean, prepared for potential recovery operations. In the Pacific, Task Force 130, led by Rear Admiral C.A. Buchanan, comprised an aircraft carrier and ten destroyers supported by strategically located aircraft across the region. Similarly robust forces under Task Force 109, commanded by Rear Admiral B.J. Semes in the Middle East, included a seaplane tender and two destroyers, backed by aircraft stationed in Aden, Nairobi, Mauritius, and Singapore.

Bioastronautic support was equally thorough. 78 medical personnel and 32 specialty team members were on standby, bolstered by nine department hospitals and over 3,400 pounds of medical equipment deployed to ensure the astronaut's health and safety throughout the mission.

During the critical recovery phase, the carrier USS Kearsarge initially saw the spacecraft, part of Task Force 130. Helicopters swiftly deployed to circle Faith 7 as it descended for its final landing. Swimmers from these helicopters secured the flotation collar and retrieved the antenna fairing, ensuring the spacecraft's stability in the water. Cooper adhered to protocol, remaining inside the spacecraft until it was safely hoisted aboard the carrier. A motor whaleboat towed the spacecraft alongside the ship, completing the meticulous recovery operation.

Gordon Cooper's Astronaut Wings

After completing his mission successfully, Gordon Cooper was bestowed with Astronaut Wings. This prestigious honor placed him among the esteemed ranks of fellow Mercury astronauts who had also earned this distinction. This recognition not only celebrated his bravery but also highlighted the flawless execution of his mission, solidifying his legacy in the chronicles of space exploration. Cooper's achievement underscored the courage and dedication required to push the boundaries of human endeavor beyond Earth's atmosphere.

Advocating for MA-10

In early June 1963, the atmosphere at the Manned Spacecraft Center crackled with anticipation and determination. On June 6 and 7, senior officials gathered to present a compelling case to NASA Administrator James E. Webb, advocating fervently for the continuation of Project Mercury through the ambitious Mercury-Atlas 10 (MA-10) mission. The room was a tableau of serious faces and hopeful expressions as the officials argued.

Central to their case was the proven capability of the Mercury spacecraft to undertake longer-duration missions. They stressed that a multi-day flight could provide invaluable insights into the effects of prolonged space exposure, data crucial for the forthcoming Gemini and Apollo programs. Importantly, they underscored the cost-effectiveness of proceeding with MA-10, given that the launch vehicle and the spacecraft were already nearing readiness.

The presentation blended meticulous detail and visionary ambition, resonating with a palpable sense of urgency tempered by cautious optimism. Each slide and statistic carried the weight of NASA's aspirations for advancing human exploration beyond Earth's bounds. As discussions unfolded, the room buzzed with the gravity of their proposal, encapsulating the high stakes and profound potential inherent in their vision for the future of space exploration.

Preparations for MA-10: June 8

Amid ongoing discussions surrounding the potential MA-10 mission, preparations surged ahead with purpose. On June 8, the Manned Spacecraft Center engineers implemented crucial upgrades to spacecraft 15B's environmental control systems. These modifications were focused primarily on enhancing the hardware and flexibility of the urine and condensate systems, a direct response to astronaut Gordon Cooper's feedback.

Cooper highlighted operational challenges with the condensate system during his post-flight press conference following the Faith 7 (MA-9) mission, noting its cumbersome and inefficient nature. The engineers made meticulous adjustments to improve functionality and efficiency in light of these insights.

Their work was driven by immediate practical needs and a broader ambition to prepare for longer and more intricate missions ahead. Every modification was subjected to rigorous testing and scrutiny, reflecting NASA's unwavering commitment to maintaining the highest standards of reliability and performance in space exploration. These efforts underscored NASA's proactive approach to refining spacecraft systems to meet the evolving demands of human spaceflight.

The End of an Era

On June 12, the fate of Project Mercury was decisively sealed as NASA Administrator James E. Webb testified before the Senate Space Committee. His words left no room for ambiguity: "There will be no further Mercury shots." Webb's testimony underscored a clear and unwavering directive—NASA's manned spaceflight efforts would now pivot exclusively towards the Gemini and Apollo programs.

Webb's declaration marked the culmination of Project Mercury, a landmark program that spanned four years, eight months, and one week. It was a moment that carried both finality and forward momentum—a conclusion to America's inaugural manned spaceflight endeavors and a catalyst for the bold missions yet to come.

The atmosphere that followed was one of reflective pride mingled with resolute determination. Project Mercury had achieved its primary objectives, demonstrating unequivocally that humans could endure and function in the harsh space environment. Its legacy, enriched by technological advancements and invaluable scientific insights, would now propel the nation towards even greater accomplishments in space exploration.

As the Gemini and Apollo programs took center stage, they would build upon the foundation laid by Mercury. The pioneering spirit and lessons learned from Mercury would guide the ambitious endeavors that aimed not just to explore but to conquer the challenges of space, culminating ultimately in the historic lunar landings and beyond.

Reflecting on Project Mercury

As the final chapter of Project Mercury drew to a close, the program's remarkable achievements were celebrated, and its profound impact was recognized. Mercury epitomized human ingenuity, courage, and an unwavering quest for knowledge, from the nerve-wracking launches to the triumphant recoveries.

Astronauts like John Glenn and Gordon Cooper had boldly pushed the boundaries of possibility, emerging as national icons and embodiments of American resilience and ambition. Their journeys into space not only captivated the nation but also yielded invaluable data and experience that would shape the trajectory of future space missions.

The insights from Mercury missions would prove pivotal in refining the design and execution of subsequent programs, notably the Gemini and Apollo projects. With NASA's focus now squarely on these ambitious endeavors, the legacy of exploration forged by Mercury was set to endure and expand.

With the prospect of landing humans on the Moon looming large, the lessons learned from Mercury would serve as a cornerstone, ensuring that the dreams of space exploration would persevere and soar to unprecedented heights. Project Mercury's legacy was a testament to past triumphs and a beacon guiding humanity's ongoing voyage into the cosmos.

Public Fascination and Legacy

Project Mercury captivated millions, with people worldwide listening to radio and television broadcasts to follow each mission. The excitement was palpable, with every launch a blend of scientific achievement and human bravery. These missions laid the groundwork for the subsequent Project Gemini, which refined space docking maneuvers crucial for the Apollo program and its goal of landing men on the Moon.

Cancelled Missions

In October 1963, NASA's plans for the Mercury-Atlas 10 mission, with Alan Shepard as pilot, were abruptly halted on June 13, 1963. This mission aimed to extend the duration of orbital flights to one day. However, NASA's decision to cancel it reflected a strategic shift towards Project Gemini. Gemini promised advancements such as longer-duration missions and spacewalks, crucial for progressing toward lunar exploration.

Similarly, Mercury-Atlas 11, intended for the fourth quarter of 1963 with Gus Grissom as pilot, was canceled in October 1962. This cancellation also pivoted towards Gemini, prioritizing its capabilities for extended missions and complex maneuvers.

The final planned Mercury mission, Mercury-Atlas 12, scheduled for the fourth quarter of 1963 with Walter Schirra as pilot, faced a similar fate when it was canceled in October 1962. Schirra and his fellow astronauts transitioned to supporting roles in subsequent programs that continued pushing the boundaries of space exploration.

Strategically, canceling these missions marked NASA's focused shift from proving manned spaceflight's feasibility under Mercury to tackling more complex challenges under Gemini. These challenges included long-duration missions, orbital rendezvous and docking maneuvers, and extravehicular activities (EVA or spacewalks). This pivot set the stage for the Apollo program, which successfully achieved President Kennedy's ambitious goal of landing humans on the Moon and safely returning them to Earth, a monumental achievement in the history of space exploration.

Epilogue: Legacy of the Mercury Seven

Memorial at Cape Canaveral Air Force Station Launch Complex 14

The Mercury Seven astronauts and the Project Mercury program profoundly influenced American culture and the course of space exploration. Their daring missions captured the nation's imagination and galvanized public support for NASA's endeavors during the Space Race.

The Mercury Seven astronauts continued to make significant contributions to space exploration and beyond following their historic Project Mercury missions, leaving a lasting impact on both scientific endeavors and popular culture.

Scott Carpenter After overcoming initial selection challenges with naval duty, Selection Challenge required intervention from Admiral Arleigh Burke to secure his release from naval duties. Scott Carpenter piloted Aurora 7, facing reentry targeting issues that led to a landing 250 miles off-course. His mission provided crucial data for future orbital missions and contributed to spacecraft navigation and reentry technology advancements.

Gordon Cooper Gordon Cooper's personal challenge involved convincing his wife to return during the rigorous astronaut training, maintaining an image of stability. He piloted Faith 7, the final Mercury flight, completing 22 orbits and demonstrating manual reentry control, which was essential for later space missions.

John Glenn John Glenn made history as the first American to orbit the Earth aboard Friendship 7. His post-NASA career included serving as a U.S. Senator, where he continued to advocate for space exploration. Glenn returned to space at 77, aboard STS-95, contributing to scientific research on aging.

Gus Grissom Gus Grissom piloted Liberty Bell 7 during the Mercury program, experiencing a premature hatch blow leading to the capsule's sinking. Despite this setback, Grissom contributed to the Gemini and Apollo programs until his tragic death in the Apollo 1 fire, leaving a legacy of dedication to space exploration.

Wally Schirra Wally Schirra achieved the unique feat of flying in Mercury, Gemini, and Apollo programs. His missions included piloting Sigma 7 in Mercury, executing the first space rendezvous with Gemini 6A, and commanding Apollo 7, the first crewed Apollo mission, which was critical for validating Apollo spacecraft systems.

Alan Shepard Alan Shepard made history as the first American in space, piloting Freedom 7 on a suborbital flight. He later commanded Apollo 14, becoming the fifth person to walk on the Moon and contributing valuable scientific data during lunar exploration missions.

Deke Slayton Initially grounded due to medical reasons, Deke Slayton later flew on the Apollo-Soyuz Test Project, showcasing his resilience and commitment to space exploration. As NASA's first Chief of the Astronaut Office, he played a pivotal role in astronaut selection and mission planning, shaping the future of American space missions.

Collectively, the Mercury Seven astronauts not only pushed the boundaries of human achievement in space but also inspired generations with their courage, dedication, and contributions to scientific discovery and exploration beyond Earth's bounds. Their enduring legacy continues to influence space programs worldwide and remains a cornerstone of space exploration history.

Cultural Icons: The Mercury Seven, comprising Scott Carpenter, Gordon Cooper, John Glenn, Gus Grissom, Wally Schirra, Alan Shepard, and Deke Slayton, became iconic figures synonymous with courage, exploration, and scientific achievement. Their journeys into space, from Alan Shepard's historic suborbital flight to John Glenn's monumental orbital mission, inspired generations and symbolized America's technological prowess during the Cold War era.

Public Engagement: Through extensive media coverage and personal appearances, the Mercury Seven engaged with the public personally, sharing their experiences, insights, and the thrill of exploration. They became national heroes, celebrated for their bravery and dedication to advancing human knowledge beyond Earth.

Technological Advancements: The Mercury program pioneered essential technologies and operational techniques critical for future space missions. From spacecraft design improvements to life support systems and orbital maneuvering capabilities, these advancements laid the groundwork for subsequent NASA programs, including Gemini and Apollo.

Educational Legacy: Beyond their immediate achievements, the Mercury Seven astronauts established the Astronaut Scholarship Foundation, which continues to support STEM education and inspire future generations of scientists, engineers, and explorers. Their commitment to education and outreach ensured their legacy would endure far beyond their time in space.

Honors and Commemorations: The contributions of the Mercury Seven were commemorated through stamps, awards, and public recognition, highlighting their role in shaping America's space legacy. The Mercury 7 monument at Launch Complex 14 stands as a testament to their achievements, with a time capsule preserving their legacy for future generations.

Legacy of Exploration: The Mercury Seven set the stage for America's ambitious lunar exploration efforts and beyond, demonstrating the feasibility of human spaceflight and paving the way for historic missions like the Moon landings. Their spirit of adventure and determination continue to inspire new frontiers in space exploration.

In conclusion, the Mercury Seven astronauts left an enduring legacy of exploration, courage, and scientific achievement that continues to inspire and captivate the world. Their contributions to space exploration and their status as cultural icons ensure their place in history as pioneers who pushed the boundaries of human endeavor beyond Earth.

About the Author

Thornton D. "TD" Barnes, author and entrepreneur, grew up on a ranch at Dalhart, Texas. He graduated from Mountain View High School in Oklahoma and embarked on a ten-year military career. Following a tour as an Army intelligence specialist in Korea, he continued his education, attending two and a half years of missile and radar electronics by day and college courses at night. Midway through the first combat Hawk missile battalion deployment, Barnes attended the Artillery Officer Candidate School, where an injury ended his military career.

Barnes first became involved with US intelligence agencies while in the Army during the Agency's Project Palladium. An ongoing relationship with the agency continues today. Barnes's career includes serving as a field engineer at the NASA High Range in Nevada for the X-15, XB-70, lifting bodies and lunar landing vehicles; working on the Nuclear Engine for Rocket Vehicle Application, NERVA project at Jackass Flats, Nevada for crewed flight to Mars; and serving in Special Projects for the CIA at Area 51. Barnes later formed a family oil and gas exploration company, drilling and producing oil and gas and mining uranium and gold.

Barnes currently serves as the president of Roadrunners Internationale, an association of Area 51 CIA, Air Force, and contractor veterans. He is the executive director of the Nevada Aerospace Hall of Fame.

Two National Geographic Channel documentaries feature Barnes: Area 51 Declassified and CIA—Secrets of Area 51. Numerous documentaries on the History Channel, the Discovery Channel, the Travel Channel, and others also feature him. Barnes authored CIA Station D Area 51 and several books approved by the CIA PRB, including the CIA Area 51 Chronicles, a three-book series about the declassified CIA U-2, A-12, MiG, and stealth projects at Area 51. Barnes remains active with oral history projects at the University of Nevada Las Vegas, the Central Intelligence Agency, the Defense Intelligence Agency, and the National Reconnaissance Office. Barnes lives in Henderson, Nevada.

www.ingramcontent.com/pod-product-compliance
Lightning Source LLC
Chambersburg PA
CBHW080900160726
48000CB00009B/2790